S0-BVN-576

Remarkable Maps

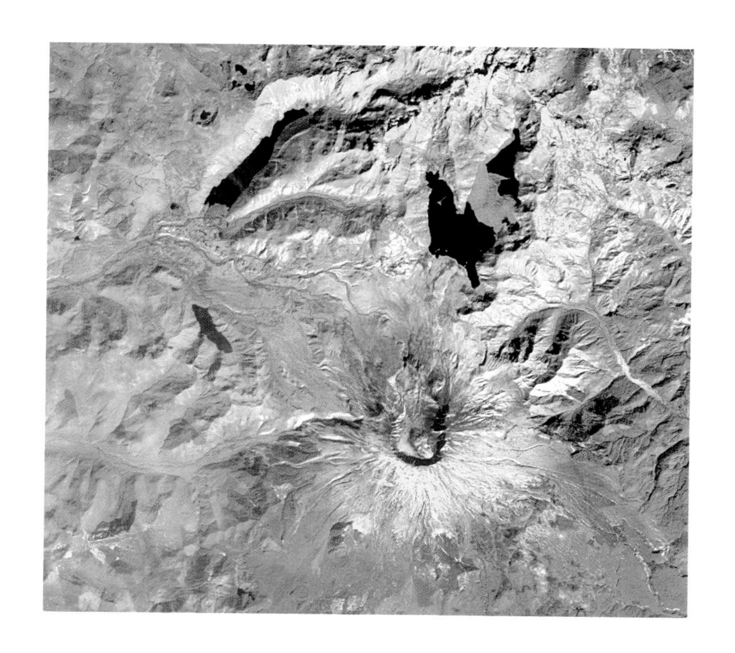

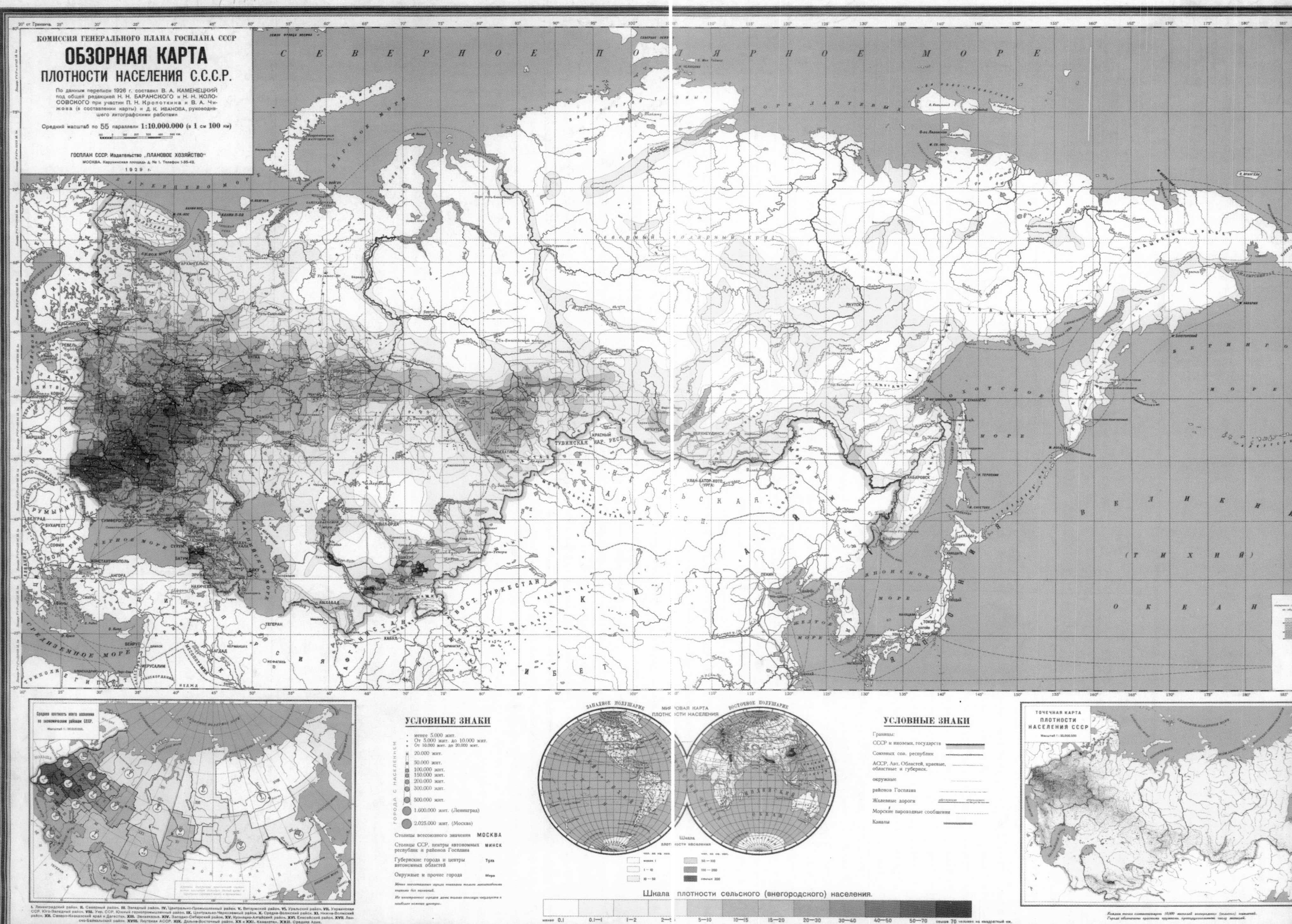

ABOVE *This is neither the most beautiful nor the most technically accomplished map in this book. But it does point to several themes in the following pages. It is a population distribution map of the Soviet Union, published in 1929. By 1927, with Trotsky expelled from the Communist Party, Stalin is the undisputed ruler. In 1928, his first Five-Year Plan is implemented. Farm collectivization entails the deportation of 1.5 million "kulaks" (land-owning peasants) to remote areas, often a death sentence. The ensuing famine of 1932-33 would kill millions. In 1929, the purges of the Commissariats and the Party had killed thousands more. This is Stalin's deathly map, a product of the 1926 census. The comparison of the Union's population and the world's is telling: for Stalin, it was a vision thing.*

Remarkable Maps

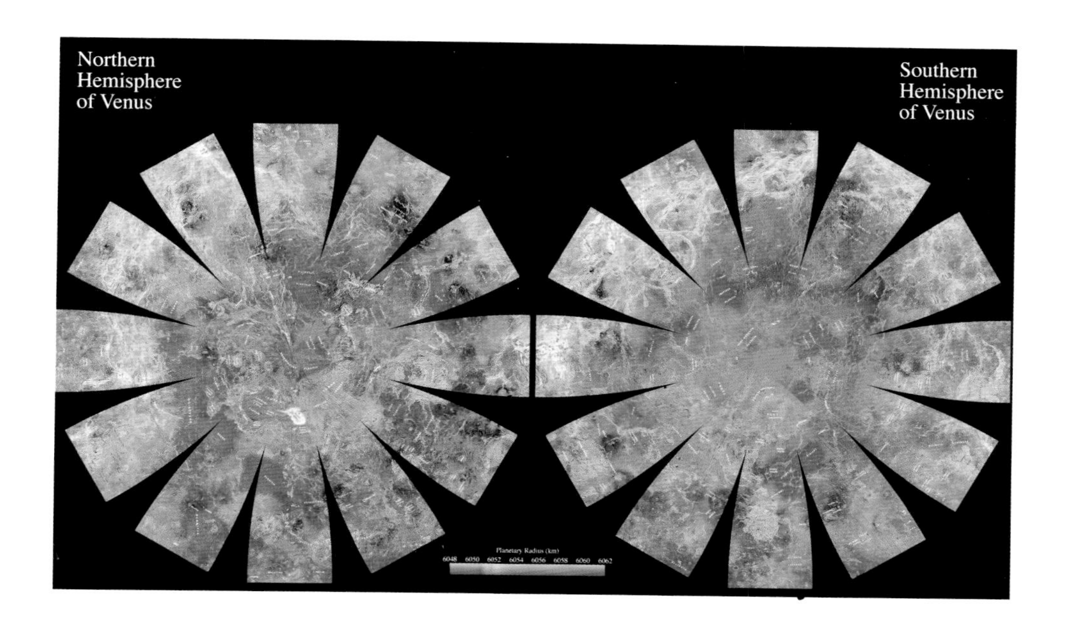

100 Examples of How Cartography Defined, Changed and Stole the World

EDITED BY JOHN O. E. CLARK WITH AN INTRODUCTION BY PROFESSOR JEREMY BLACK

CONWAY

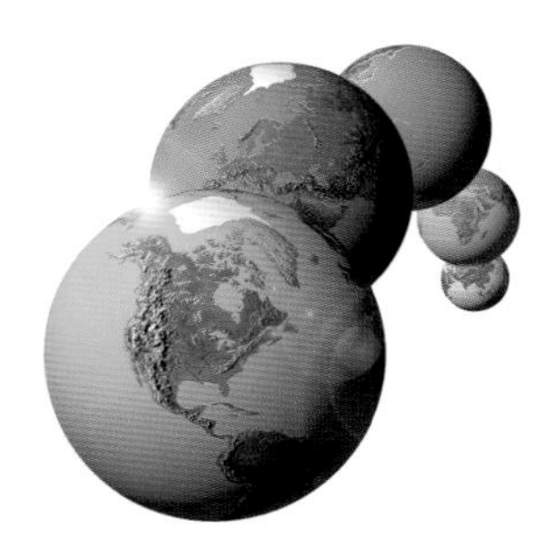

First published in Great Britain in 2005 by Conway Maritime Press

An imprint of Chrysalis Books Group plc

The Chrysalis Building, Bramley Road,
London W10 6SP
www.chrysalisbooks.co.uk

Project Editor: Shaun Barrington
Designer: John Heritage
Production: Alice Reeves
Picture research: Rebecca Sondegren, Shaun Barrington
Index: Diana LeCore
Colour reproduction: Classicscan

British Library Cataloguing in Publication Data
A catalogue record for this book is available
from the British Library

Library of Congress Cataloging in Publication Data available

ISBN 1 84486 027 2

Printed in China

Illustration on page 1: Advanced Spaceborne Thermal Emission Reflection Radiometer (ASTER) image of Mount St. Helens, Washington, one week after the March 8th, 2005 eruption. ASTER was launched in 1999 on NASA's Terra satellite and can map the Earth's surface with a spatial resolution of 50 to 300 feet (15 to 90 meters). Infrared has captured hot spots from incandescent lava.

Page 3: The northern and southern hemispheres of Venus, computer-generated not from single images or even single missions, but from ten years of radar investigations. Even the projection required a specially written computer program to create the "flower petal" globe effect, combining two different projections, the Transverse Mercator and Lambert Azimuthal Equal-Area. While not explaining the calculations, it is hoped that this book will indicate the advantages of such an approach to one of the mapmaker's ultimately insoluble problems: how to wrap an orange neatly in a single sheet of paper.

Contents

Introduction by Professor Jeremy Black

Maps have the capacity to open worlds of reality and imagination, to depict, in their lines, points, and spaces, both hopes and fears, to urge the wanderings and wonderings of the mind. This collection exemplifies how cartography is both a science and an art. Maps existed before the written word and today exploit the most up-to-date computer technology and imaging systems. Many inform us about the cosmological beliefs of the people who made them, as as much as they do about geophysical reality. Maps are also repsesentations of social and political aspiration and power, making statements about the ownership and control of territory. The history of the development of cartography, both east and west, is indivisible from the history of grasping space, both imaginatively and in reality, from depictions of human relations with the heavens to the worlds of invasion and conquest. Cartography also responds to technological developments, from cuneiform to CADCAM. It is hoped that this collection of varied maps, and the stories of their creation, will illuminate the ever-changing relationship between cultures and their graphic representation through cartography.

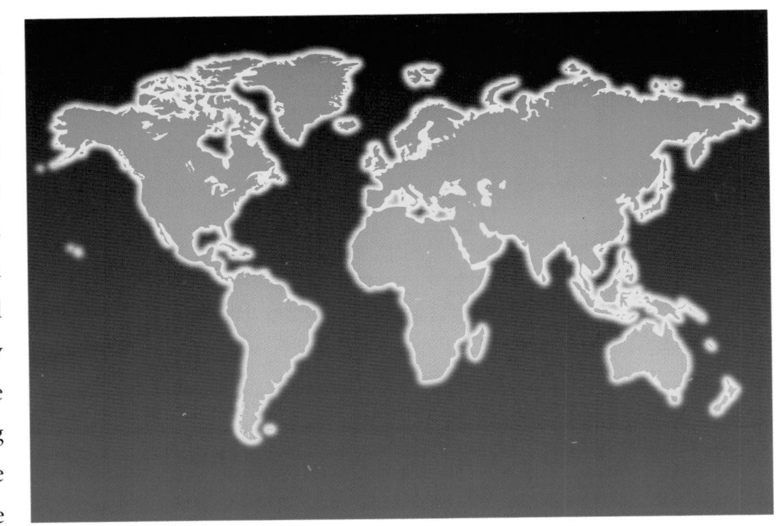

This cannot be a complete history of maps. As some kind of indication of just why not, *The History of Cartography*, published by the University of Chicago Press, had its first volume (of six) *Cartography in Prehistoric, Ancient, and Medieval Europe and the Mediterranean*, published in 1987. The latest installment at the time of writing, which arrived in 1999, was volume two, book three—and they are still looking for financial donations to finish this huge project. This book is emphatically not an academic treatise, but the maps featured should enthuse and intrigue, and it is written for the non-expert.

Many maps are simply beautiful, and though it would be desirable to understand the science behind, say, von Humboldt's maps of the earth's magnetic field, or NASA's survey of Venus, it's not essential to enjoy them visually. On the other hand, for those who see cartography as more of an aesthetic endeavor than a scientific one, here is a timely caveat: $x = \tan^{-1}(\tan\text{Ø}\cos\text{Ø}_p + \sin\text{Ø}_p\sin(z\text{-}z_0))$ over $\cos(z\text{-}z_0)$ where z is longitude and Ø is latitude is a simple coordinate calculation for a simple oblique Mercator projection. So don't get above yourself.

INSET *The basic Mercator projection (this is not actually standard Mercator, but close) is no longer a map, it's a kind of world logo. You could stretch it, or maybe add a half-inch strip in the Pacific, and you might accidentally create something more "accurate," particularly with regards to area representation.*

RIGHT *This stupendously flamboyant 1786 map by Dane Charles Louis Desnos (1725-1812), does not follow the dictum "less is more." The hemispheres show the voyages of Captain Cook. Thr historical notes on the countries of the first world inform us that Africa is the hottest continent, Asia the least known and richest.*

MAPPE-MONDE CELESTE
AVROY
TERRESTRE ET HISTORIQUE
NOVVEAV MONDE
ANCIEN MONDE
CIVILES
SYSTHEME
DE LA LVNE
HEVRES
HEVRES
ANNEES
CIVILES
SYSTHEMES.
DV. MONDE
DIVERSITE

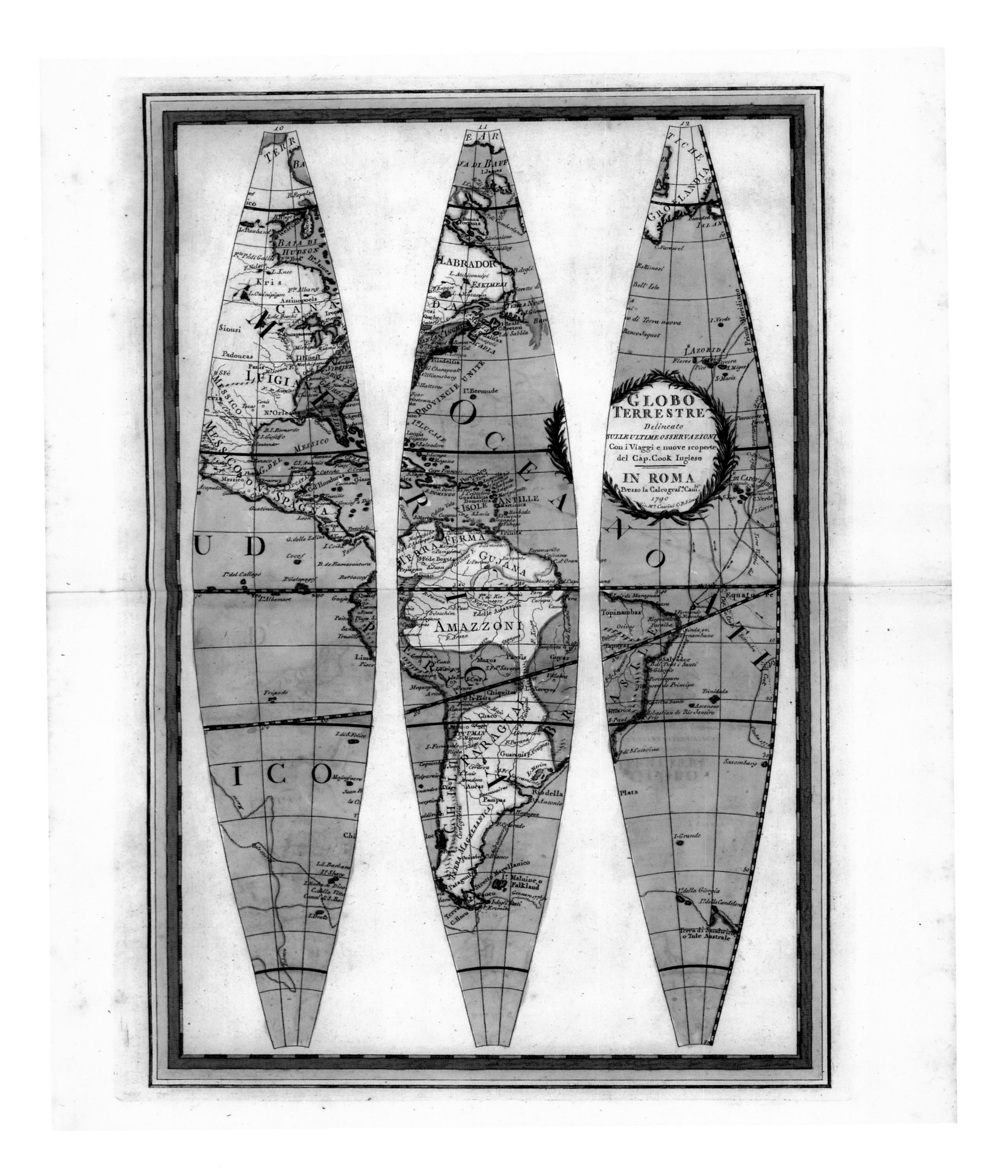
GLOBO
TERRESTRE
Delineato
SULLE ULTIME OSSERVAZIONI
Con i Viaggi e nuove scoperte
del Cap. Cook Inglese
IN ROMA
1790
BAIA DI HUDSON
LABRADOR
ESKIMEAUX
PROVINCIE UNITE
AMAZZONI
GUJANA
PARAGUAI
TERRA MAGELLANICA
ANTILLE
Rio della Plata
Falkland
Tropico
Equatore
Primo Meridiano
L AZORIDI
GROENLANDIA

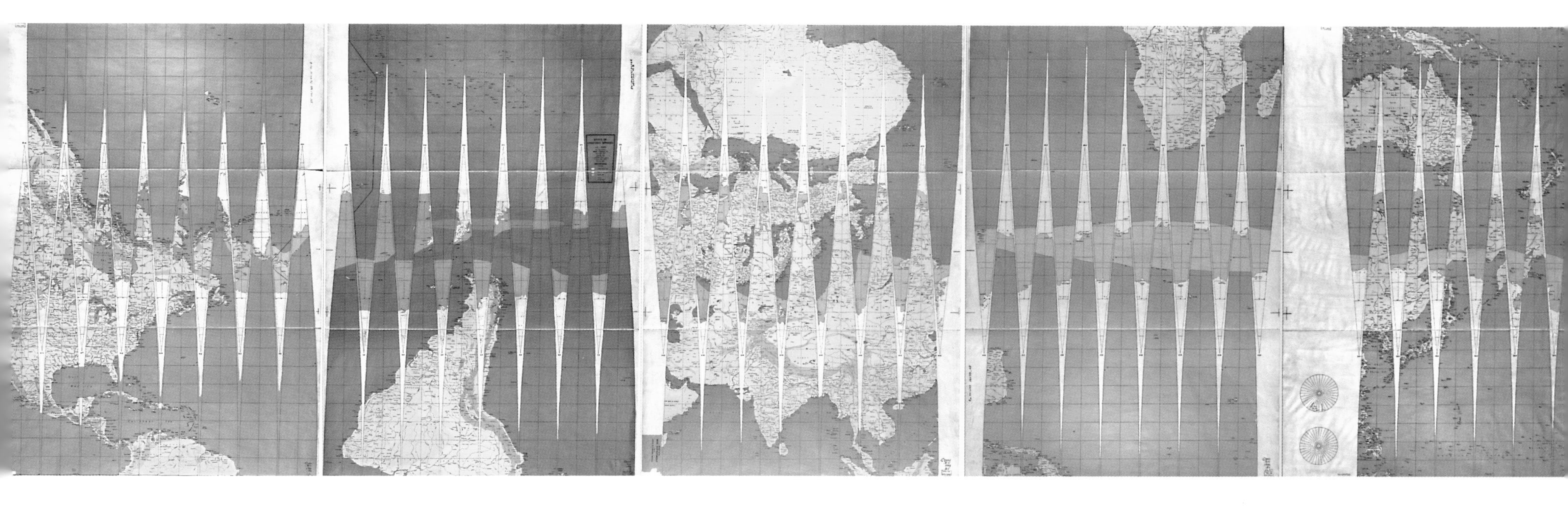

LEFT *These gores for a 13-inch (34-centimeter) globe (1790) by Giovanni Maria Cassini (1754-1824), printed in Rome, also show Captain Cook's voyages. To labor the point a little, this map and the Desnos map on page 7 are created at the same time, on the same —evidently salable—subject, yet they could not be more different. Cassini's globes enjoyed widespread success, as did his* Nuovo Atlante geografico delineato sulle ultime osservazioni, *in which the gores were printed.*

ABOVE *The same approach to depicting the world 150 years later, in 1942. These 50-inch (127-cm) globe gores were prepared for the US War Department by the Office of Strategic Services. "Copyright by Weber Costello Co. except as to modifications made and materials added by the Government of the United States." It would be interesting to know just what were the "materials added." The OSS was established in that year to collect and analyze strategic information.*

Some of the maps assessed are neither beautiful, nor technically or scientifically accomplished. They are included for other reasons. Examples of propaganda mapping are a reminder that the language of cartography demands careful reading. Drafting a "General map of that part of North America which has been the seat of war wherein is distinguished the roads ... rivers ... new forts," Colonel James Montressor in 1760 was certain that "it will be very acceptable, as well to the ministry as the military," reminding us that maps of exploration were designed to serve the needs of trade and territorial gains. Social issues have also been depicted here, as in John Snow's cholera map.

Maps have always posed a series of graphic challenges. The accurate depiction of a large sphere in two dimensions on a small scale entails formidable problems, and means that mapmaking is about compromises. Aside from this basic point, there are major issues for particular types of map: in finding the relevant information, in locating it spatially, so that it can be reproduced at a different scale, and in depicting it accurately. This varies by type of map. It is easier to show rail routes than to indicate landscapes of fear: which parts of cities people are reluctant to enter; and yet the latter are as much part of their spatial world as the former. It is possible to show majority religious affiliation—Italy as Catholic, Israel as Jewish and so on—but far harder to depict degree of religious commitment.

Maps are a tactile form of spatial perception, and the shifting, unfixed quality of the latter greatly affects how the maps themselves are understood. The extent to which maps are capable of multiple meanings adds to their fascination, complexity, and importance. The range of meanings is dramatized in the issue of "which way up" the world should appear.

The idea that the northern hemisphere should appear at the top of the map has been challenged, not least by "McArthur's Universal Corrective Map of the World" (Artarom, 1979), which carries a caption ending "Long live AUSTRALIA – RULER OF THE UNIVERSE." There is no reason why the map should center on the Greenwich meridian, with Europe in the middle. Indeed, many early maps did not. Many American maps put the western hemisphere at the center.

Many of the maps in this collection were originally in atlases; the works of Ortelius and the Blaeu family, for example, feature strongly, the beauty of their creations reason enough. Historical atlases—that is, atlases about history, rather than old atlases—provide an interesting insight into how cartography has changed. Until the 20th century,

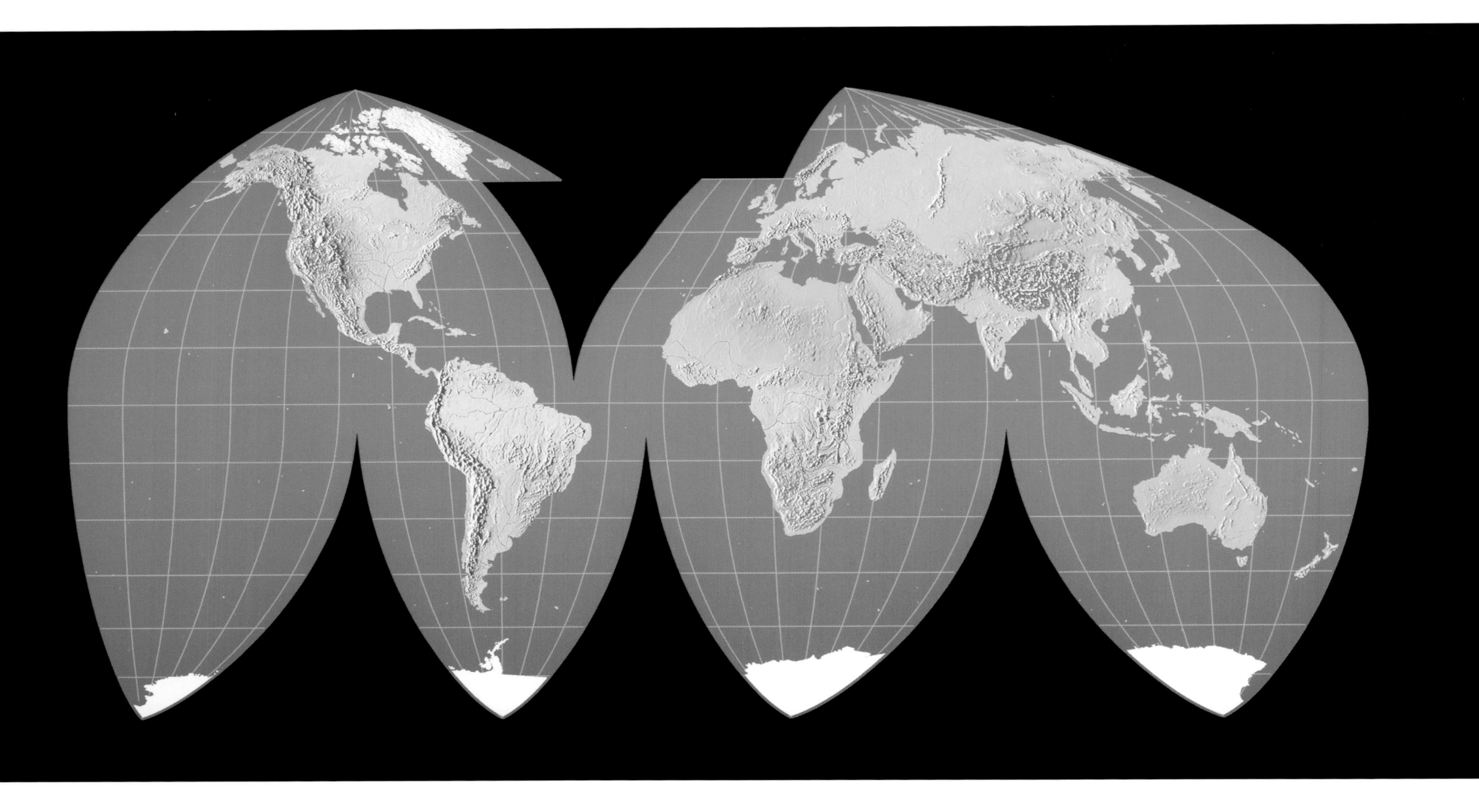

ABOVE *The Goode projection minimizes surface area and shape distortion. To give it its full technical description, the Interrupted Goode Homolosine projection is an interrupted, pseudocylindrical, equal-area, composite map.*

historical atlases' content was predominantly defined in terms of international relations—particularly warfare and shifts in control over territory. The state was assumed to be the crucial unit (and objective) in the historical process, and atlases were accordingly concerned with changing state boundaries, especially the rise and fall of empires. The apparently cyclical character of the rise and fall of empires, especially the Roman empire, gave historical atlases the character of morality tales, mirroring historical works of this period, such as Edward Gibbon's *Decline and Fall of the Roman Empire* (London, 1776–88). In the 19th century (and before) expansion of European power served to organize and rank the rest of the world. In the introduction to his *Historical Atlas in a Series of Maps of the World, as known at Different Periods* (London, 1830), London barrister Edward Quin employed color to depict "civilization," in Eurocentric terms. "We have covered alike in all the periods with a flat olive shading … barbarous and uncivilized countries," he wrote, "such as the interior of Africa at the present moment."

Post-1945, there was evidence of a crisis of confidence in the map. Reduced emphasis on physical geography in historical atlases reflected a number of factors, including globalization and analytical shifts away from materialistic explanations. The net effect has been that as maps have become more innovative in design, there has been less confidence in their ability to *explain* on their own, as opposed to *describe*.

RIGHT *Frederick de Wit's early 18th-century map of Ireland is a good example of a problem that no longer vexes the cartographer. The scale bars are provided in German, French, English, and "Hibernian" miles! Though the "statute mile" had been defined at the end of the 16th century, the length of a mile varied even within countries. A "Lieue de Bourgogne," for example, was much longer than a "Lieue de Paris." An Irish mile was 2,240 yards.* *(See page 122 for more on de Wit.)*

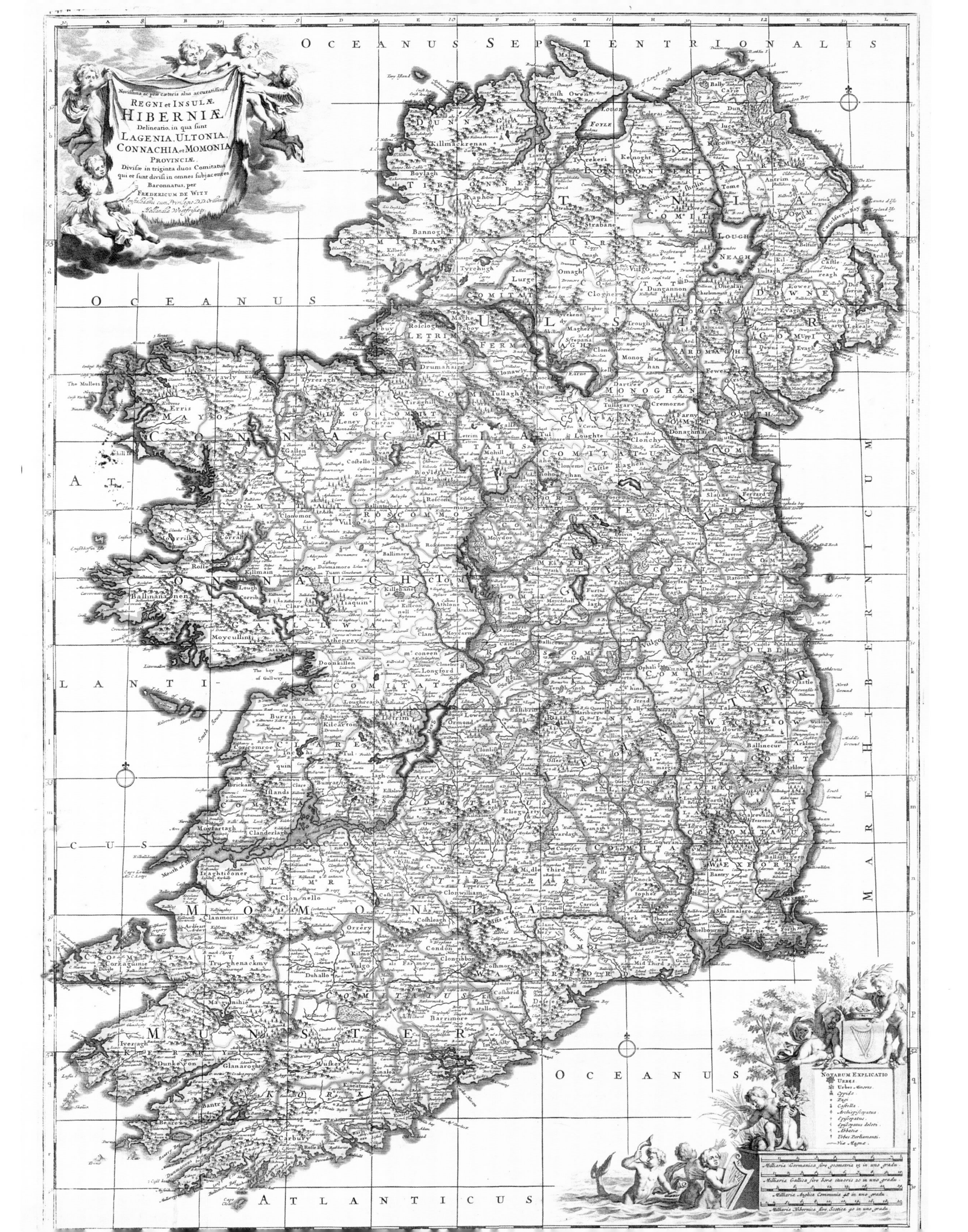

Novissima ac præ cæteris alus accuratissima
REGNI et INSULÆ
HIBERNIÆ
Delineatio, in qua sunt
LAGENIA, ULTONIA,
CONNACHIA et MOMONIA
PROVINCIÆ,
Divisæ in triginta duos Comitatus
qui et sunt divisi in omnes subjacentes
Baronnatus, per
FREDERICUM DE WITT
Amstelodami cum Privilegio D.D. Ordinum
Hollandiæ Westfrisiæq.
OCEANUS SEPTENTRIONALIS
OCEANUS ATLANTICUS
MARE HIBERNICUM
OCEANUS
ULTONIA
CONNACHIA
LAGENIA
MOMONIA
LONDONDERI
ANTRIM
DOWNE
ARDMAGH
MONOGHAN
CAVAN
LOUTH
LETRIM
SLIGO
MAYO
ROSCOMMON
GALLWAY
DUBLIN
WICKLOW
CATHERLAGH
WEXFORD
KILKENNY
TIPPERARY
WATERFORD
LIMRICK
CLARE
CORKE
KERRY
LOUGH FOYLE
LOUGH NEAGH
The bay of Gallway
Mouth of Shannon
NOTARUM EXPLICATIO
URBES.
Urbes Minores.
Oppida.
Pagi.
Castella.
Archiepiscopatus.
Episcopatus.
Episcopatus deleti.
Abbatiæ.
Urbes Parliamenti.
Viæ Magnæ.
Milliaria Germanica sive geometria 15 in uno gradu.
Milliaria Gallica sive horæ itineris 20 in uno gradu.
Milliaria Anglica Communia 48 in uno gradu.
Milliaria Hibernica sive Scotica 40 in uno gradu.

How autonomous is the cartographer? As pointed out in relation to the great Ortelius (see page 111), even his productions were the result of teamwork, and driven by commercial considerations. I would question the notion of the cartographer as bringing the panoptic eye of scholarship, or of his or her own views and suppositions. It is clear that, in contrast to most other books, the name on the cover of an atlas—whether of author, cartographer, editor, or all three categories—tells one only so much. This is because the framework of the atlas is set by the publishers. At the most basic, they decide how long the work should be and how many maps it should contain. The last is a crucial point because maps, both today and in the past, are expensive, certainly more so than text or pictures. (In earlier times, the cartographer sometimes *was* the publisher, which simplified matters.)

By way of example, for an atlas on the history of warfare some years back I produced a map of India in the 18th century designed to show the peripheral nature of the European impact in the first six decades of empire. The standard north-south map places a premium on European penetration, making the relationship between India and the surrounding seas central: India appears primarily as a peninsula. Eye lines focus on Delhi from European coastal positions, such as Bombay, Calcutta, Goa, and Madras. The customary maps also indicate only European victories, such as those of Clive at the Battles of Arcot and Plassey, and organize space and time in terms of British annexations. A totally different narrative of Indian history could be told focusing on European defeats, such as the Convention of Wadgaon in 1779, or Haidar Ali of Mysore's victory over the British at the Battle of Perumbakam in 1780, or Tipu Sultan's victory on the Coleroon river in 1782, or the unsuccessful British campaigns against Mysore in 1790 and the summer of 1791.

The major theme of my draft map was the contested succession to the Mughal Empire by a number of expansionist powers: Britain of course, but also the Maratha Confederation, the Nizam of Hyderabad, the Nawabs of Bengal and the Carnatic, and the Sultan of Mysore. To do this, one would adopt a perspective in which India opens up from the Khyber Pass, with a central alignment thence via Delhi. Marketing considerations prevented the map's publication. Such a perspective simply was not expected, and giving the public what they want is as good advice for publishers as it is for any other manufacturer. When maps are made by governments or public bodies, the pressures on the cartographer—or rather, the team—are different, but no weaker. If an academic, historical map like mine of India can't make it, imagine the leverage on a cartographer working for, say, Stalin.

Space and distance look the same on a map, established and measured by the scale. But they are not the same. The very notion of distance has changed over time. The rate of change in perception is not constant. Journeys and concepts of space and time in 1776 were more similar to the situation 223 years earlier than 223 years later. The menace of the dark (as unknown areas were often depicted), when space shrank to the shadowy spots lit by flickering lights, cannot be captured by maps. Similarly, the sense of direct providential intervention, of a daily interaction of the human world and wider spheres of good and evil, of heaven and hell, of sacred places, is heavily constricted today by secularism and science. Those earlier concerns were important aspects of cartography, as the maps of the Australian Aborigines, of the Music of the Spheres, and even of the mythical isle of Avalon, all included in the this book, are intended to show.

I became interested in maps as a child. I liked it if the books I read, whether *Swallows and Amazons* or *The Hobbit*, contained a map. It made the story solid, comprehensible, real. Growing up in outer London, my world was also defined by maps: the subway map devised by Harry Beck that showed different routes into town, structuring the sprawl of the city with its clear symmetry, and the far more inchoate street maps of my suburb, depicted in the A-Z street atlas, a great help when my paper deliveries took me farther afield than the immediate streets. As a child, I wrote an imaginary history of an invented land, which required maps, and also drew maps for an interpretation of the history of at least one real state. (Readers may have similar memories evoked when considering the last chapter in this book, "Fantasies, Follies, and Fabrications.") School meant geography as well as history, and meant "where bananas came from," not the modern "where bananas would come from it they read analyses of locational geography." Maps, therefore, unlocked the real world, excitingly so in my teens when I was the route-planner for family driving trips on the Continent and for walking vacations in England. Still, to this day, I find the maps in airplanes—the paper ones in the in-flight magazines, and their electronic counterparts on the monitors—seductive. Their difference is also arresting, showing how the same routes can be presented in contrasting ways.

When I was first thinking about writing this introduction, the British newspapers were full of election maps, which clarified, and yet, also misleadingly simplified: constituencies in which more than sixty percent of the voting electors voted against the winning candidate were defined by the color of the latter's party, thanks to the British "first-past-the-post" voting system. At once, this is an accurate account of the result, in terms of who was elected, and a misleading portrayal of voter preferences: they would be better shown in a color-coded dot map, with the number and color of dots per constituency proportionate to the votes cast; but that would sacrifice comprehension for the sake of accuracy, a balancing act performed in all the following maps.

My comments will strike echoes with some readers; all will have their own history of understanding and appreciating maps. If this understanding and appreciation varies and is often distinctively personal, that helps unlock the issue of how maps can be presented in very different ways. That this has been exploited for propaganda purposes, often brilliantly so, does not mean that maps are without value, or that they are simply systems to control territory, by allocating it or by manipulating views. Instead, it is necessary to understand the nuances of perception, and therefore representation, at the same time as appreciating the inherent problems of mapmaking.

This book will encourage people to look to the future. In a world in which the visual is increasingly dominant over the literary, maps will play a prominent role. This will in part be because the systems that need depicting, whether natural, such as the human brain, or artificial, such as microchip mechanisms, are ones with which people are unfamiliar, or that cannot be understood in familiar terms. In appreciating the widening world of maps, it will be important to understand the insights that considering the invigorating and splendid history of cartography offers.

BELOW *A bird's-eye map of the battle fought near Lake George on September 8th 1755. The British with their Mohawk allies defeated a greater French and Indian force. Cartography is a vital tool of warfare, as is made clear in this book in the section on military maps; but some maps are simply post-battle memorials. This is of course an English creation. To the victor, the map.*

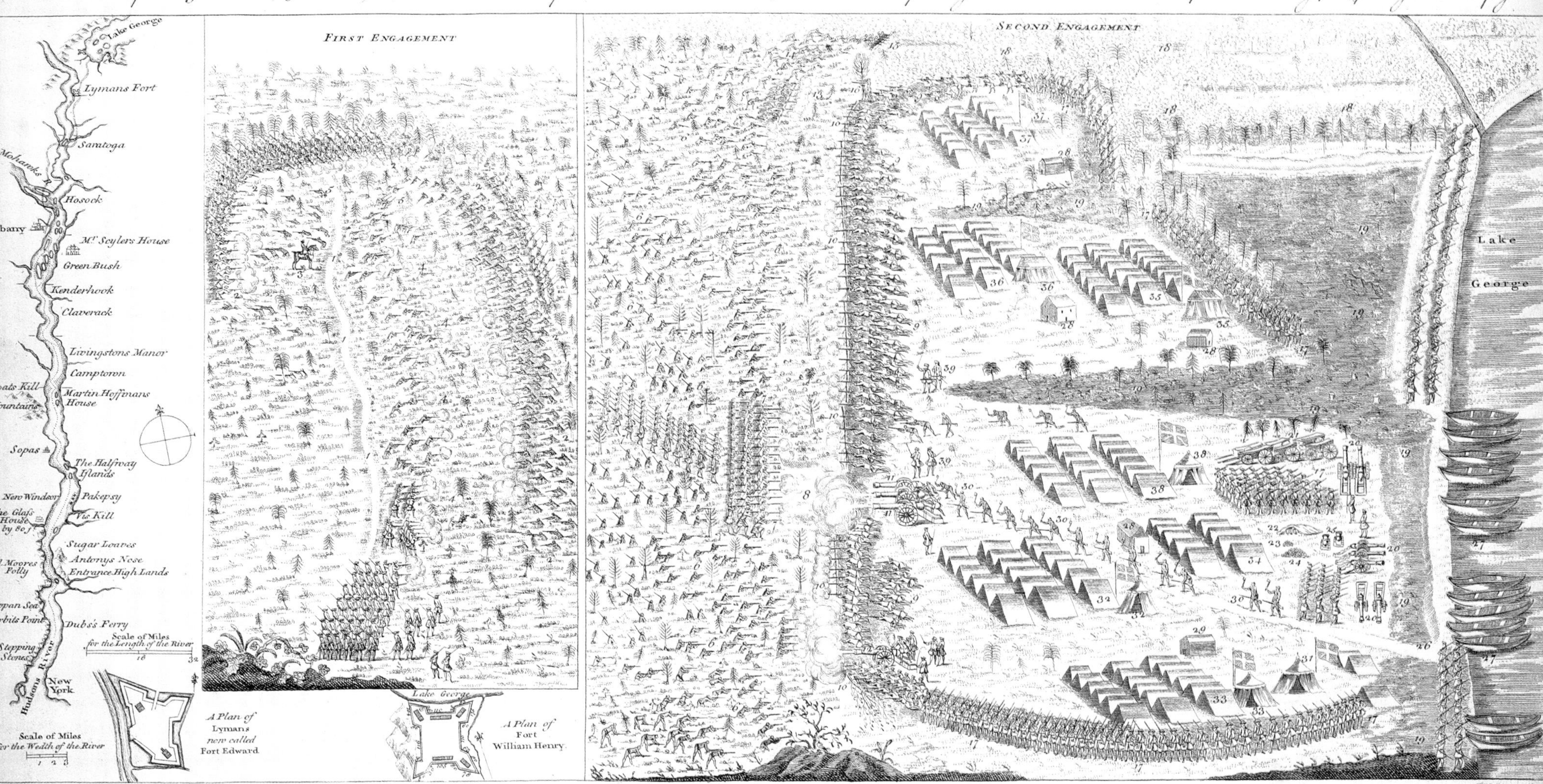

The Choice of Maps

As Pofessor Black has pointed out in the Introduction, a single volume like this cannot hope to address the whole history of cartography. The authors and the publishers are acutely conscious that anguished cries will rend the air from cartographic historians when they consider the selection of maps that follows. Where are the indigenous maps of South America? where is César-François Cassini's *Carte Géométrique de France*? Where is the map of the human genome? The defense to such charges is twofold. The first is that the maps have been chosen to highlight specific cartographic themes; this may be a cartographic innovation, such as Harry Beck's orthogonal schematic map of the London subway system, or Hermann Bollmann's axonometric city guides. The map may be included simply because its creator is an important figure in the history of cartography: so the maps of Christopher Saxton, Philip Apian, Ortelius—and of course Ptolemy—are featured. Some maps deserve their place as examples of specific mapping techniques or approaches, such as the computer-generated maps of the 2004 tsunami and the statistical mapping of world poverty. Some of the maps symbolize or encapsulate an entire cosmology, like the Australian Aborigine map and the map of the nine worlds of the Norsemen.

The second defense is rather more nebulous. As should become clear, almost all maps are partial, prejudiced representations of the world, compromises among the competing demands of accurate area, shape, distance, and direction even if the sole aim is geographical, physical. If the aim is more complex or even deliberately obscured, as in maps designed to glorify leaders or nations, to claim territory, and in propaganda mapping, then it is even harder for the mapreader to interpret and decode the image confidently. *Therefore* we claim our map selection is beyond criticism, because not only beauty, but significance, is in the eye of the beholder! It must be admitted that beauty has played a part: there is a preponderance of Dutch 17th-century maps because they are such marvelous images.

The first section considers the beginnings of mapmaking and indicates some of the early transfers of cartographic knowledge, East to West and West to East. The second considers cartographic breakthroughs chronologically, most, but not all, associated with individual genius. The third section, inevitably, analyzes the maps of exploration. This section is called *The Ages of Exploration*, simply because the standard definition of the Golden Age from the late 15th to the 19th century seems too broad. Christopher Columbus and Captain Scott do not inhabit the same world. The selection of military maps in the following section is consciously circumscribed. So many maps can loosely be defined as military in purpose. Many 19th-century European maps of Africa, for example, if not military, are certainly for

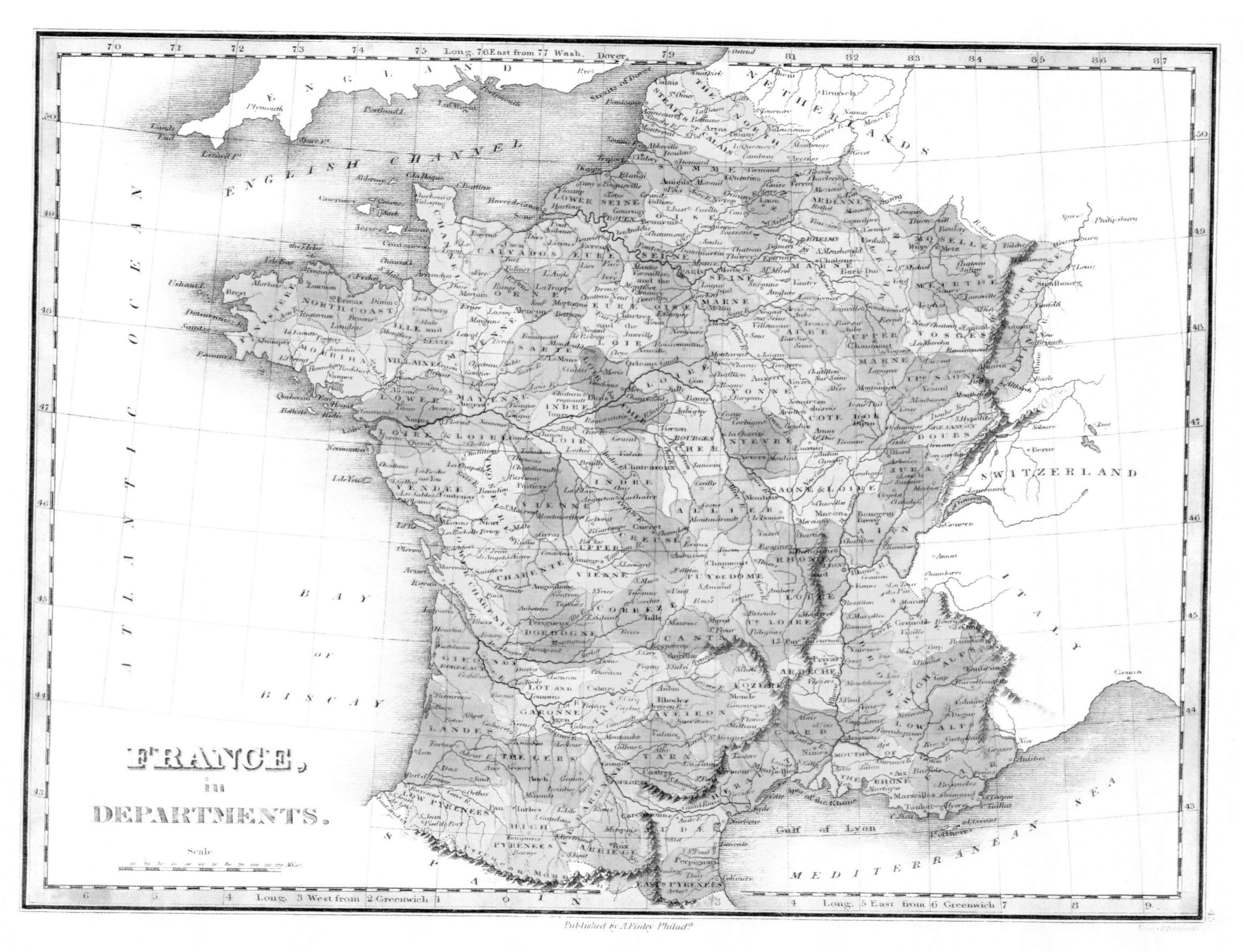

the purpose of annexation. Most of the military maps chosen are battle maps, maps that had an actual effect on military outcomes, or reveal the military decision-making process. In *Drawing the Line* are just those maps that were expressions of territorial annexation. The final section considers map controversies and works of the imagination.

ABOVE *Why this map? Joseph Perkins' 1826 map of France shows district boundaries, or departments. It also demonstrates rather beautifully the "rule of four." For any collection of contiguous shapes on a flat plane to be differentiated, you only need four colors. Try it for yourself: and no, starting with the Vatican inside Italy makes no difference. Obvious to a mathematician, mystifying to most.*

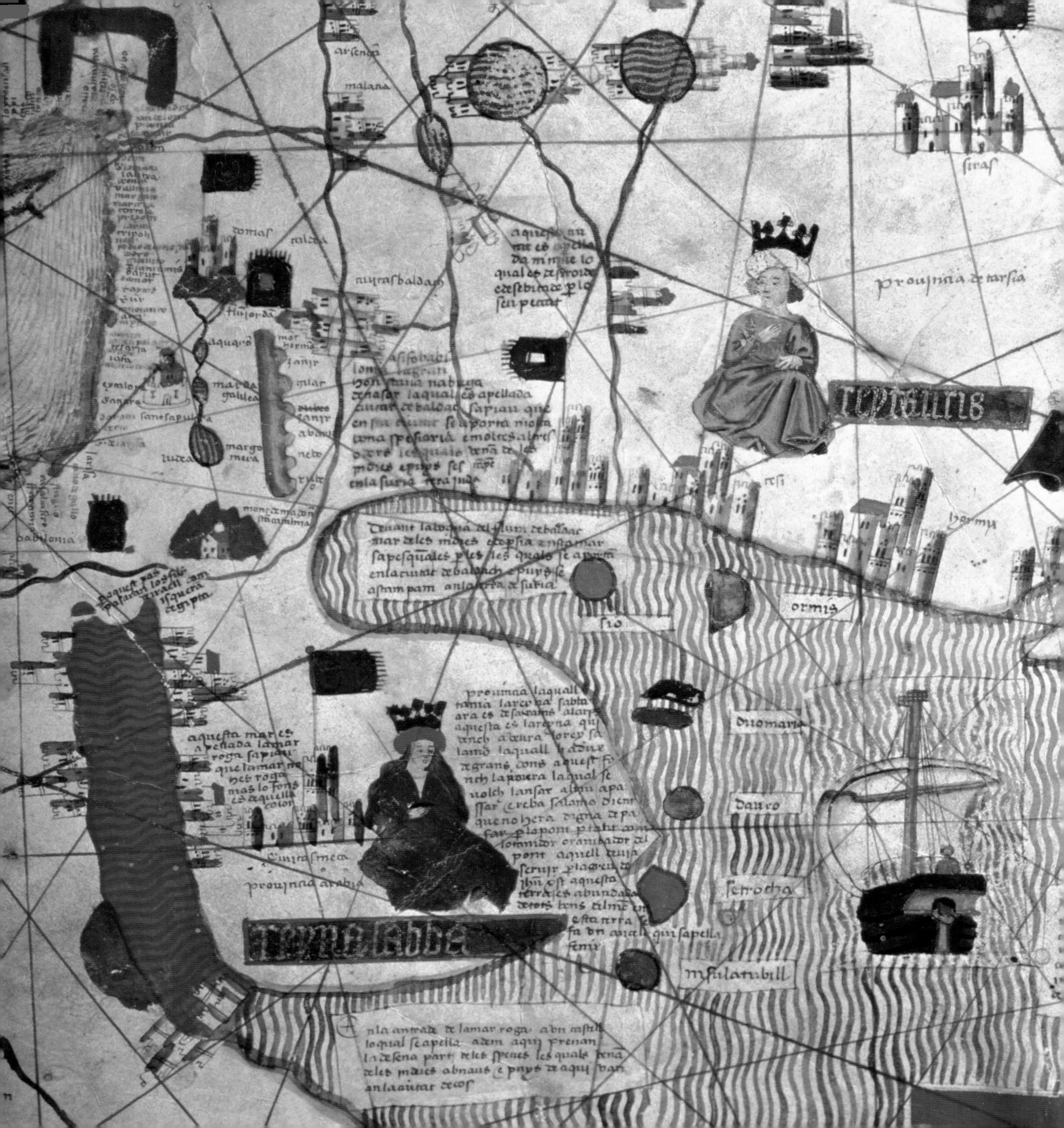

ormis
siras
prouincia de tarsia
babilonia
dauro

The Earliest Maps

The Middle East, Persian Gulf, and Red Sea, from the Catalan map by an unknown cartographer, commissioned by (or acquired by) ambassador Alberto Cantino for the Duke of Ferrara, manufactured and illuminated in Portugal in 1502. Cantino smuggled the portolan-style map out of Portugal to Italy. It informed the Duke of the discoveries of Columbus on the northern coast of South America and the Caribbean islands, of Perdo Alvars Cabral (Brazil, 1500), and of the Corte Real brothers Miguel and Gaspar of the "North Cape of Asia" (probably Labrador, Newfoundland). One question for the Duke was which monarchs would rule the new lands? Would it be the same ones shown here in the first world? This portolan represents something of a cartographic endpoint; the first printed world map to include the New World, the Contarini map of 1506, recognized that a regular projection was necessary to depict the vast discoveries.

It Was Always About Ownership:

Ancient Clay Maps

Like the wheel and the plough, the map was probably invented (if maps can be considered an invention in any meaningful way) in Mesopotamia, the ancient Middle-Eastern land that lay between the rivers Tigris and Euphrates. Arguably the oldest existing map was made in that region well over 4,000 years ago. Though there are other claimants, such as the Bedolina petroglyph map in northen Italy (ca 2000-1500 B.C.), with its apparently abstract symbols for fields, walls, and waterways.

In about 2350 B.C. Sargon I founded the Akkad kingdom based on Agade in northern Mesopotamia. The people's language, Akkadian, used the cuneiform writing of the Sumerians, in which combinations of wedge-shaped marks formed the characters. For permanent records, scribes used a triangular "pen" to impress cuneiform script into damp clay tablets, which were then sun-dried or baked in an oven to preserve them. Thousands of such tablets have been found dating from the whole of the Babylonian period. In 1930-31 archeologists digging at the ruined city of Ga-Sur at Nuzi (present-day Yorghan Tepe in Iraq) unearthed a clay tablet with a difference—instead of the usual cuneiform characters it depicted a map. The site where it was found lies 200 miles (320 kilometers) north of Babylon near the modern towns of Kirkuk and Harran. The archeologists dated the tablet to 2300 B.C., making it the oldest known map. It measures 3 by 2.7 inches (7.6 by 6.8 centimeters), slightly smaller than a man's palm.

The lines inscribed in the clay are difficult to decipher, but experts agree that they represent an area of land, probably the plan of an estate, complete with the owner's name: Azala. The estate lies between two ranges of hills, which are denoted by overlapping semicircles, and has a waterway running through the middle, either a river or a canal. The central plot has a stated area of 354 iku, which equals about 30 acres (12 hectares). Three inscribed circles probably indicate the cardinal points north, east and west, although some archeologists suggest that these represent cities. The map's function is not known, although it may have been some sort of title deed establishing ownership.

Smaller-scale maps or plans of towns and cities on clay tablets also occur from later in the Babylonian period. One tablet from about 1500 B.C. depicts the city of Nippur, south of Babylon and capital of the Sumerian civilization. It shows major buildings, including the Temple of Enlil in its own square. Among the nearby fields, lines indicate the boundaries of the estates of some of the wealthy landowners, separated by irrigation channels that run down from a curving river. Later still the town of Sippar, on the banks of the River Euphrates just north of Babylon, was the subject of a clay tablet made in about 500 B.C. The town is depicted as a rectangular shape, surrounded by an intricate system of canals. The street plan of Babylon itself was also mapped on clay tablets. One of these locates the Temple of Marduk and the route of a processional path through the Ishtar Gate to a smaller temple sited outside the city's walls.

The "city" interpretation of the circles on the Nuzi map is given credence by another later clay tablet map, which was made in about 600 B.C. It is a Babylonian map of the world as it was then known, with the Earth represented as a small disc surrounded by a ring of water (the "Bitter River"). Near the center is a rectangular box denoting Babylon, straddling the Euphrates River. Circular impressions indicate the locations of neighboring cities. This is the first known attempt to show the Earth as a whole and predates similar interpretations by the Greeks Anaximander and Heccataeus.

Notwithstanding the extant world maps, most ancient clay maps are large-scale representations of small areas, showing irrigation, fields, and most surely, ownership. They are title deeds, necessary records for a new kind of urban society that has abandoned a hunter-gathering existence. Of course these are not the oldest maps, only the oldest to survive.

RIGHT *Babylon's view of the world was recorded on this clay tablet made in about 600 B.C. The city occupies the rectangle just above the center, with mountains to the north and a south-flowing Euphrates that debouches at the Persian Gulf into the surrounding annular "Bitter River," representing the world's oceans.*

82
7-14
509
92687

Alcheringa:

The Mapping of the Dream Time

The Australian Aboriginal word "Alcheringa" is commonly translated as "Dream Time" and usually signifies a sacred primordial time during the creation of the world when the totemic ancestors of all living creatures roamed the earth. It was believed all-powerful ancestors of caterpillars and kangaroos and scorpions–among the many others–created the world by dreaming it into existence. These entities shaped the Earth and made it habitable for all the animals and humans that came after.

Traditional Aboriginal religion maintains that these immortal entities still inhabit (in a "hibernating" dreaming state) sacred rocks and features of the landscape today. Visions of these totemic ancestors and their deeds have traditionally been recorded in bark drawings or on the walls of sacred caves. They are also recited in songs and re-enacted through dramatic dances and rituals. In all cases these illustrations or aural tales have served as maps to teach and guide the souls of initiates through the world in the Dream Time. However, as these supernatural beings inhabit geographically recognizable features of the land; they are also maps that are also practical guides to the natural world.

The concept of Dream Time is common to all the Aboriginal people of Australia, however, there are many variations on this system. "Alcheringa" is actually an anglicization from the Aranda language, just one of the many Aboriginal languages. More correctly transliterated as "Altjiranga," it is more accurately translated "Eternal Dream Time" or "The Dreaming." However it is translated, it denotes not so much a "time" as an eternal "dimension" in the Australian Aboriginal tradition.

Alcheringa describes a dimension common to most mythologies known as *in illa tempore*—"before history began"—wherein gods or ancestral beings establish the laws and taboos for the human race. To the Aborigines, this was the dimension of the supernatural ancestors who were *altjiranga ngambakala* or "born to eternity" and shaped the land and filled it with life. These spirits or beings remain in the land and can be conjured up so they can teach us how to survive.

According to Australian Aboriginal traditions, mortals are still able to communicate with these entities. This is, in part, because of the Aborigines' perceived relationship to the land. In Australian Aboriginal society, everyone has two souls, one mortal and one immortal. The mortal soul inherits its natural powers through his human ancestry. The immortal soul inherits its supernatural powers through it totemic ancestors. One's immortal totemic ancestor may be that of a wallaby or a caterpillar, depending on the circumstances of one's birth.

These immortal totemic beings sleep within the land: in rocks, in trees, in springs. However, these supernatural beings are able to enter the wombs of mortal women if these women pass by their sanctuaries during certain stages of their pregnancies. Each totemic ancestor bestows certain supernatural gifts upon each child. This second soul dwells as an immortal twin alongside his mortal soul; and lives within the human body so long as it survives. This totemic being's spirits may become animated within human forms. However, when the mortal body and soul perish, the immortal soul returns to the ancestral being that resides forever within the land.

In traditional Aboriginal societies, every individual knows how to identify and communicate with these ancestral spirits in the land. Failure to do so could be fatal. To learn the myths and rituals relating to these ancestors was also the means by which one could safely travel from one waterhole to another in the proper seasons; or one hunting ground to another in time with animal migrations. The mapping of these eternal patterns as set down in the traditions of Alcheringa was more than a set of spiritual beliefs. It was also a practical means by which all the Aboriginal tribes managed to survive.

RIGHT *In Gumatj lore the power of the crocodile is linked to fire. The crocodile Baru brought fire to Biranybirany on Caledon Bay, and from there fire blazed across the country, represented in the diamond designs. The Gumatj —or "crocodile"— people are a clan of the Yolngu from the northeast coast of the Northern Territory. This image represents both the ancestral figure of Baru and the Biranybirany area. Where tail meets body is the river mouth.*

Space Maps:

The Nazca Enigma

The Nazca were a South American people who flourished between about 200 B.C. and A.D. 600 in the southern part of what is now Peru. They lived in an inhospitable region of arid desert dominated by a windswept plateau, and they left behind two main reminders of their existence–fine, multicolored ceramics and huge figures drawn on the desert ground.

These so-called Nazca Lines consist mainly of outlines that to our eyes resemble animals—birds, a monkey, a whale, and even a giant spider—as well as trees and flowers. A 160-foot (50-meter) hummingbird sits next to a 935-foot (285-meter) pelican. Many of them are more than 300 feet (91 meters) across and can really be seen only from the air, and that is how they were first discovered, in about 1927. In addition there are regular geometric figures including spirals, triangles, rectangles, and arrays of apparently arbitrary dead straight lines. Together the collection of desert designs occupies an area of about 400 square miles (1,036 kilometers).

Anthropomorphic figures are found on the slopes. They include a bizarre being with two enormous hands, one normal and the other with only four fingers. Also represented are objects such as yarn, looms, and ornamental clasps. All these figures seem to have clear "entrances," which suggest paths to be followed and places to line up.

The Nazca made the huge designs in the desert by removing the weathered brown stone from the surface of the Pampa Colorada ("Colored Plain") to expose the much lighter sand beneath. The persistent winds keep scouring any accumulated sand from the grooves, which may have been deepened originally by the countless feet of the Native Americans as they tramped around the lines. The Nazca plain is ideal to preserve the markings, owing to the the climate (one of the driest on Earth, with only twenty minutes of rainfall per year) and the flat, stony ground, which minimizes the effect of the wind at ground level. But if that is how they were constructed, what were they for?

There have been various theories about the function of the Nazca Lines. It has been suggested that they are plans—large-scale maps—for (now defunct) irrigation systems. Some rivers from the high Andes do intersect the desiccated Nazca territory, and ensuring the water supply must have been high on the agenda of its water-starved inhabitants. Some of the giant "pictures" resemble the designs on Nazca pottery and may, like some of the pots, have had a religious significance. The image of worshippers or pilgrims ritually walking around the patterns in the desert, perhaps while offering sacrifices to propitiate the gods to guarantee the crucial water supply, adds weight to this theory. Perhaps, in this case, the figures and lines are all that remain of a huge outdoor temple. The long straight lines, some diverging from a central point, are supposed to lead to particularly sacred positions on

INSET *The Nazca lines are found in southern Peru in the Pampa region on the coast in the province of Nasca, 250 miles (400 kilometers) south of Lima. They cover nearly 400 square miles (1,036 square kilometers) of desert. Chile and Bolivia also have such geoglyphs, but nothing to compare with the variety and huge scale of the Nazca lines.*

ABOVE *The most famous of the Nazca "human" figures is the 105-foot (32-meter) "Astronaut," only recently discovered in 1982 by Eduardo Herran. Is this the representation of a human figure, or of a god? For Erich von Däniken it was of course the representation of a visitor from outer space. The people were trying to call back the space beings by drawing the lines, some of which copied the scorch marks of rocket propulsion!*

the plain. One ingenious suggestion is that the priests took to the air in hot-air balloons so that they could look down on the proceedings; this would account for the scale of the Nazca figures and why they can be appreciated only from high above the ground.

Another possibility, put forward by American historian Paul Kosok in 1941, is that the designs were made from the people's observations of the heavens. With the eye of faith the shapes of some of the animals do correspond to groupings of stars, rather like the fanciful animals associated with the constellations by early first-world astronomers. According to this theory, the whole site might be a vast celestial map or calendar of some sort. But American astrophysicist Gerald Hawkins poured cold water on this idea in 1967 when he could find no match between the Nazca patterns and the constellations, even allowing for changes in star positions over the last 2,000 years. Six years later Dr. Hawkins studied 186 lines with a computer program and found that only 20 percent had any astronomical orientation; no more than by pure chance. Nevertheless, if the Nazca figures and lines are the remnants of some ancient celestial map, they must constitute the biggest map ever constructed by human beings. The Nazca enigma is insoluble: we cannot even be sure that the Nazca people made the lines, since the lines cannot be radiocarbon dated. They have been associated with cat cults, zodiacal symbols, running races—and this comes as no surprise (see above)—visitors from outer space.

Al-Idrisi:

Islamic Guardians of Knowledge

Much of the great body of knowledge generated by the Classical world has only survived into modern times thanks to the efforts of Islamic scholars in the early Middle Ages. Most famously, the 11th-century Persian philosopher and physician Avicenna (Ibn Sina; 980–1037), and the 12th-century philosopher Averröes Ibn Rushd (1126–98) from Moorish Spain brought key writings of Aristotle to the attention of their Christian counterparts. Similarly, Arab scholars played a vital role in handing down the cartographical works of Ptolemy.

One of the most notable contributions from Islamic learning to cartography was made by the geographer Abu Abdallah Mohammed al-Sharif al-Idrisi. Born in the Moroccan city of Ceuta on the straits of Gibraltar in 1100, al-Idrisi went to study in the major cultural center of Córdoba in al-Andalus (Spain) before undertaking a wide-ranging series of journeys that took in Northern Europe, North Africa, and Asia Minor. These travels lasted for some 15 years, and laid the practical groundwork for the theoretical writings and mapmaking that occupied his later years.

Al-Idrisi's great aptitude in cartography eventually came to the attention of the Norman King of Sicily, Roger II (1095–1154). Roger's court at Palermo was renowned not only for its splendor, but also for the dialog that this enlightened ruler fostered there between Christian and Muslim scholars. Although the Normans had only recently wrested the island from Arab rule, at the end of the 11th century (1072–91), Roger had no thought of purging Islamic influence. Being especially interested in geography, he commissioned a number of important works from the Moroccan cartographer, from around 1140 onward. The most renowned of these was a globe of the world, partitioned into sections and containing all the most up-to-date geographical information that was available. Engraved on the surface of a ball of silver weighing some 880 pounds (400 kilograms) al-Idrisi's magnificent creation—sadly now lost—depicted the seven continents and also included detailed information such as rivers and lakes, major cities, and trade routes.

Accompanying the globe (and how we come to know about its existence in the first place) was a compendium of diverse information on countries of the world, such as their religion, languages, and customs. This work, which has survived to the present day, was called the *Al-Kitab al-Rujari* (Latin Tabula Rogeriana, or "Book of Roger"), and included 70 maps of different regions. Although these maps, for example the zonal map reproduced left, display some inevitable inaccuracies, like major islands scattered throughout the Atlantic Ocean or the portrayal of Scandinavia as an island, they are decidedly superior to contemporary maps of the same kind produced in Europe. The text of the Tabula Rogeriana is at pains to point out that all the distances given, along with the positions, height, and length of certain topographical features, are as accurate as possible, having been meticulously based on data gathered by al-Idrisi and other recent (mostly fellow Muslim) travelers. Al-Idrisi's mapping shows a certain influence from Ptolemy, whose works had been translated into Arabic in the 9th century, but is otherwise strikingly original.

Al-Idrisi's considerable learning was not confined to cartography. He wrote a book on botany and one on zoology. This remarkable Muslim polymath died ca 1166, and his renown quickly spread around Europe as a result of Sicily's nodal position on several trade routes. Though we should not exaggerate his influence on western Medieval Letters: no translation of his *Geography* is found before 1619.

RIGHT AND INSET *Al-Idrisi's 1154 world map (inset)—this a 1553 copy—is superior to other maps of the time in its use of curved parallels. The map is based in part on Ancient Greek* periploi, *or sailing instructions. The course of the Tigris and Euphrates Rivers are shown (right) in the 11th-century "atlas" of al-Istalhry.*

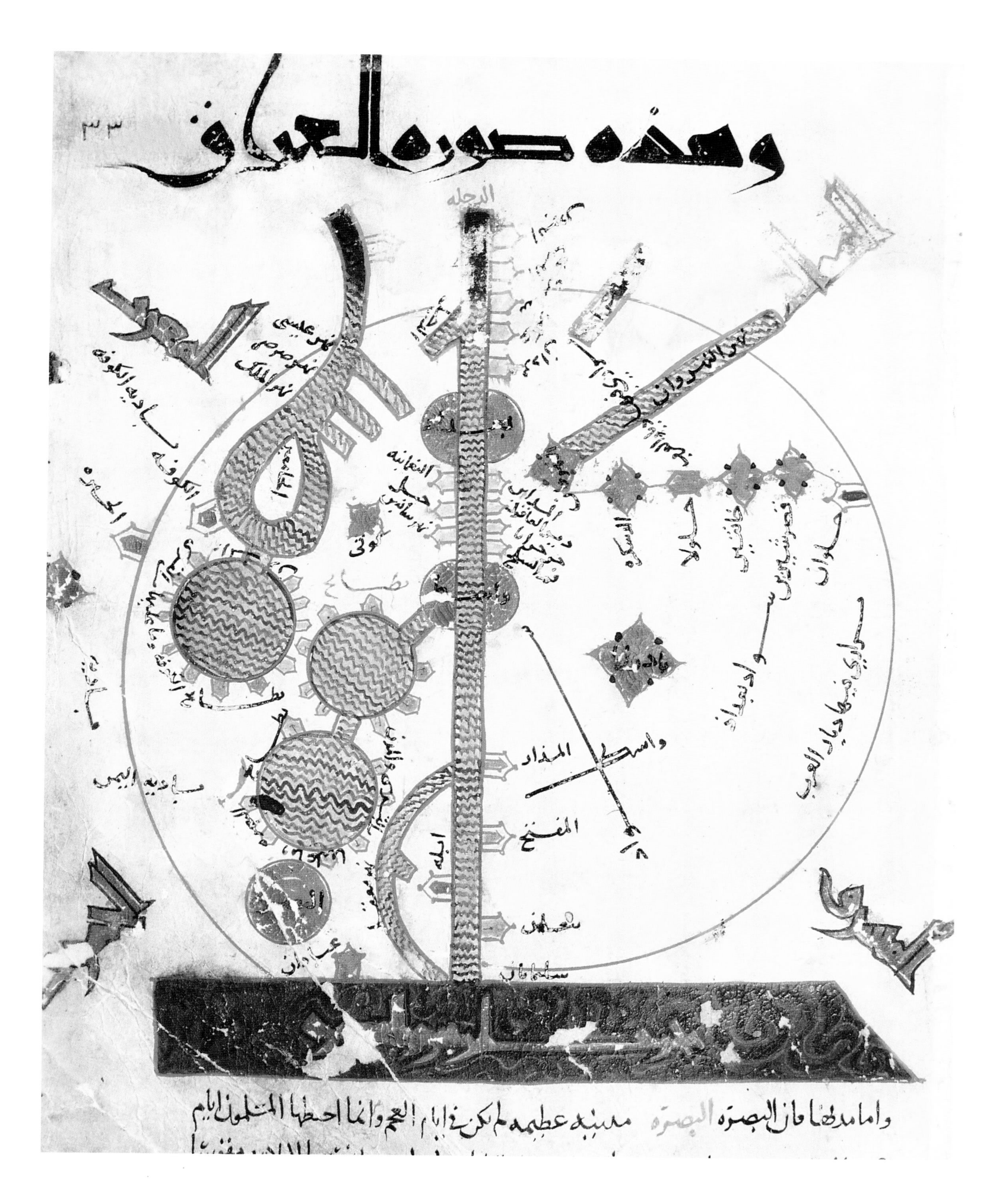
وهذه صورة العراق

Left-Right Hemispheres—East-West Parallels:

Chinese Cartography

In a book of this size, it is of course impossible to survey the history of Chinese or Asian maps meaningfully, (just as it is beyond reach to summarize the western tradition); but certain key breakthroughs must be mentioned. And one fascinating aspect that even the non-expert can respond to is "coincidence," or parallel development. It is as if cartographic endeavor reaches beyond cultures, as if it is a "natural" mental achievement, an inevitable human way of looking at the world.

As Norman J. W. Thrower points out in his marvelous book, *Maps & Civilization*, "The earliest survey of China is approximately contemporaneous with the earliest mapmaking activites of the Greeks, that of Anaximander (sixth-century B.C.)." He continues:

"In the centuries following, there are remarkable parallels between the geographical literature of China and that of Greece and the Latin West (especially later Roman writers), indicating more than casual contacts beteen these cultures."

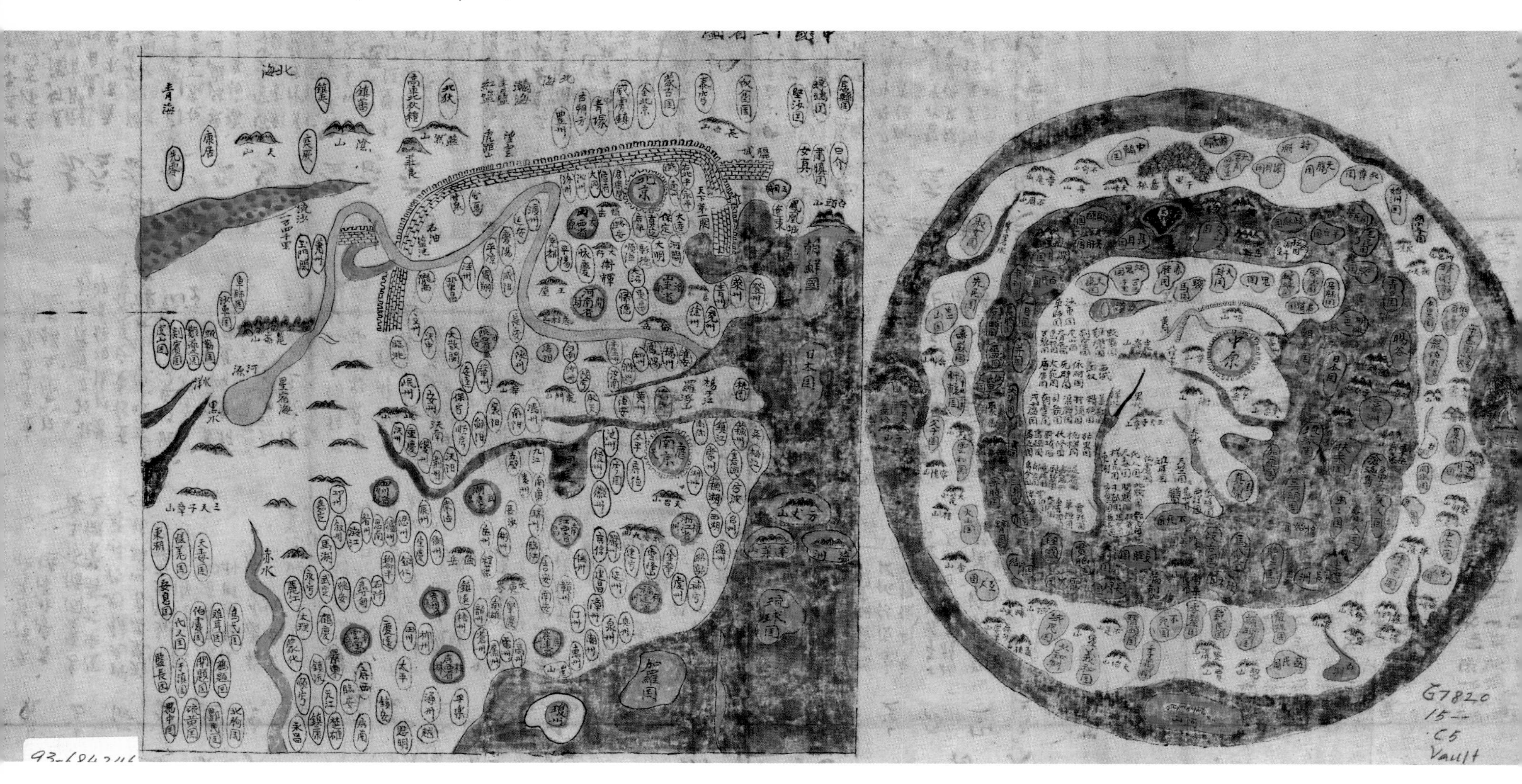

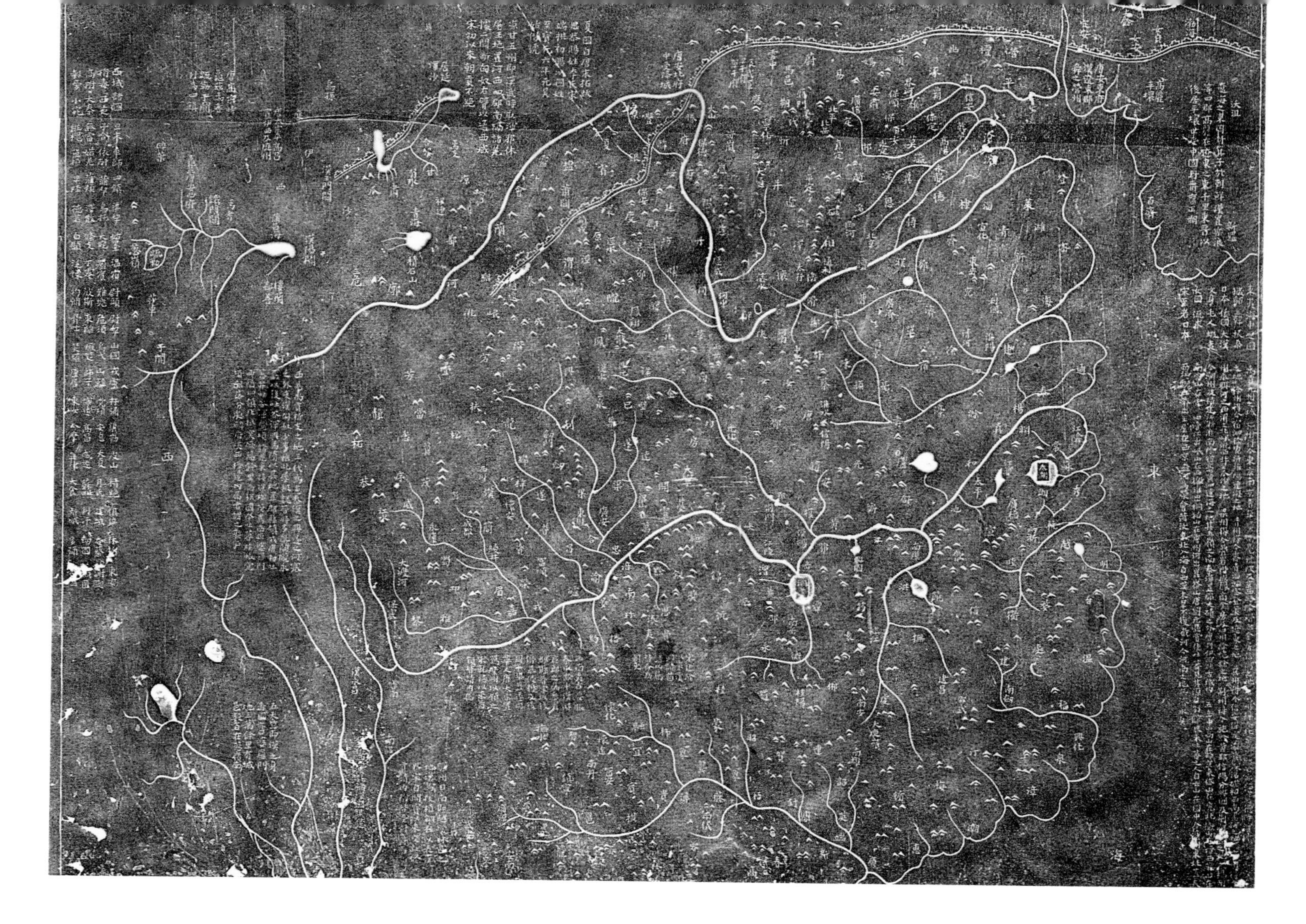

LEFT *The "Ch'onha Chungguk" map, created in pen and ink and watercolor in about 1800, shows the World and (left) the Ming Empire that lasted from the mid-14th century to the mid-17th. "Chungguk" is Korean for China: Chinese cartography strongly influenced that of Korea and Manchuria in particular.*

ABOVE *This 1136 stone map shows China during the Nan Song Dynasty (1127-1279), including part of Korea. The Yangtse and other rivers are accurately shown and more than 400 place names given. To indicate its sophistication, Magna Carta will be signed 80 years later, and Marco Polo will visit China 135 years later.*

Perhaps more than "casual contacts" is occasionally at work here: the Chinese astronomer Chang Heng, who introduced the rectangular grid to eastern cartography, was a contemporary of Ptolemy. They are not in contact with one another, but they are both dividing up the World with "geometrical" lines. Maybe it was just an idea whose time had come!

China as a discrete nation was unified for the first time in 221 B.C., when the western frontier state of Qin, the most aggressive of the Warring States, subjugated the last of its rival states. The repressive centralization initiated by the first emperor did not ensure the longevity of the first dynasty, but the imperial system and its bureaucratic control would set the pattern for two millenia. As is hopefully made clear elsewhere in this book, bureaucracy needs maps. Cadastral maps—those produced to define the extent, value, and ownership of land for the purposes of taxation—are found in early Chinese maps (as they are in Ancient Egypt). This need, and the pressing need for military mapping of the constantly threatened Chinese state, produced cartography with a clearly understood viewpoint, scale, and symbolism; plus the square and rectangular grid. As early as the third century A.D., minister Pei Hsiu set down the rules or principles for offical mapmaking, including altitude measurement, having considered the problems of mapping uneven terrain on a plane surface. To these accomplishments we must add the invention of paper in the previous century, and the earliest known printed map, of western China and dated about 1155. The compass is in use in the 11th century, (and probably earlier), finding its way west within the next 100 years. China also gave us the first celestial globe in 440 A.D.; but their mapping of *the* globe was Sinocentric even after the arrival of European cartographic sciences from the 16th century. The planets and stars were of more interest than Africa or the Americas. One final characteristic of Chinese cartography (characteristic as in non-Western), linked to this Sinocentrisim, is that despite its great practical value, mapmaking was not divorced as a graphic practice from the literary and visual arts until late in the 19th century. It didn't become its own "science."

Viking Mythology:

The Nine Worlds of the Norsemen

The Viking cosmos was shaped by the supernatural traditions of shamanism, a set of beliefs common to most tribal peoples, and practiced since the dawn of history. The shaman is a magician, mystic, healer, and poet. At the heart of his multi-leveled universe is the unifying concept of a cosmic pillar, a towering mountain, or a great tree. The World Tree is the most vital and powerful. It is a life force in itself that binds and nourishes all elements and levels in the shaman's universe. Here we consider not an actual old map, but one of the oldest psychic "maps."

In Norse mythology we have Nine Worlds bound together by Yggdrasill the Great Ash. Beyond its purpose as central pillar of support for the Nine Worlds, Yggdrasil is the means by which shaman spirits ascend to the world of the gods; or descend to the world of the dead.

Yggdrasil, literally means "the steed of Ygg (Odin)." Odin was the Norsemen's King of the Gods. He was also the supreme Shaman. For just as the shaman's spirit climbs his tree in a trance, so Odin rode Yggdrasill through all the Nine Worlds to become the lord and master of each. For a long time, Odin was a wanderer and seeker of wisdom and visions. He traveled through all the Nine Worlds and questioned every living thing: Giants, Elves, Dwarfs, Nymphs and spirits of the air, water, earth, and wood.

He questioned the trees, plants, and the very stones themselves. Odin endured many trials and dangerous adventures, but from each he wrung what wisdom there was from all things he encountered. It was, however, on Yggdrasill that Odin underwent his most harrowing rites of passage. Like the crucified Christ, Odin was wounded by a spear and hung from the sacred tree for nine days and nine nights. Hanging from the tree in great pain, Odin maintained a state of meditation on the markings cut in the stone by Yggdrasill's roots. By the ninth night Odin discovered the secret power of the runes, and brought about his own resurrection. From Yggdrasill he cut the limb and made his magician's staff. And by his magician's wisdom, Odin learned to cure the sick, make the dead speak, render weapons powerless, gain women's love, and calm storms by land and sea.

Ever thirsting for more knowledge and power, Odin went to the spring of Mimir—the fountain of wisdom and inspiration—but for this too there was a price. For one deep draft from the spring, Odin must sacrifice an eye. Without hesitation, he drank, and from that time he was always the one-eyed god. Resurrected as King of the Gods, Odin was a fearful god to look on. He was stern, one-eyed, gray-bearded, and gigantic. He wore a gray cloak with a broad blue mantle and a warrior's eagle-winged helmet. Upon his golden throne, Hlidskialf ("the watchtower of the gods") Odin's one eye could see all that happened in all the Nine Worlds in a single terrible glance. At his feet crouched two fierce wolves (Ravener and Greed) and on one shoulder perched two ravens (Thought and Memory).

The poet and author, Kevin Crossley-Holland, in his definitive *The Norse Myths*, has given the most lucid and instructive map of the Norse cosmos. His World Tree reaches the heavens, and upon its topmost branches is a Great Eagle with a Hawk perched between its eyes. Its leaves drip with honeydew, and deer and goats chew at its bark and shoots. The squirrel Ratatosk is the messenger who passes insults from the Great Eagle upon the highest branch down to the Dragon gnawing on its deepest root.

Crossley-Holland describes the Norse cosmos as a tricentric structure—like three plates set one above another. Yggsdrasil the World Tree has a root sunk in a well or spring on each of these levels. The first root is sunk in the Well of Urd (Fate), the second is the Spring of Mimir (Wisdom), and the third is the Spring of Hvergelmir (Underworld). A tree. A column, or a mountain at the center of the world (or universe) can be found in many world mythologies; and the division of three cosmic regions fed by, or feeding, a tree is found in Vedic Indian and Chinese cosmologies.

The simplest way of roughly categorizing these three levels would be: heaven, earth, and hell. However, the Norse cosmos is much more complex. The first level encompassed three separate worlds: Asgard, Vanaheim, and Alfheim.

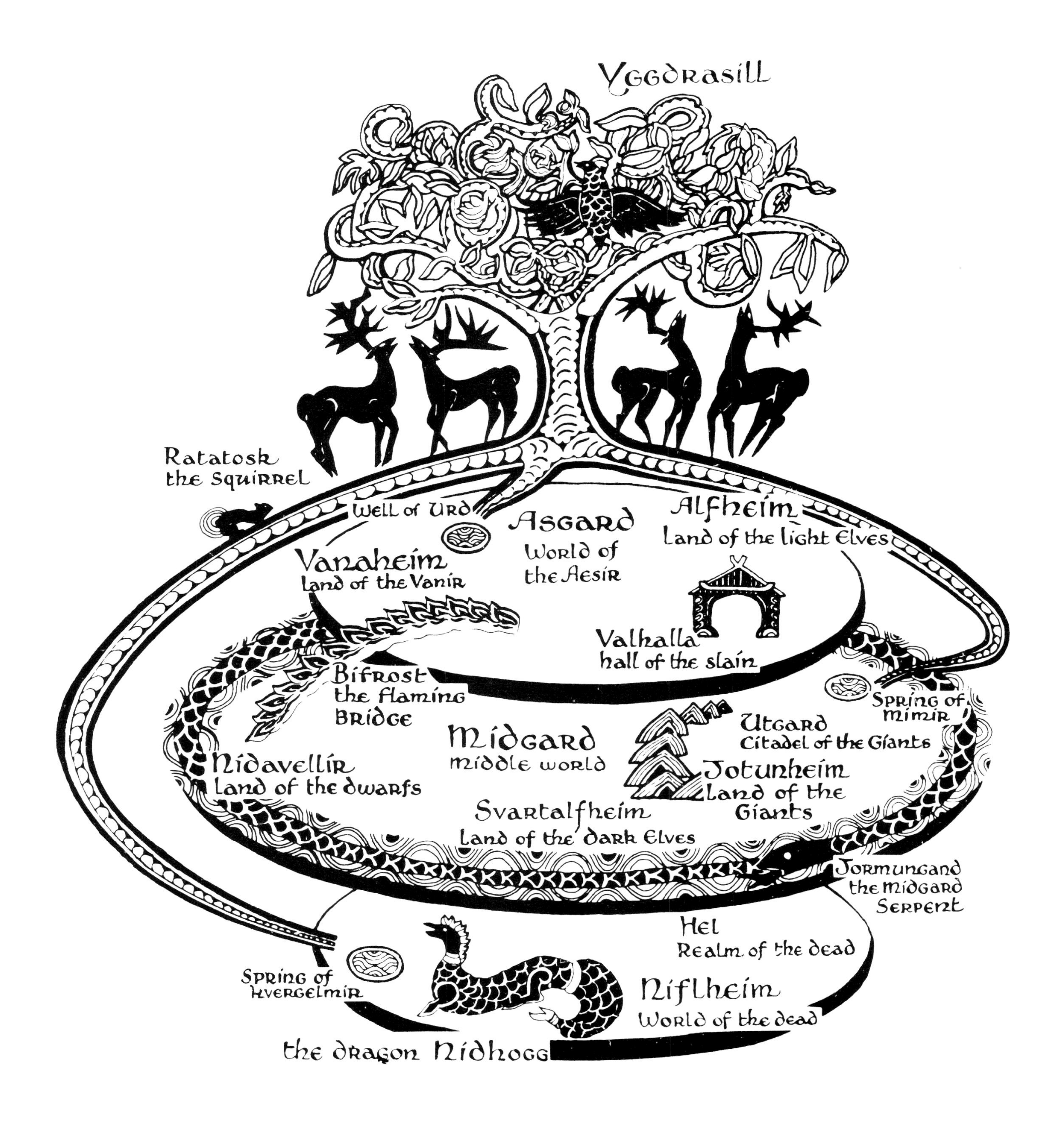

ABOVE *Kevin Crossley-Holland's map of the Viking cosmology does what all maps should do: it simplifies information. The three levels of life and death and their interconnections are captured rather beautifully. He points out that a familiar sight in Iceland, outside the few towns, is a lone farmhouse with a single tree growing right up against it. These trees "must be a 20th-century echo of the traditional guardian tree … the first and greatest of such trees was Yggdrasill."*

Asgard is the World of the Aesir, where the most powerful warrior gods and goddesses have their great halls. The greatest was Valhalla, Odin's "hall of the slain," and heaven of warriors. Vanaheim is the World of the Vanir, the home of fertility gods and goddesses. And the third world is Alfheim, the World of the Light Elves.

The second level is encircled by the monster Jormungand. This is

BILDSTEN
TJÄNGVIDE,
700

LEFT *An eighth-century Viking stele, with depictions of the Norse god Odin riding Sleipnir, his eight-legged horse, and Valkyries guarding the gates of Valhalla. The eight legs symbolize the directions of the compass.*

the world serpent who holds his tail in his jaws and sleeps upon the ocean-river bed. The Gods of Asgard descended to the worlds on this second level by passing over the fiery rainbow bridge, known as Bifrost, "the Trembling Roadway." This was the preferred route for Thor when he rode his goat-drawn chariot down to the land of mortals.

On this second level there were four separate worlds. The first was called Midgard, the World of Men. The second world was to be found in the eastern mountains of Jotunheim, the World of Giants, with its citadel of Utgard. The third and fourth were subterranean realms to the north and south of Midgard: Nidavellir is the World of the Dwarfs and Svartalfheim is the World of the Dark Elves.

On the third and deepest level are two final worlds: Hel and Niflheim. Both are Worlds of the Dead. The first world of the dead was Hel. It is a massively walled world with a great gate guarded by a demonic monster known as Garm the Hound. This realm of the dead was ruled by the hideous female monster, half black and half white, also called Hel. Evil men pass through the gates of Hel, and seem to die again, and travel into an even more terrible world of the dead. This place is called Niflheim. It is a misty land that is bitterly cold and endlessly dark. It is also the dwelling place of slithering serpents, and the evil dragon, Nidhogg.

In the cosmos of the Norsemen, even the gods do not live forever. The Gods knew that all the inhabitants of the Nine Worlds were eventually doomed. Fittingly, it would be a warrior's end. There will be one last mighty final battle between the Gods and the Giants. It would end in universal slaughter and a fiery conflagration that will consume all the Nine Worlds.

This was the day of doom the Norsemen called Ragnarok. It was foretold in an ancient prophecy. On that fatal day, the dark celestial wolf Skoll would devour the sun; and his brother Hati would devour the moon. Darkness would fall and the mountain would shake and the sea would surge upon the lands. Fenrir the Wolf would break loose from his chains and Jormungand the World Serpent would arise in wrath. All these with the Hell Hound Garm and the Dragon Nidhogg would join the Giant Legions in their war with the Gods. And in the end, all would be destroyed. But then too, so would Odin, Thor, and all other almighty Gods. None would survive but the demons of fire whose flames leapt across all the barriers between the worlds. The Nine Worlds would then become one vast inferno and all life be consumed in the blaze.

Can we relate this cosmological map to the physical world in any meaningful way? Firstly, the second level, Midgard, was surrounded by a vast and seemingly uncrossable ocean. If anyone knew the truth of that from the 9th to the 11th centuries, it was the greatest seafarers, the Vikings (though of course the Viking cosmology described here is far older). Secondly, Niflheim was specifically described as nine days' ride from Midgard, (just as the vast plain where the last battle is fought, Vigrid, stretches 120 leagues in every direction from Valhalla), suggesting that the Norsemen really did have a geographical, topographical sense of the three regions. And which direction would hell be? North of course; into darkness, into the wasteland.

RIGHT *The Old Stone Mill at Touro Park, Newport, RI; some claim the structure is of Norse origin, either a lighthouse or church. No English unit of measurement seems to have been used in its construction; tenuous grounds for such a claim, and an example of the innate desire to claim antiquity, longevity, in the "New World."*

Musurgia Universalis:

The Music of the Spheres

The Music of the Spheres is the oldest continuous intellectual tradition in the history of western civilization. As the Renaissance Jesuit scholar Athanasius Kircher wrote in his **Musurgia Universalis**: "The ancient philosophers assumed the world consisted of a perfect harmony, namely, from earth to the starry heavens is a perfect octave." This was no twilight, pagan superstition; it would exercise the minds of the Enlightenment. What is shown here is not therefore the oldest map: but rather, the oldest Western idea of a cosmological map.

In the Music of the Spheres the seven steps of the musical octave were believed to contain the structure of the universe in a single all-embracing synthesis of geometry, astronomy, mathematics, harmonics, philosophy, and aesthetics. It is a system that conceives of a universe governed by the laws of musical harmony.

The codification of the arithmetic basis of music by Pythagoras in Greece in the 6th century B.C. is considered the true beginning of the science. It was Pythagoras who discovered that exact numerical ratios determined the intervals of the musical scale within the structure of the octave (as in the"'Doh-ray-me" seven-tone scale.) It was also Pythagoras who saw the universe as a musical structure. It was by means of this musical-arithmetic structure of the octave scale that man found a precise means of measuring and exploring the universe. Later, the philosophers Plato and Socrates elaborated on what became known as Music of the Spheres.

In the Music of the Spheres, there were three types of music. The first type was *musica instrumentalis:* the mathematically precise means of measuring musical time, interval, tone, and pitch in musical instruments; including voice, percussion, wind, and string instruments. The second was *musica humana:* the rhythm of the human body's pulse and the breath of the human organism. The third was *musica mundana:* the belief in musical tones produced by the movement of the earth and the other planets as they travel through the universe.

Musica mundana is understood in terms of mathematical ratios and the annual cycle of the planets moving around the Sun as compared to the cycle of the planet Earth. Just as the rotation of the Earth around the Sun controls our daily and yearly rhythms, and the moon rotation around the earth controls the tides and our weekly and monthly rhythms, so the movement of other planets was believed to affect other aspects of our lives. These influences were the result of algorithms arrived at through mathematical calculations of movements within the solar system.

The ancient Greeks conceived of the Music of the Spheres through intense speculation on the nature of the cosmos. However, when Christianity became the official religion of the Roman Empire, the same system proved to be an

RIGHT AND INSET, ABOVE *Johannes Kepler (above) discovered the elliptical, rather than circular, orbits of the planets. Far from undermining the concept of celestial harmony, the equivalence he found between the changing velocities of the planets and precise musical intervals suggested a new, more subtle, music of the spheres. Johann Bayer's star atlas* Uranometria *(1603) was the first to illustrate the whole of the celestial sphere (right).*

ΟΥΔΕΙΣ ΕΙΣΙΤΩ
ΑΓΕΩΜΕΤΡΗΤΟΣ
AETERNITATI.
IOANNIS BAYERI
RHAINANI I. C.
VRANO-
METRIA,
OMNIVM ASTERISMORVM
CONTINENS SCHEMATA,
NOVA METHODO
DELINEATA,
AEREIS LAMINIS EXPRESSA.
MDCIII.
ATLANTI
VETVSTISS.
ASTRONOM.
MAGISTRO.
HERCVLI
VETVSTISS.
ASTRONOM.
DISCIPVLO.

extremely useful structural model for the status quo (now sanctified by God and the Church) within the new social religious orders that came into being throughout European history.

The concept of the Music of the Spheres not only survived the transition from pagan philosophy to Christian acceptance, it became one of the most powerful doctrines used to maintain the order and authority of the Church. In the 15th century the English jurist Sir John Fortescue saw in the Music of the Spheres an ancient theme of celestial harmonic order that duplicated the social and religious hierarchy of the medieval world.

> "In this order angel is set over angel, rank upon rank in the kingdom of heaven; man is set over man, beast over beast, bird over bird, and fish over fish, on the earth, in the air, and in the sea: so that there is no worm that crawls upon the ground, no bird that flies on high, no fish that swims in the depths, which the chain of this order does not bind in the most harmonious accord."

Throughout the Middle Ages and the Renaissance a key masterpiece of Latin prose was a fable called known as "Scipio's Dream" by Cicero. It was also the most famous instructive text to explain the cosmos in terms of the Music of the Spheres. In "Scipio's Dream," Cicero explains that the Roman hero Scipio Africanus is transported to the heavens where his ears are filled with the celestial, transcendent Music of the Spheres.

With a sense of wonder and delight, Scipio asks: "What is that sound, so loud and sweet, that fills my ears?" Unfortunately, the reply to this question comes from one of the world's most boring teachers:

> " It is the sound which, connected at spaces which are unequal but rationally divided in a particular ratio, is caused by the vibration and motion of the spheres … [blah… blah… blah …] to produce seven different sounds, a number that is the key to almost everything."

Two millennia later, a far more satisfying description of this vision of the Spheres was described by the metaphysical poet, Henry Vaughan: "I saw Eternity the other night. Like a great Ring of pure and endless light. All calm, as it was bright; and round beneath it, Time in hours, days, years driv'n by the Spheres."

Henry Vaughan was a near contemporary of Johannes Kepler, the greatest astronomer of his time and the author of the *Harmony of the Universe.* Kepler's refinement of this ancient system has been described as "a mathematician's Song of Songs." And yet his eloquent study of the solar system is a near duplication of the two-thousand-year-old Music of the Spheres, except that Kepler was careful to explain that this "music" is primarily meant as an intellectual concept. Kepler insisted that "the movement of the heavens are nothing except a certain everlasting polyphony, perceived by the intellect, not by the ear."

In his *Harmony of the Universe,* Kepler's "celestial music" is a metaphor for a deeper mathematical reality. After all, in any kind of harmony, the ear and the eye require the intellect to perceive perfect geometric shapes, precise numbers, sequential spaces or units of time. This was also the point the ancient Greeks wished to make in their conception of the Music of the Spheres.

In *The Republic,* Plato has Socrates describe the visible movement of celestial bodies through the heavens: "These intricate traceries in the sky are, no doubt, the loveliest and most perfect of material things." However, these traceries in the sky were considered irrelevant to the real concerns of the Music of the Spheres. In Socrates' view, they were "part of the visible world, and therefore they fall far short of the true realities—the real relative velocities, in a world of pure number and all perfect geometric figures … These can only be conceived by reason and thought, not seen by the eye."

We can see from the commentaries on the Music of the Spheres—from Socrates to Kepler—that this system was primarily developed to explore the elemental realities of the world. In the age of Bach and Newton, the mathematician Liebniz wrote: "Music is the hidden arithmetic exercise of the soul unconscious that it is calculating." Nothing could more precisely express the Pythagorean principles.

However naive we may think this doctrine of a musical universe to be now, this combination of music theory, mathematics, cosmology and philosophy was the basis of virtually every modern science. The Music of the Spheres conception of a harmonic cosmos led directly to the development of the mathematical languages of algebra, projective geometry, and calculus. And for over two thousand years, the Music of the Spheres was the most important and advanced system for the mapping out of our multi-dimensional universe.

RIGHT *The Constellation of Pisces by Johann Bayer (1572-1625), engraved by Alexander Mair (1559-ca 1616) for the* Uranometria, *hand-colored copper. Johann Bayer's preeminence in the history of astronomy is assured, as his use of Greek letters to label stars visible to the naked eye is still in use. The atlas consists of 51 maps. The 49th map displays 12 new southern constellations; and maps 50 and 51 show* Synopsis coeli superioris borea *and* Synopsis coeli inferioris austrina *respectively.*

羅定州城
西寧縣城
晉康司
封門所
羅鏡所
蟠龍汛
古榕汛
都城汛
羅旁汛
冷水汛
南江口汛
雙埇汛
通逕汛
思約汛
五里塘
封門塘
麻地汛
金鎖汛
封門口汛
小崗汛
大甘汛
東壩汛
東安頭塘
中伙汛
泗倫汛
大陂汛
林菁汛
白石汛
河口汛
楓木汛
旺久汛
河六汛
雲致汛
石橋頭塘
石根汛
函口汛
林崗汛
界牌汛
楓崗汛
良村汛
龍虎塘
金山逕汛
夾崗汛
牛路汛
雲廉汛
龍崗汛

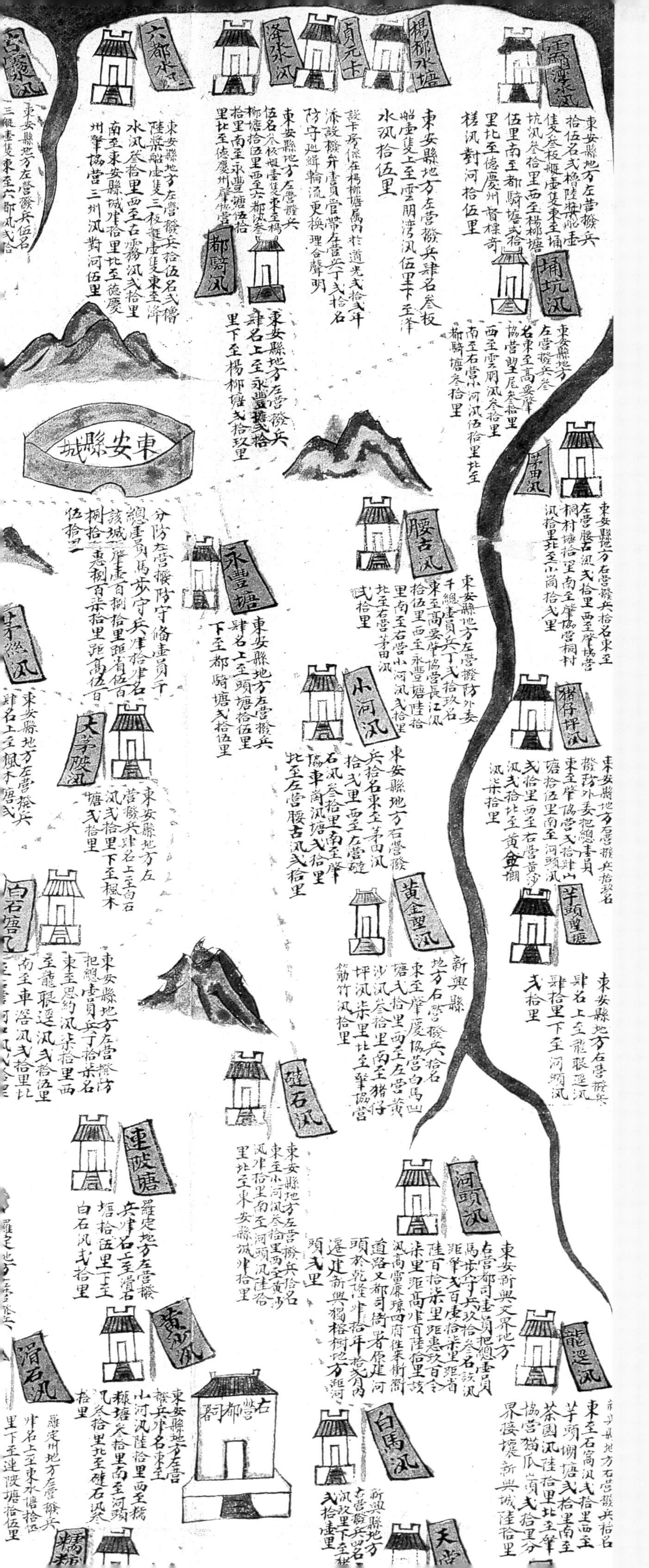

Cartographic Breakthroughs

From the Kwangtung Provincial Archives, a Chinese military map from the 1850s showing the coast and sea of Chin-chou from Ho-p'u to the Vietnamese border, the area garrisoned by the Lung-men regiment. In 1838 the Vietnamese successfully rebelled against occupying Chinese forces to become an independent nation—but only until 1858, when the French invaded to establish colonial control. By the mid-19th century, this looks like a primitive map, though it does include the military essentials: topography, unit location. It also serves as a reminder of one or two earlier Chinese contributions to cartography—pictograms, the compass, paper … add printing and the city plan with graticule to the list.

Mapping Heaven and Earth:

Ptolemy

Every author dreams of having a book printed and reprinted for a whole lifetime—just think of the royalty earnings. Consider a book that remained "in print" for around 1,600 years. That was the achievement of the **Guide to Geography** (*Geographike Hyphygesis* in Greek), written in A.D. 127-155 by the Egyptian-born Greek astronomer and geographer Ptolemy.

Claudius Ptolemaeus, usually known as Claudius Ptolemy, was probably born in Egypt in about A.D. 90, possibly in Ptolemais Hermii on the banks of the River Nile. Practically nothing is known about his life, except that he became Hellenized and worked in the Great Library in Alexandria, which by that time was coming under Roman domination. It was there that he produced his *Geography*, which consisted of eight volumes. Volume 1, perhaps the most significant, includes a map of the world and deals with locating places in terms of latitude and longitude. In this he drew on the work of earlier Greek writers such as Eratosthenes of Cyrene (c.276-194 B.C.) and Hipparchos (or Hipparchus) of Nicaea (c.180-125 B.C.), who had suggested using imaginary lines drawn on the surface of the Earth as references for giving locations. Ptolemy proposed that lines of latitude (running east-west) should be drawn parallel to the equator and calibrated in degrees, with 0 degrees at the equator rising to 90 degrees at the North Pole. He divided lines of longitude into 180 degrees east and west of the "prime meridian" (assigned 0 degrees), which he positioned in the Canary Islands, which were then known as the Fortunate Islands. Unfortunately, his location was 7 degrees too far east of their true position. Even so, any place on Earth could now be defined in terms of its latitude and longitude. He sited the places using information provided by merchants and Roman officials who traveled widely (not by using astronomical data); critics even of the time probably considered some of them merely travelers' tales.

To represent the curved surface of the Earth, Ptolemy used two main map projections. In the conic projection (obtained by projecting a spherical shape onto a cone) the lines of latitude take the form of concentric circular arcs. Lines of longitude become radii of these circles, fanning out from the North Pole. In the second, pseudoconic projection, lines of longitude are also circular arcs, rather than straight lines.

Volumes 2 to 7 of the *Geography* consist of tables of world places with their latitude and longitude coordinates. They encompass the whole of the known world, including Africa, Asia, and Europe. We are told that Volume 8 contained individual area maps, with four maps for Africa, twelve for Asia, and ten for Europe, probably drawing on the earlier work of Marinus of Tyre (fl.150 A.D.). Together they constituted an atlas of the known world, although none of these maps has survived. Ptolemy's world map may look strange to modern eyes, but it was good enough—along with some even more inaccurate, not to say preposterous, reasons—for Christopher Columbus (1451-1506) to decide that it ought to be possible to teach Asia by sailing westward from Europe (especially as Ptolemy's map had considerably telescoped the Europe-Asia distance). See page 98 for more on Columbus.

What kept Ptolemy's *Geography* top of the bestseller list for 16 centuries? The answer lies in the fact that it was not merely reproduced, but it was added to, or "improved," as the successive editors would claim. As long as Ptolemy's name remained prominent on the title page, they sold well. Of course, before the advent of printing, all copies had to be made by hand. But then it was easy to add. For example, in 1427 the Canon of Rheims, Guillaume Fillastre, had his scribe add a map of northern Europe that had been produced by Danish geographer Claudius Clavus. Twenty years later the Duke of the Italian town of Ferrara received a copy of the book that contained additional maps and a method of denoting physical features and boundaries. Contemporary maps of France, of Italy and of Spain, along with views of major

RIGHT *A gorgeous illumination of Ptolemy in one of the many editions of the* Geography *of the first half of the 15th century. He is accompanied on the right by the tools of his trade, and appears to be taking measurements with an astrolabe.*

RIGHT *This 15th-century example of the Ptolemaic world map employs the cloak-shaped* (chlamys) *projection. The lines of of latitude and longitude are numbered. The Indian Ocean is enclosed and there is no sea route around the Cape. The "inhabited" (Old) World is massively inflated.*

Mediterranean cities, were added in manuscript copies produced by the Florentine painter Piero del Massaio in 1469 and 1472.

Ptolemy's *Geography* really gained its preeminence during the rapid spread of printing in the 1450s. Printed editions appeared thick and fast, particularly in Italy: from Vicenza (1475), Rome (1478) and Florence (1480). In 1482 one of the most beautiful versions was published in Ulm, Germany, with colored woodcuts showing blue seas with yellow margins. The discovery of the Americas was acknowledged in the maps of editions produced after about 1508. The number of maps also increased, from 20 in Martin Waldseemüller's 1513 edition to 33 in that produced by Giacomo Gastadi in 1548. One of the "non-improvers" was Flemish cartographer Gerardus Mercator (1512-94), whose 1578 edition reproduced 27 of Ptolemy's own maps, together with an accurate version of the original text. It is probably the most authentic surviving edition. By 1730 more than 50 editions had been published.

Ptolemy also wrote a seminal book on optics, in which he took account of both physical and geometric principles. But his best-known work remains the *Almagest*, a book that probably was even more influential than—if not so quite long-lived as—his *Geography*. Ptolemy called his collection of astronomical works *Megale Syntaxis*, but it has come down the ages under its Arabic name, which means simply The Greatest. Central to the account in the *Almagest* is the theory that the Earth is the center of the Universe, as taught by Greek philosophers Plato and Aristotle and now called the Ptolemaic system. It required some tricky mathematics to explain why all the planets move in circles (they actually move in ellipses), and whereas the explanations for individual planets could be made to agree with observation, the system fell down when the whole of the Solar System was considered together. Even so, Ptolemy's reputation was beyond reproach and this is probably why the Sun-centered Universe remained entrenched for so long, until overturned in the 15th century by Polish astronomer Nicolas Copernicus (1473-1543) and his followers such as Galileo Galilei (1564-1642); his reputation, allied to the immense theological bouleversement that a non-Earth-centered Universe represented.

In his fourth major work, the *Tetrabiblos* (Four Books), Ptolemy gathered together information about the Earth in the *Guide to Geography* and about the heavens in the *Almagest* to produce a treatise on astrology, a legitimate area of scholarship at the time.

SEPTEM
MARE INDICUM
SINUS BARBARICUS
Terra incognita secundum Tholomeum
SCHITHIA
ASIA MINOR
GERMANIA
ARABIA FELIX
Fora Regio

All roads Lead to Rome:

The Peutinger Table

Although not the oldest, most detailed, or most accurate map, the so-called Peutinger Table is certainly one of the longest. Painted in color on a roll of parchment nearly 23 feet (7 meters) long, it is a traveler's map that indicates landmarks and distances along all the major roads that branched out from Rome. The Peutinger map is also in a sense one of the most unreliable of all types of maps: a copy of a copy of a copy.

The original was made in the second half of the 1st century B.C. by a team directed by Roman general and statesman Marcus Agrippa (c.63-12 B.C.). They surveyed the Roman Empire's major roads from Britain and Spain in the west, across Europe and the Middle East as far as India (beyond the Empire). They carved the finished map, which located Rome at the center, onto slabs of marble that were erected near the Forum in Rome.

The map was copied and updated over the years, and in the 3rd century a monk in Colmar (now Alsace) made a copy on 12 sheets of parchment. In 1494 Conrad Pickel, librarian to Holy Roman Emperor Maximillian I, discovered this copy in a Benedictine monastery in Tegernsee, Bavaria. Pickel, or Conradus Celtius (1459-1508) as he preferred to be known, was a German humanist and Latin poet. He was created the first German poet laureate in 1487 and championed the study of German antiquities. When he died, he left the map in his will to his friend Conrad Peutinger (1467-1547), who was an official administrator, antiquary, and coin collector in Augsberg. When he in turn died, the map was part of his library and eventually found its way into the State Library in Vienna, where it was referred to as the *Tabula Peutingeriana,* or the Peutinger Table.

The Peutinger map is very long, 22 feet 4 inches (6.8 meters) and narrow, 13. 4 inches (34 centimeters). Originally drawn on 12 sheets, 11 of them have survived in the Vienna Library. On the map roads are represented as more-or-less straight lines running from one end of the sheet to the other, joining named towns. It is not drawn to scale; the distances between towns are given at the sides of the road, so that 15 miles (25 kilometers) is represented as no longer than 3 miles (5 kilometers) to scale. But this does not matter to the traveler, who merely wants to know how to get from A to B. Prominent landmarks—again important reference points for travelers—are also indicated alongside the roads. For example a castle, church, lighthouse, tower, or grove of trees has its own characteristic symbol. Towns are represented as collections of buildings. One of the most traveled sections was the *Via Francigena,* which was the main communications artery between central Europe and Italy, followed each year by thousands of pilgrims going to and from the Holy City of Rome. A recent Dutch study checked some of the distances on the Peutinger map and found that, where the route is unchanged, they are essentially correct.

Similar linear, or strip, maps were produced in more recent times. In 1675 John Ogilby (1600-76) produced a strip map of England's roads drawn to scale, with up to six strips to a page. Roadside features—mainly churches and windmills—were included as landmarks to aid the traveler. In Britain in the 1930s the Great Western Railway published maps of its most popular routes, such as the line from Paddington (London) to Penzance (Cornwall). In a slim book of rectangular pages bound at the top, like a stenographer's notebook, the line ran vertically with landmarks that could be seen from the train on either side. The travelers turned over the pages one way for the outward journey, and reversed the procedure for the trip back. The Automobile Association produced similar maps for its members on request. The AA maps also showed the required route from top to bottom, with road numbers and distances indicated. A similar presentation is used today on an in-car satellite position indicating system. The Peutinger map is the forefather of all these eminently practical approaches to traveler's cartography.

RIGHT *The Peutinger map appears to have been mostly created in the 4th century, with emendations from as late as the 16th century. East-west is the dominant axis, as it always was for ancient "Mediterranean," Western civilization.*

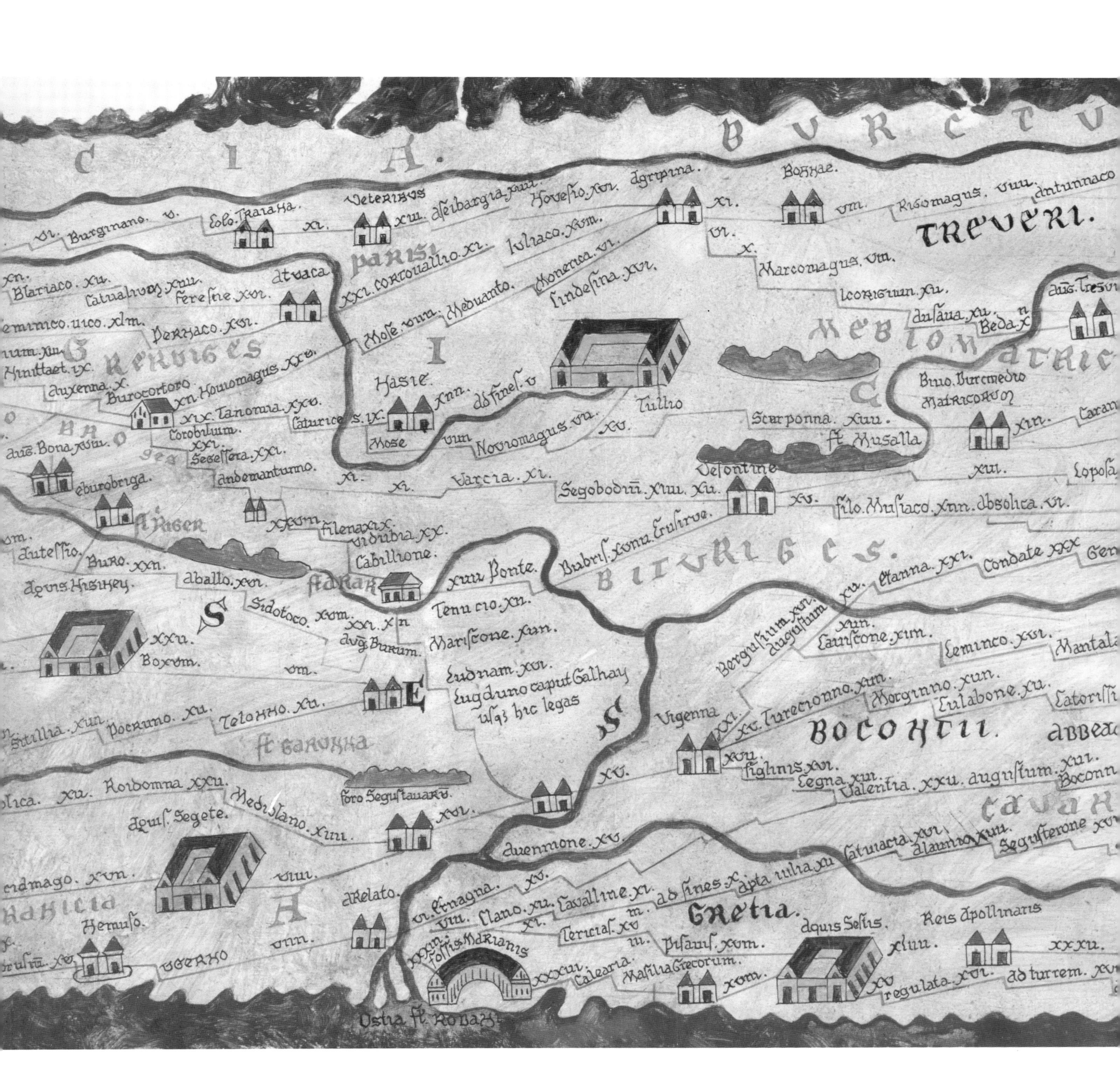

TREVERI.
Marcomagus. viii.
Agripina.
Bonnae.
Risomagus. viii.
Antunnaco
Veteribus
Beda.
Tullio
Scarponna. xiiii.
Vesontine
Segobodiu. xiiii. xii.
Cabillione.
Filenaxxx.
Vidubia. xx.
Tenu cio. xii.
Mariscone. xiiii.
Vigenna
Valentia. xxii. augustum. xiii.
Avennone. xv.
Arelato.
Fossis Marianis
Calcaria.
Masilia Grecorum.
Aquis Sestis.
Tericias. xv
Cavallinc. xi.
Ad fines. x.
Apta iulia. xii
Ad turrem.
Regulata. xvi.
Aquis Segete.
Foro Segustavarû.
Mediolano. xiiii.
Roidomna xxii.
Telonno. xii.
Pocrino. xii.
Sidoloco. xviii.
Aballo. xvi.
Autessio.
Eburobriga.
Andemantunno.
Burocortoro
Blariaco. xii.
Augustonemeto
Ostia fl. Rodani

Some of Us Are Looking at the Stars:

Celestial Charts

Some of the most beautiful and technically complex maps to emerge during the heyday of cartography in early modern Europe had as their subject not terrestrial features, but the heavens. Particularly since the invention of the telescope in around 1610, both the scientific and the popular imagination had been captivated by astronomy, creating a ready market for works that graphically displayed and summarized the current state of knowledge. This form of map reached its zenith with Andreas Cellarius' magnificent **Harmonia Macrocosmica** of 1660.

The star atlas that set the standard for others to follow in the 17th century was the *Uranometria* (see page 32) produced by the Bavarian lawyer and publisher Johann Bayer (1572-1625) in 1603. Bayer's seminal work was based on the star catalog of the Danish astronomer Tycho Brahe (1546-1601). In this period before the arrival of the telescope, the calculations for the projections used in these charts were made using such instruments as the astrolabe and the quadrant. The *Uranometria* was a key scientific work, intended primarily for use by astronomers to follow and chart the motions of planets, the moon, and other heavenly bodies; it also introduced the system of identifying individual stars in a constellation, ranked according to their brightness, by letters of the Greek alphabet (e.g. Alpha Centauri). Yet for all its functionality, Bayer's Atlas was also an object of great beauty, presenting its information in 51 skillfully engraved plates.

Celestial maps are mainly familiar nowadays in the form of glossy reproductions as *objets d'art* on calendars. Their highly graphic execution has instant appeal—especially the depiction of each constellation not simply as a pattern of individual stars but also in the shape of an elaborate image of the animal or mythological figure for which it is named. Such vignettes of Cancer, Ursa Major, or Taurus may strike the modern observer as quaint and arcane, but this is an anachronistic reaction. All star charts of this period, whether produced by commercial cartographers or by established scientists such as John Flamsteed, England's first Astronomer Royal at the Greenwich Observatory (established 1675), portray the constellations in this way. A very different mindset informed humanity's view of celestial bodies at that time, and found its expression in contemporary cartography.

The heavens in all their diverse aspects played a far more prominent role in the lives of all people in 17th-century Europe than today, irrespective of their social standing. Ordinary citizens anxiously consulted horoscopes to try and foretell their fate. Sales of almanacs that purported to predict the future on the basis of zodiacal data boomed; during the devastating English Civil War (1642-51), fully one-third of the country's populace is thought to have regularly bought such publications. The dread of events such as solar eclipses or the passage of comets was still widespread. Rulers and other prominent figures were equally convinced of the influence of planetary powers over their lives. Albrecht von Wallenstein, the most successful general of the Thirty Years' War (1618-48) took an astrologer on campaign with him to advise on favorable moments to undertake military action. At the Prague court of the Holy Roman Emperor Rudolf II (1552-1612), the astronomers Tycho Brahe and Johannes Kepler were, successively, retained not just as mathematicians but were also required to prognosticate on the likely course of future events. Records show that Kepler, for instance, advised Rudolf on the outcome of a conflict between Papal Forces and the Republic of Venice, and also drafted hundreds of horoscopes. Astronomy and astrology were by no means the mutually exclusive pursuits that they became after the rise of the empirical sciences.

The growing acceptance of the heliocentric model of the universe—the view, first put forward in the late 16th century by

RIGHT *Cellarius' engraving of an armillary sphere, from the 1708 edition of the* Harmonia Macrocosmica. *Already outmoded by the time Cellarius published his famous work, the geocentric model of the universe shown here, in which all the planets orbit around the Earth, dated back to Aristotle. Typical of Cellarius' comprehensive treatment is his inclusion, bottom left and bottom right insets, of Ptolemy's and Tycho Brahe's cosmological systems.*

ORBIUM PLANETARUM TERRAM COMPLECTENTIUM SCENOGRAPHIA.
SPHÆRA ZODIACI
SPHÆRA SATVRNI
SPHÆRA IOVIS
SPHÆRA MARTIS
SPHÆRA SOLIS
SPHÆRA VENERIS
SPHÆRA MERCVRII
SPHÆRA LVNÆ
CIRCVLVS ÆQVINOCTIALIS
HYPOTHESIS PTOLEMAICA, in qua Terra totius Universi centrum.
HYPOTHESIS BRAHEA, in qua centrum Lunæ et Firmamenti est Terra, reliquorum quinq. Planetarum Sol.
FIRMAMENTUM STELLARUM FIXARUM SEDES.

Nicolaus Copernicus, that all the planets, including the Earth, orbit around the Sun—became a keystone of the new rational, empirical science. Famously, Galileo was put on trial and forced by the Inquisition to recant his support for Copernicus' position. Yet even this insight did not bring a clean break with astrology. No less a figure than Isaac Newton (often held up as the very epitome of the Age of Reason) claimed that, since astrology dealt with the relationships between planets, it was irrelevant what heavenly body lay at the center of the system, and it therefore remained a valid subject of inquiry.

It is with this background in mind that we should approach the most celebrated celestial atlas of the 17th century, the *Atlas Coelestis* (1660) of the German mapmaker Andreas Cellarius (c.1596-1665), who spent most of his life in the Netherlands. Also known as the *Harmonia Macrocosmica,* this monumental compendium of astronomical knowledge comprised 29 double-folio hand-colored plates, plus over 200 pages of Latin commentary (organized into essays of four to nine pages following each plate). The lavish illustrations in Cellarius' celestial atlas cover a wide range of topics, including the orbits of the planets around the Earth, planispheres (projections of spheres onto flat surfaces) of various astronomers' different systems, a diagram of lunar phases, the constellations of both the northern and southern hemispheres, and the relative sizes of the celestial bodies. Intriguingly, it differs qualitatively from Bayer's earlier work in presenting a complete survey of astronomical systems down the ages, as opposed to only the current mode of thought. Thus, Cellarius presents the reader with the 2nd-century Ptolemy's geocentric view of the universe and his ingenious, fallacious, hypothesis concerning planetary motion, alongside Brahe's and Copernicus' more modern systems. Similarly, different maps show both the classical (and still accepted) form of the constellations and a Christian conception of them, with mythological figures supplanted by Biblical ones, which had been advanced by the Augustinian friar Julius Schiller in 1627.

For all the undoubted accuracy of its mapping of stars and other celestial bodies, certain aspects of the *Harmonia Macrocosmica* are decidedly unscientific, and appear to be aimed at a wider readership than just the astronomical community. For example, on the plate depicting the sky of the Southern Hemisphere the Pacific and Antarctic regions of the Earth are pictured, as though from a distant point in space, beneath the constellations. This imaginative use of perspective certainly enhances the aesthetic appeal of the map, but it is also so confusing as to render it useless to an astronomer. The popular appeal of the work was great, and it was reissued in 1708 minus the weighty commentaries. We can reasonably regard the overview offered by Cellarius' atlas as itself embodying a new *Zeitgeist.* It seems to stand beyond the polemical disputes of half a century before, and rather to celebrate the sheer diversity of conflicting views of the heavens.

The dedicatee of the first edition of the atlas was the recently crowned English king, Charles II. Under his reign, in 1662, the Royal Society was founded in London, the first of a number of similar institutions across Europe that officially promoted the new experimental method of scientific enquiry and the dynamic exchange of ideas. Cellarius' famous work seems very much in accord with the spirit of the coming age.

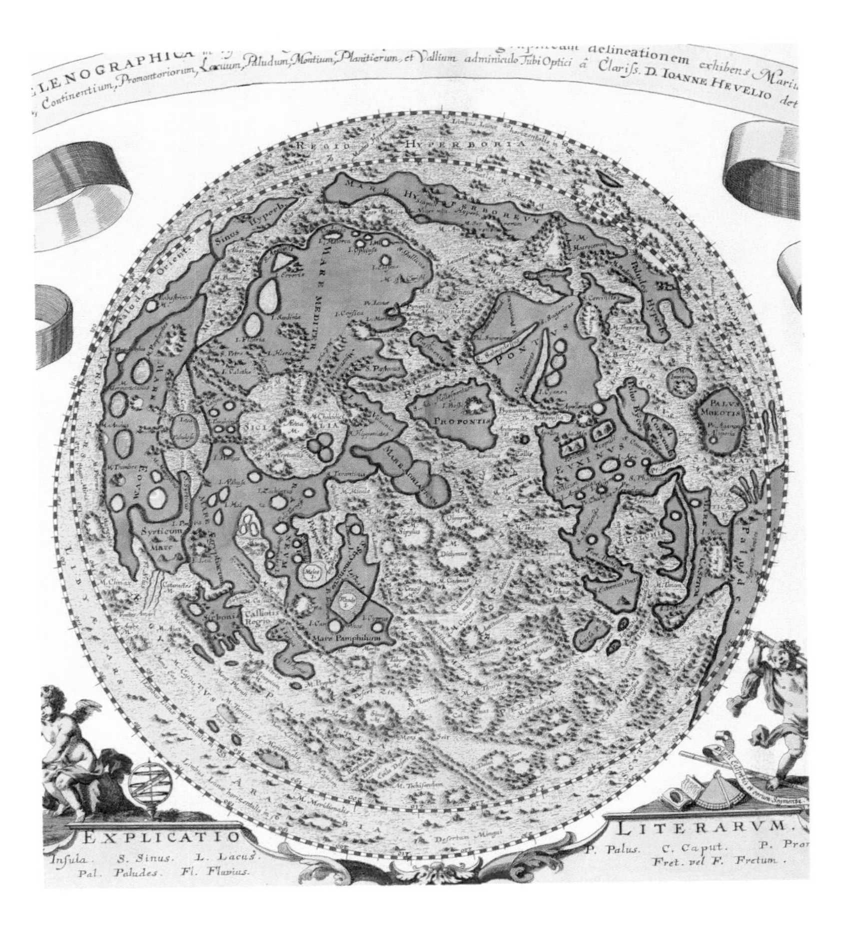

ABOVE *Johannes Hevel (Hevelius) produced this "selenographia," or map of the moon, first published in 1647. He built his observatory in Danzig on the upper floors of four houses, and installed a 130-foot focal length telescope. His ten years of lunar observations led him to estimate the height of lunar mountains, and more accurately, produce diagrams of lunar phases.*

RIGHT *Celestial mapping was a vital part of Islamic science. The twelve signs of the zodiac represent an unbroken body of knowledge that passed from the East to the Greeks. There were 48 ancient constellations, but post-Copernicus, many "modern" constellations were quickly described.*

Mapping the Nation-State:

Saxton's Elizabethan England

The publication of early national atlases in England and elsewhere was inextricably bound up with power politics. For example, John Speed's lavish Atlas of the British Isles–The Theatre of the Empire of Great Britaine (1611-12), which contained maps of England, Scotland, and Ireland, was a graphic embodiment of the monarch Charles I's desire to reign supreme over a united kingdom. No less politically inspired was the work of Speed's English predecessor Christopher Saxton, whose seminal 1579 Atlas was the first such work to appear in Europe.

It is thought that Christopher Saxton (ca1543-1610) received an early training as a cartographer and draughtsman from John Rudd, the vicar of Dewsbury in Yorkshire, in the north of England, near Saxton's birthplace. Records certainly show him to be in Rudd's employ as a surveyor by 1570. This profession would likely have provided Saxton with a steady, indeed probably extremely comfortable living at this time. In the wake of Henry VIII's dissolution of the monasteries in the late 1530s, following his split with the Church of Rome, extensive ecclesiastical properties were parceled out for sale to private buyers, and disputes were constantly arising over the rights to land enclosures, making work for surveyors and agents.

By around 1573, Saxton had moved to London, where his work came to the attention of Thomas Seckford, Master of the Court of Requests, whose master William Cecil, Lord Burghley, as longstanding Chief Secretary of State to Queen Elizabeth I, occupied one of the most powerful posts in England. Seckford commissioned Saxton for the huge task of surveying and mapping all the counties of England and Wales. This undertaking was both time-consuming and expensive, with Seckford meeting the considerable finances. In all, Saxton was engaged on the task for a total of five years.

Clearly, this commission was regarded as a matter of major national importance. Seckford's political superiors were in no doubt about the vital significance of mapping; Burghley is documented as having himself sketched maps of areas of tension and potential future conflict, such as the borders with Scotland under its troublesome Catholic monarch Mary. Elizabeth's reign had been under grave threat since her accession in 1558, and fears were rife of Popish plots for invasion or insurrection. It is difficult for the non-historian to comprehend the desperate *realpolitik* that characterized the reign of the Virgin Queen, simply because it lasted so long! That longevity suggests a stability that never was. It was not for reasons of national prestige and pride that the mapping of Britain was embarked upon, but primarily from strategic considerations, to reveal such key information as possible landing sites that a seaborne invasion force might use. With this in mind, the cartographer was instructed to begin his work in the east and south of England, the shores that faced England's most dangerous foes (most notably Spain, which would indeed attempt to overrun England with their formidable Armada, only nine years after publication of the Atlas).

The first county map to be completed was that of Norfolk, in southeast England, in 1574. It displayed the general accuracy that characterized Saxton's work throughout the entire series. However, what is shown in detail and what is omitted is significant.

Key rivers and other bodies of water are depicted, as are large and discrete areas of vegetation, while major towns are shown by groups of buildings, with notable landmarks accurately portrayed. Mountains and hills are also indicated pictorially; although precise or relative altitudes are not given. This landscape relief was nevertheless a vital and telling feature, as these high points had been designated by Elizabeth's strategists as the site of a chain of warning beacons to raise the alarm in the event of invasion.

We may also surmise that the absence of roads suggests that Saxton was under orders to survey and map only the most salient features, so as to ensure the quickest possible publication of the Atlas. Comprising fully 34 double-page maps of individual counties or groups of counties, plus a general map of England and Wales, the work was achieved in a remarkably short time. Just five seasons of surveying in total meant that he could only devote, on average, less than a month to

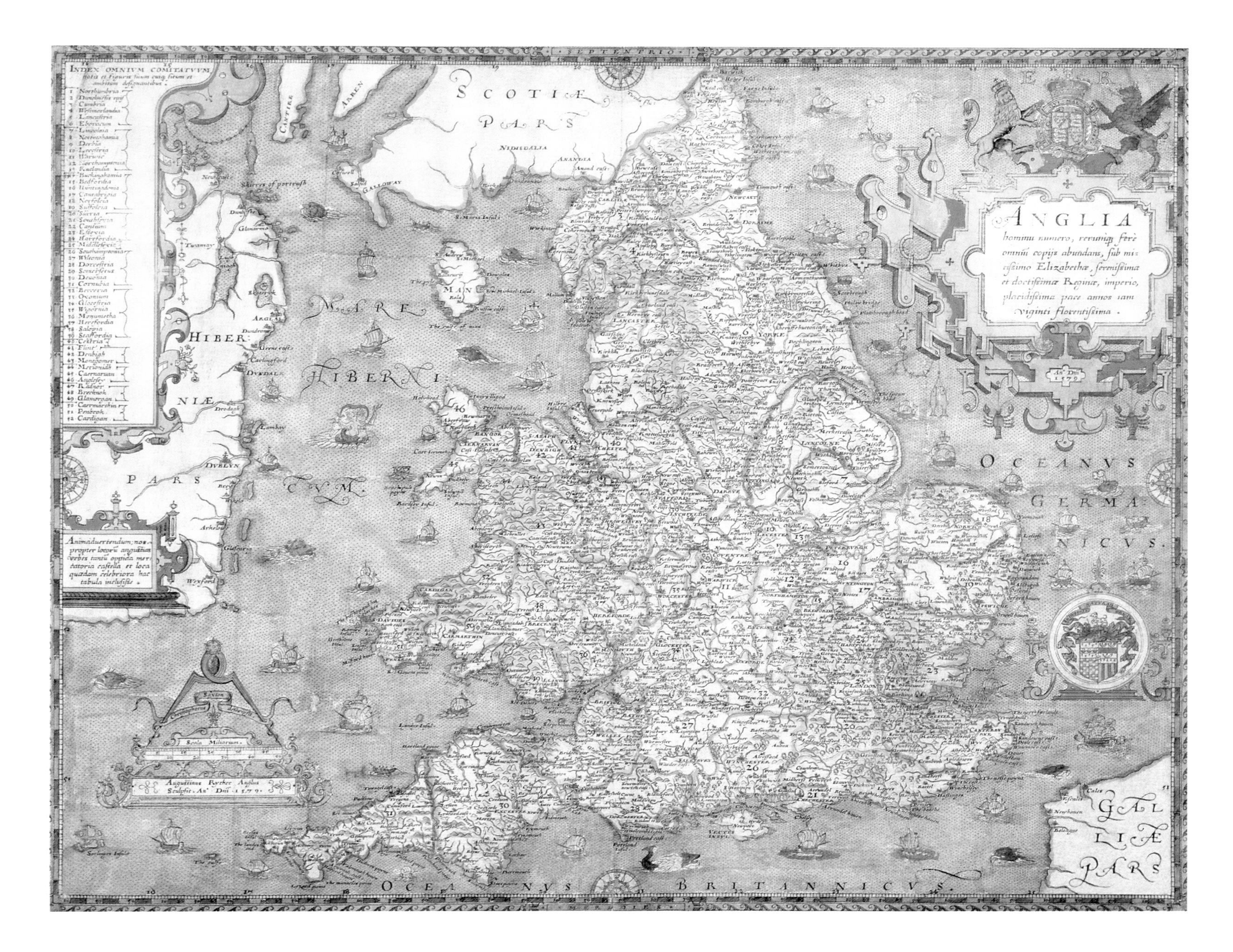

each county! Moreover, nowhere in the scant documentation on Saxton is there any mention of his being helped by assistants. Rather, he simply relied on the local populace of each county to aid him in his task. To this end, he carried with him a letter from the Privy Council, requiring local officials "to see him conducted unto any tower, castle, high place or hill to view that country … accompanied with two or three honest men such as do best know that country." Little is known about his exact method of surveying, but he is thought to have used a rudimentary form of theodolite and worked by triangulation.

Even so, elements that entailed no extra work on the part of the cartographer—such as galleons on the sea, fishes and the occasional marine monster, and highly flamboyant cartouches containing the title of each map—do appear, indicating that Saxton's creations were not intended, at least by him, to be purely functional. These decorative flourishes doubtless contributed to their later widespread popularity.

The highest official backing for Saxton's undertaking is clear from the frontispiece of the published Atlas—a portrait of Elizabeth I. She is

ABOVE *Saxton's map of England and Wales published by Augustin Ryther in 1579. The list of counties on the left uses latinized names—"Northamptonia," "Buckinghamia"—and surely there is some satisfaction to be gained by the inhabitants of those wild northern climes that "Cumbria" and "Northumbria" are still current terms. The names elsewhere are not Latinized: this a map of a nation, not a northern island of the world, and employs its own language.*

RIGHT *John Speed's atlas,* The Theatre of the Empire of Great Britaine *was also a county atlas, first published in its entirety (including Ireland and Scotland) in 1612. This extravagantly decorated version of "Britannia" is from the large-format Basset and Chiswell edition of 1676. The cartography owes a huge debt to Saxton. The genealogist and historian Speed produced his atlas to complement his enormous* History Of Great Britaine, *published around 1600. The (almost subconscious) aim of such cartography is in part to establish the historic "reality" of the country, its venerable antiquity.*

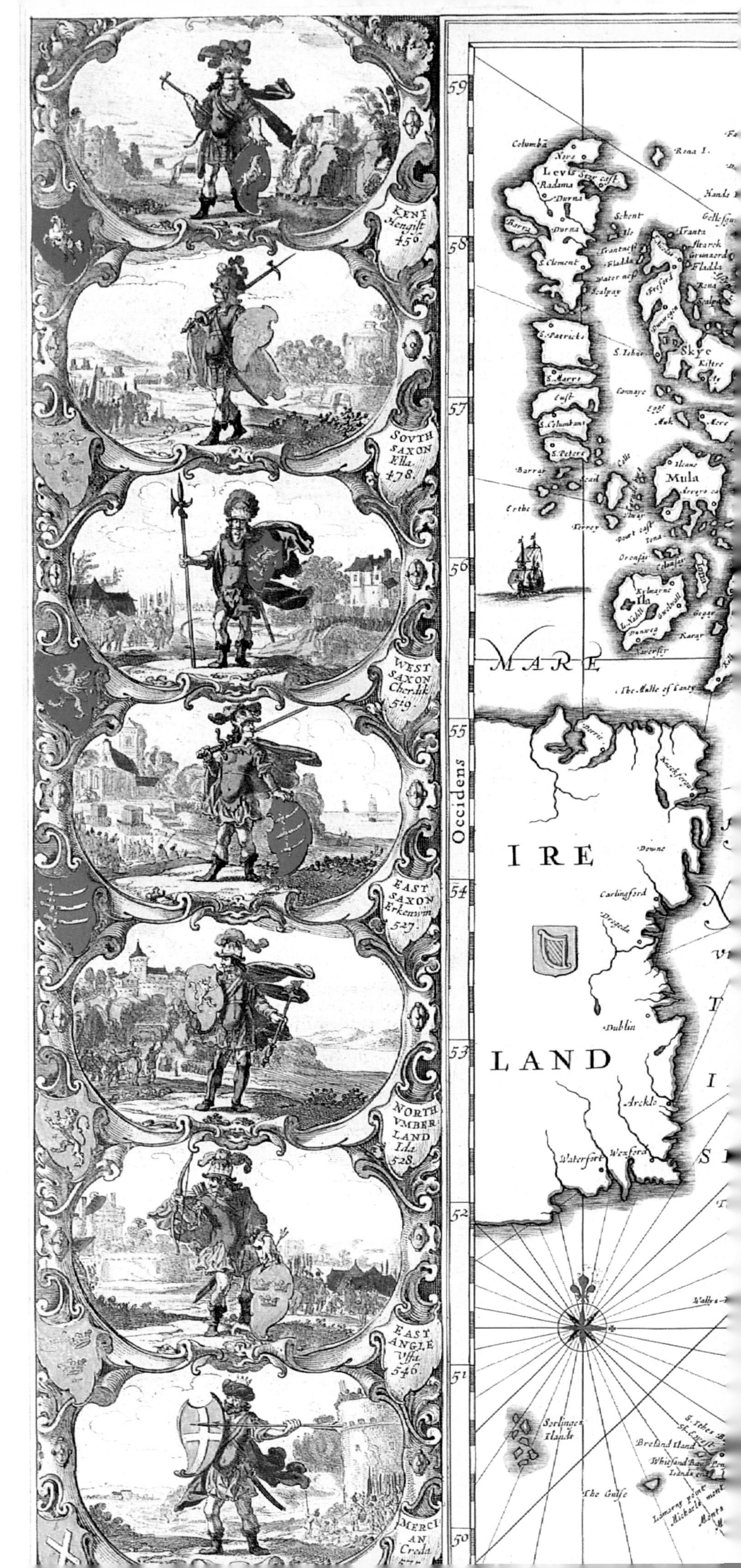

shown seated as the patron of the sciences of geography and astronomy, represented by the two bearded figures that flank her. England's redoubtable Virgin Queen is also the dedicatee of the work. She showed her gratitude to Saxton for his diligence, even while the survey was still in progress, by granting him the lease of Grigston Manor in Suffolk in 1574: " … for certain good causes, grand charges and expenses lately had, and sustained, in the survey of divers parts of England." Further evidence of the keen interest that the country's ruling elite took in the project comes from the papers Lord Burghley left behind (now in the possession of the British Museum). As each map plate was engraved on copper plates from Saxton's drawings—by leading English and Dutch engravers of the age, such as Augustine Ryther, Remigius Hogenberg, and Cornelis de Hooghe—Burghley gathered together proof copies carefully in a special book, bound with other documents relating to the Atlas. As a final accolade, the Queen granted the mapmaker the exclusive rights to publication of his maps for a ten-year period from 1577 onward.

Saxton's achievement in being the first person accurately to render all the constituent divisions of a nation-state can scarcely be overstated. His maps were so well-regarded that they remained the basis of all subsequent county maps for over one hundred years. His original engravings were adapted and reissued constantly until the end of the 18th century. The modern English novelist and cultural observer Peter Ackroyd conveys vividly the great impact that these strikingly beautiful, novel artifacts must have had on those who first saw them, in his marvelous book *Albion – Origins of the English Imagination*:

> "Saxton's county maps … provided the first complete set of visual images [of the country of England], as fresh and illuminating to his first audience as photographs of the outer universe to a more recent generation … the four thousand names inscribed upon Saxton's wall-map of England are a holy litany … It is a highly localized vision like that of Blake or Langland."

Septentrio
OCEANVS GERMANICVS,
BRITANNIA
prout divisa fuit temporibus
ANGLO-SAXONVM,
præsertim durante illorum
HEPTARCHIA.
Anglis
THE GERMAN SEA.
THE BRITISH OCEAN
GALLIA
Oriens
KINGDOM OF THE PICTS
THE SCOTS
WALES
THE KINGDOME OF MERCIA
EAST ANGLES KINGDOME
EAST SAXONS KING DOME
THE KINGDOME OF WEST SAXONS
THE SOUTH SAXONS KINGDOME
THE KINGDOM OF KENT
KINGDOME OF NORTHUMBERLAND
THE PICTES WAL
KENT Ethelbert 595
EAST SAXON Sebert 604
EAST ANGLE Erpenwald 624
NORTH VMBER: LAND Edwin 627
WEST SAXON Kengils 635
MERCIA Peada 650
SOUTH SAXON
59
58
57
56
55
54
53
52
51
50

Comets and Compasses:

Edmond Halley

In any name association test, "Newton" would almost certainly elicit the response "gravity," just as the name "Halley" would invariably lead to "comet." But there was a lot more to Edmond Halley than the comet that bears his name. Not least of his achievements was a map of the Atlantic Ocean that showed for the first time how the magnetic variation–the difference between true north and the direction indicated by a compass–itself varies from place to place.

Edmond (sometimes spelled Edmund) Halley was born at Haggerton in Shoreditch, near London, in 1656, the son of a wealthy soap-maker. He was educated privately at home and then at St Paul's school, before going to Queen's College, Oxford, in 1673. While he was still at school he measured the magnetic variation in London to be 2° 30' (magnetic variation is the difference between true north and the direction indicated by a compass; it varies from year to year as the position of the north magnetic pole moves gradually around true north). As a student he made many astronomical observations, using various expensive instruments bought by his father, finding a new way of calculating the orbits of planets. At the age of 19 Halley published three papers on sunspots, planetary orbits, and the occultation of Mars.

Halley left Oxford in 1676 without a degree and, encouraged by the Astronomer Royal John Falmsteed (1646-1719), went to the island of St Helena in the southern Atlantic Ocean—then Britain's southernmost possession—to plot the positions of the stars of the southern hemisphere, which hitherto had remained uncharted. After 18 month's work in poor viewing conditions (because of the cloudy weather) he listed the coordinates of 341 stars in his *Catalogus Stellarum Australium* of 1679, and with its publication his reputation was made. Flamsteed christened him the "Southern Tycho," reeferring to the celebrated Danish astronomer Tycho Brahe (1546-1601). Halley was saved further undergraduate drudgery by being granted his M.A. by royal mandate of King Charles II. He was elected a Fellow of the Royal Society when he was still only 22 years old.

After traveling around Europe for two years and visiting the Paris Observatory, where he and Giovanni Cassini (1625-1712) observed a new comet, he took a house in Islington in 1682 and began a series of lunar observations to help with the problem of finding longitude at sea.

He consulted Robert Hooke and Christopher Wren about problems with gravity, but receiving no satisfactory answers decided to consult the great man himself and went to Cambridge to talk to Isaac Newton. The meeting led to a firm and lasting friendship between the two men, which was later (1687) to be instrumental in the publication of Newton's *Principia*, which Halley largely financed and edited. He was clerk of the Royal Society by then and had inherited a lot of money in 1684 after his father was found dead, presumably murdered.

In 1686 Halley published a map that showed the directions of the prevailing winds across the world's oceans, using data obtained mainly from the reports of returning sea captains. It was the first large-scale meteorological map to be published. In return for his earlier help, in 1696 Newton secured Halley the lucrative job of Deputy Comptroller of Chester Mint; Newton himself was at this time Warden of the Royal Mint in London.

In 1698 King William III gave Halley command of a Royal Navy warship with the unlikely name of the *Paramour*, a pink, which was a small, square-rigged ship with an overhanging stern. He spent most of the next few years at sea.

William had asked him to try to discover what land lies to the south of the Western Ocean. He voyaged extensively in the Atlantic as far as Barbados, plotting the magnetic variation, which he hoped could be

RIGHT *Halley's 1700 chart of the western Atlantic and the Americas, showing lines of magnetic variation. The journey Halley undertook to make these measurements was arguably the first purely scientific sea voyage. He would later produce a map with the same information including Europe, Africa, and Asia. The note in the center of this map admits "I durst not presume to describe the like curves in the South Sea wanting accounts thereof."*

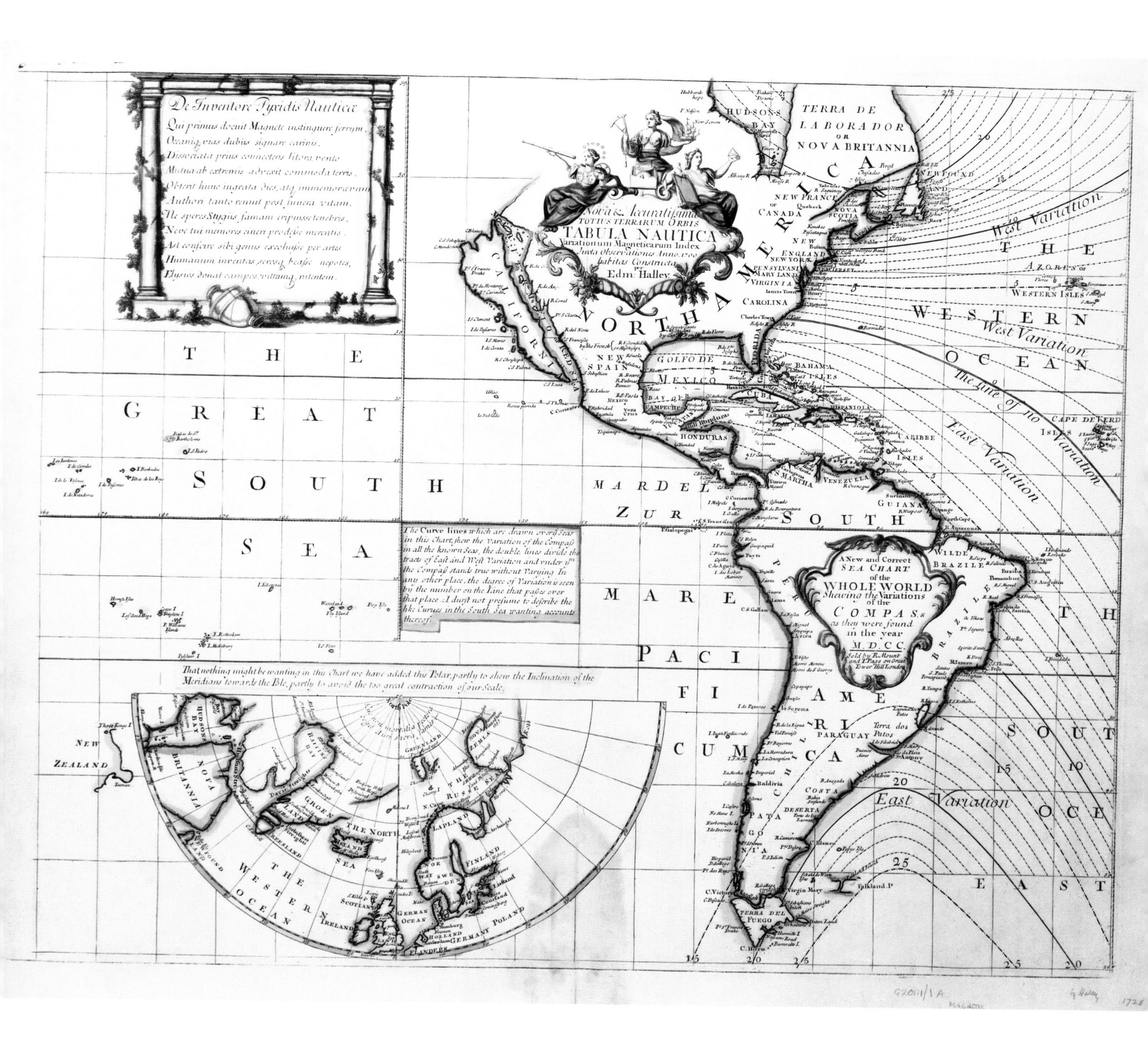
De Inventore Pyxidis Nauticæ
Qui primus docuit Magnete instinguere ferrum,
Oceaniq vias dubiis signare carinis,
Dissociata prius connectens litora, vento
Mutua ab extremis advexit commoda terris.
Obtexit hunc ingrata dies, atq immemor ærum
Authori tanto renuit post funera vitam.
Ne speres Stygiis famam eripuisse tenebris,
Neve tui memores cineri prodesse merentis.
Ast conferre sibi genus excoluisse per artes
Humanum inventas, serosq beasse nepotes,
Elysios donat campos, vittamq nitentem.
Nova & Accuratissima
TOTIUS TERRARUM ORBIS
TABULA NAUTICA
Variationum Magneticarum Index
Juxta Observationes Anno 1700
habitas Constructa
per
Edm: Halley
A New and Correct
SEA CHART
of the
WHOLE WORLD
Shewing the Variations
of the
COMPAS.s
as they were found
in the year
M.D.CC.
Sold by R. Mount
and T. Page on Great
Tower Hill London
The Curve lines which are drawn over ye Seas in this Chart, shew the Variation of the Compass in all the known Seas, the double lines divide the tracts of East and West Variation and under ym the Compass stands true without Varying. In any other place, the degree of Variation is seen by the number on the Line that passes over that place: I durst not presume to describe the like Curves in the South Sea wanting accounts thereof.
That nothing might be wanting in this Chart we have added this Polar, partly to shew the Inclination of the Meridians towards the Pole, partly to avoid the too great contraction of our Scale.
THE GREAT SOUTH SEA
MAR DEL ZUR
MARE PACIFICUM
NORTH AMERICA
SOUTH AMERICA
TERRA DE LABORADOR OR NOVA BRITANNIA
HUDSONS BAY
NEW FRANCE OR CANADA
NEW ENGLAND
NEW YORK
PENSYLVANIA
MARY LAND
VIRGINIA
CAROLINA
CALIFORNIA
NEW SPAIN
GOLFO DE MEXICO
HONDURAS
BAHAMA ISLES
CUBA
JAMAICA
HISPANIOLA
CARIBBE ISLES
VENEZUELA
GUIANA
PERU
WILDE BRAZILE
BRAZILE
PARAGUAY
Terra dos Patos
CHILI
PATAGONIA
COSTA DESERTA
TERRA DEL FUEGO
Falkland I.
THE WESTERN OCEAN
WESTERN ISLES
West Variation
The Line of no Variation
East Variation
CAPE DE VERD ISLES
SOUTH OCEAN
EAST
NEW ZEALAND
NOVA BRITANNIA
GREENLAND
THE NORTH SEA
LAPLAND
NOVA ZEMLA
FINLAND
SCOTLAND
IRELAND
GERMAN OCEAN
GERMANY
POLAND
HOLLAND
FLANDERS
THE WESTERN OCEAN

ABOVE *Hermann Moll's map of South America; this version published in about 1715. The note reads "The World is in nothing more scandalously imposed upon, than by maps put out by ignorant Pretenders, who most falsly and impudently assume ye titles of ye Queen's geographers." Moll, a Dutch émigré, became England's most prominent map publisher. He was even cited in Congress in 1811 when the possibility of Washington D.C. as the prime meridian was considered, because he had used not only London but also Edinburgh Castle on his maps.*

used for determining longitude. He prepared maps or nautical charts with lines joining places of equal declination, that is, the angle between the magnetic meridian and the geographic meridian, called isogones—(from the same Greek word as isobars and isotherms, *isos*, equal).

Halley also traveled along the coasts of the English Channel to survey the tides and to measure the variation in terrestrial magnetism. At the behest of Queen Anne, he inspected harbors at the ports around

the Adriatic Sea, and on another trip went to Trieste in order to study the fortifications.

Such was Halley's cartographic fame by now that imitators—described as "ignorant pretenders"—attached his name to spurious maps of their own. For example, the Hermann Moll map of 1715 shown above, based upon Halley's work, complains that a contemporary French map, dedicated to Halley, locates:

"C. Horn in Lat. 56 and makes ye Long. Between C. Horn and C. St. Augustin 17 Deg [instead of 57.30 degrees and 45 degrees, respectively] … These false maps differ from Dr Halley's and all other late Observations … and consequently make our Sailing to ye South Seas less by about a Thousand Miles than it really is. Every Body may easily judge what a Dangerous consequence these maps may produce, if ever they should be us'd at Sea, and ye wrong notions they must give others at land are no less apparent. NB that ye Projection of these Maps is also notoriously false."

Other genuine maps also acknowledged their indebtedness to Halley, including one from 1768 and authorized by Benjamin Franklin and Timothy Folger of the North and South Atlantic, showing the Gulf Stream as determined by Halley's measurements.

No description of Halley would be complete without an account of the comet that bears his name. In 1703 he was appointed professor of geometry at Oxford University even though, according to Flamsteed, he "drinks brandy like a sea captain" and would "corrupt the youth of the university" (by now Flamsteed was his archenemy). Halley built an observatory on the roof of his Oxford house, which is still there, to begin a study of comets. With Newton's help, he charted the known positions of as many comets as possible. Comets are visible only for a part of their orbits when they are near the Sun, and nobody really knew where they came from or where they then went. Halley suggested that the bright comets seen in 1456, 1531, 1607, and 1682 were, in fact, all the same comet that had traveled in an elongated ellipse around the Sun and back out beyond Saturn (then the most distant planet known). By taking into account the gravitational effect of Jupiter on the comet's orbit, he predicted that it would return at the end of 1758.

Among Halley's many other non-astronomical achievements he discovered the magnetic character of the aurora borealis, demonstrated that atmospheric pressure decreases with increasing altitude, studied monsoons and trade winds, invented a practical diving bell, produced solutions to polynomial equations, and compiled sets of mortality tables (for the city of Breslau) that actuaries could use to calculate life insurance premiums.

Halley succeeded Flamsteed as Astronomer Royal in 1720, even though he had again earned his wrath by colluding with Newton to publish some of Flamsteed's observations before they were finished. Halley held the position until his death in 1742, and had been dead for 16 years when the comet turned up again on Christmas Day 1758, just as predicted. Ever since the comet has been named after him.

Alexander Von Humboldt:

German Polymath

German scientists were key players in the formative years of physics, chemistry, and biology. But Germans took little or no part in the great 16th-century voyages of discovery. Those fell to navigators from Portugal, Spain, and Britain. Even the Italian Christopher Columbus (1451-1506) was adopted by Spain. In the following centuries Dutchmen such as Abel Tasman (1603-59), the Dane Vitus Bering (1680-1741), and the English James Cook (1728-79) put their names on the map. In 1800 a German polymath and explorer finally made a massive contribution.

His work would influence many who came after him, including British naturalists Charles Darwin (1809-82) and Alfred Russel Wallace (1823-1913). That man was Alexander von Humboldt. Darwin himself described him as "the greatest scientific traveler who ever lived." Baron Friedrich Wilhelm Heinrich Alexander von Humboldt was born in Berlin in 1769, the son of a Prussian army officer. He went to study at universities in Frankfurt-an-der-Oder, Berlin and Göttingen. While at Göttingen in 1790 he met British naturalist J.G.A. Forster (1754-94), who had accompanied Captain Cook on his second voyage round the world (1772-5). The acquaintance gave von Humboldt what he later described as an "ardent desire held from my earliest youth" to explore exotic places. In 1793 he published *Flora Subterranea Fribergensis* while he was studying under German mineralogist Abraham Werner (1750-1817) at the Mining Academy in Freiberg. For two years he took a post as an engineer in Upper Franconia's mining department and became inspector of mines in the Bavarian town of Beyreuth.

For six years von Humboldt traveled with Forster to various parts of Europe, while during the same time he continued his studies in botany, geology and geophysics. His plan to accompany a French expedition to the South Pacific was thwarted because of the war with Italy. But with the botanist of the aborted expedition, Aimé Bonpland (1773-1858), von Humboldt went to Madrid and was presented to King Carlos. The king gave the pair passports to go to Spanish America, and in 1799 the intrepid travelers set out from Corunna. Calling off at the Canary Islands to study their botany and geology, they finally arrived at Cumaná, Venezuela, more than a month later, having successfully evaded British warships (the Napoleonic Wars were just beginning).

They began an extensive series of trips to explore Latin America. On a journey that covered 6,000 miles (9,600 kilometers), they traveled up the River Orinoco, visited the source of the Amazon, journeyed up the River Magdalena in present-day Colombia, and over the Cordilleras to Quito in Ecuador and Lima in Peru. They stayed for a while in Cuba and spent a whole year in Mexico. Von Humboldt even climbed the 19,278-foot (5,876-meter) Chimorazo volcano in the Andes, setting an altitude record that remained unbeaten for 30

RIGHT AND ABOVE *The director of the Royal School of Mines in Mexico City, M. d'Elhuyer, persuaded von Humboldt (above) to take the information he had collected regarding "New Spain's" national industry and produce a map of the 37 mining districts and major mines. Maps like this of 1804, based on trigonometrical and astronomical observations, were later corrected and published in Paris.*

General Chart
of the
Kingdom of New Spain
Betn parallels of 16 & 38° N.
From materials in Mexico at commencement
of year 1804
by
Humboldt
PROVINCIA DE SONORA
PROVINCIA DE NUEVA BISCAYA
PROVINCIA DE TEXAS
PROVINCIA DE COHAHUILA
BOLSON DE MAPIMI
NUEVO REYNO DE LEON
PROVINCIA DE SINALOA
CULIACAN
CINALOA
DURANGO
NOMBRE DE DIOS
ZACATECAS
S. LUIS
INTENDENCIA DE GUANAXUATO
INTENDENCIA
VALLADOLID
GUADALAXARA
QUERETARO
CHIHUAHUA
MONTEREY
SANTA FE
ISLES DE REVILLAGIGEDO
Pacific Ocean
Gulf of Mexico

BELOW *Illustration of Humboldt's expedition at the volcano of Jorullo, published in the mid-1820s. Humboldt included an elevation (or cutaway) of the volcano in the* Atlas géographique et physique des régions équinoxiales du nouveau continent *(1814). He used barometric readings to determine altitudes. El Jorullo is a small volcano in the Sierra Madre mountain range of Mexico. Because of his interest in geology, Humboldt was fascinated by volcanoes. Not the most important of his observations, but fascinating nonetheless, he was the first person to realize mountain sickness was caused by the lack of oxygen at high altitudes.*

years. Everywhere they made surveys and maps, recorded the climate and magnetic field, and collected samples of plants, animals, and minerals so that when they returned to Europe that had more than 60,000 specimens, including many plants never seen before. Presciently, von Humboldt also suggested building a canal across the Isthmus of Panama to connect the Atlantic and Pacific oceans, and suggested that guano (mineralized bird droppings) from Peru could be imported to Europe for use as a fertilizer.

Back in Berlin von Humboldt began organizing and publishing his findings. He moved to Paris in 1808 and gave himself two years to complete the work. In the event it took a further 20 years. The resulting 23 volumes of *Voyage de Humbolt et Bonpland aux Régions Équinoxiales* covered such a wide range of subjects that he gained enormous prestige throughout the world. It was among the five or six books that Charles Darwin took with him on his epochal voyage aboard HMS *Beagle*. In his book von Humboldt presented statistical information in diagrammatic form using cartograms. Maps of mountains and volcanoes included information about the vegetation and how it varied with altitude. He also demonstrated that many volcanoes are associated with underlying geological fault lines, especial in Central America and the southern United States. He diagrammized temperature variations using for the first time isothermal maps (an isotherm is a line on a map that joins places of equal temperature). He also introduced isobars (lines joining places of equal atmospheric pressure) on his weather maps. His book *Relation Historique* was a personal account of the trip and one of the first travel books, and it is still read.

Von Humboldt then went to Paris, where he carried out experiments with French chemist Joseph Gay-Lussac (1778-1850) on the chemical composition of air. The Prussian King Frederick William III occasionally employed von Humboldt on various duties, and after he returned to Berlin in 1827 he undertook diplomatic missions to France and became friends with King Louis Philippe.

It was in Paris that he wrote and published *Géographie du Nouveau Continent* (1835-8). Later he proposed a scheme for collecting meteorological and magnetic data throughout Russia, Asia, and the British colonies. He went to Russia himself in 1829 at the invitation of Tsar Nicholas I. He traveled with German naturalists Christian Ehrenberg (1795-1876) and Gustav Rose (1798-1873), and published the results of their meteorological and geological findings on the 9,000-mile (14,500-kilometer) journey through the Ural Mountains and Siberia in *Asie Centrale* of 1843.

Von Humboldt summed up his philosophy and experiences in 1847 in the two-volume *Kosmos*, which is an illustrated history of the physical world. In it he tried to show that, although nature is so complex, there is an underlying unity. His stated ambition was to discover "the laws which wind a uniting bond round a multitude of isolated fact."

Humboldt believed that the Earth's history provides the clues to physical geography, and that the rotation of Earth and its magnetism were responsible for many of these physical characteristics. The work also related the physical environment to indigenous plants and animals, and as such is commonly viewed as the first approach to what was to emerge as the science of ecology.

He added further volumes to *Kosmos* over the years until his death in 1859. His name was commemorated in the Humboldt Current, the cold current that runs northwards from the Southern Ocean along the Pacific coast of Chile and Peru.

At last, a German had his name on the world map; unfortunately, it has since been renamed the Peru Current.

Strata Smith:

A Single Map Creates a New Science

William Smith earned himself the remarkable nickname "Strata Smith," although he would probably have preferred his more formal sobriquet Father of English Geology. More remarkable than his nickname, however, is his achievement. Single-handedly and over a period of 20 years, he traveled around the country, making a complete survey of all the rocks and fossils he found.

William Smith published his findings in 1815 in the form of maps entitled *A Delineation of the Strata of England and Wales, with Part of Scotland*. The first geological map of England and Wales, it occupied 15 sheets at a scale of 5 miles to the inch.

Smith was born in Churchill, Oxfordshire, in 1769, the son of the village blacksmith. His father died when Smith was only eight years old and he was left largely to his own resources. He was taught at the village school and studied surveying, also collecting fossils on his walks in the local countryside. This was a time of a boom in canal building and in 1787, when he was 18, Smith went to work for surveyor Edward Webb at Stow-on-the-Wold who was planning the routes of canals. In 1791 he moved to northern Somerset to survey the coal deposits and from 1794 to 1799 was engineer to the Somerset Coal Canal. He also became involved in overseeing the construction of canals, and in 1794 made a six-week tour of northern England by carriage to look at the methods being used for excavations. During all this time he noticed how the rocks consisted of layers, or strata, and he began keeping notes of his observations, drawing sections of rock formations and labelling them by their geological names. His friends called him Strata Smith.

Encouraged by associates at Bath, Smith drew his first detailed geological map. The first map of its kind ever, it showed in different colors the rocks of the region around Bath. He then set out in 1805 to complete his mapping of the whole of England and Wales. But as well as these unique maps, he established two key principles of geology. From his observations of rock strata, he stated that whenever one type of rock lies above another, the younger rock is always on top (unless major Earth movements have scrambled the order). This is known formally as the theory of superposition. And from his observation of fossils, he noticed that wherever a certain type of sedimentary rock occurred, it always held the same invertebrate fossils. Even if the rock stratum dipped down and vanished underground to reappear a mile away, it contained the same fossils. The age of the fossils had to be the same as the age of the rock, so that if you know the rock's age you also know that of the fossils, and vice versa.

Smith was no writer of textbooks—possibly because of his lack of schooling—but he published his findings in two pamphlets: *Strata Identified by Organized Fossils* of 1816, which illustrated typical fossils from each rock formation, and *Stratigraphical System of Organized Fossils* of 1824, which included geological maps of 21 counties as well as sectional drawings of rock strata. He regarded the map as the best way of showing a region's stratigraphy and he enlisted the help of his nephew John Phillips (1800-74) in drawing them. Some of the maps were still being printed in 1911. During most of this time Smith was based in London but in 1819 he moved north to Yorkshire. The importance of his work went largely unrecognized, however, and he became short of money. He had to sell most of his fossil collection to the British Museum to raise funds. In 1831 the Geological Society belatedly awarded him the first Wollaston Medal, named after the British chemist and physicist William Wollaston (1766-1828) who discovered the platinum-like metals palladium and rhodium. In 1835, four years before he died, the University of Dublin awarded the blacksmith's son an honorary doctor's degree.

RIGHT *Representing the geology of the whole of England and Wales at a scale of 5 miles to the inch created a huge map nearly 9 feet (2.6 meters) tall by just over 6 feet (1.8 meters) wide. The base map alone, prepared by John Cary, required 15 separate copperplate engravings. Publication was financed by public subscriptions, initiated by Royal Society president Joseph Banks, who contributed £50.*

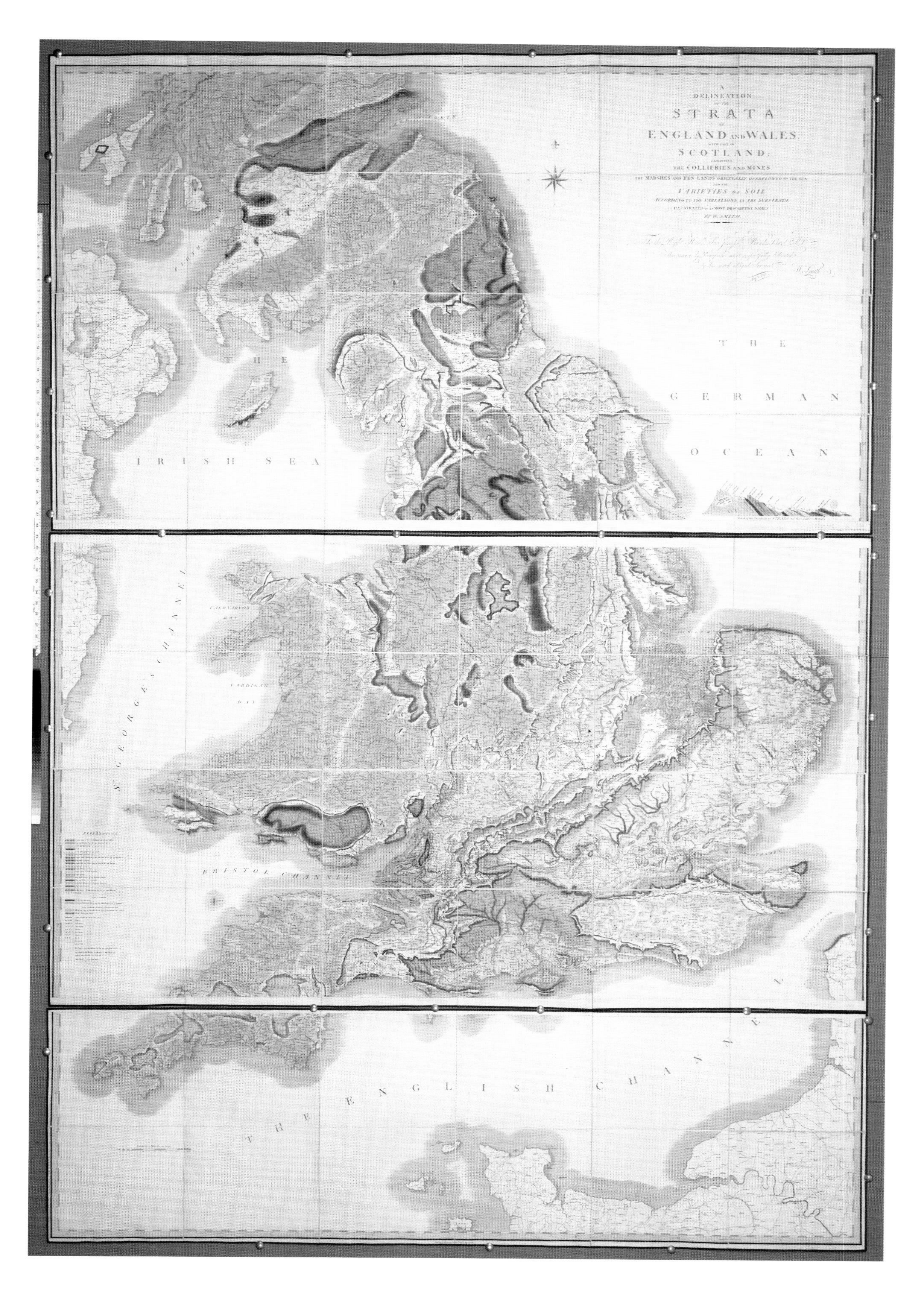
A
DELINEATION
STRATA
ENGLAND AND WALES,
SCOTLAND:
THE COLLIERIES AND MINES
VARIETIES OF SOIL
BY W. SMITH
THE GERMAN OCEAN
IRISH SEA
BRISTOL CHANNEL
THE ENGLISH CHANNEL

Matthew F. Maury:

"Pathfinder of the Seas"

Born on a Virginia farm in 1806, Matthew Fontaine Maury had few prospects in life but to till the soil. After the family moved to Tennessee, he badgered his father for an education and eventually enrolled in Harpeth Academy. He diligently searched for a better life and in 1825 entered the U.S. Navy as a midshipman. For the next eleven years he spent much of his time at sea and became engrossed in the intricacies of navigation.

In 1829 Maury enjoyed the unique opportunity to sail around the world in the USS *Vincennes*, the first U.S. Navy ship to do so. The voyage opened his eyes to the massiveness and the mysteries of the oceans. He observed changes in the winds and currents as the ship passed from the northern latitudes deep into the far southern latitudes. The captain sailed without a chronometer, so when on watch Maury practiced his navigation by chalking equations on the surface of cannonballs. When entering the belts of Cancer, Capricorn, and the equator, he joined the struggle to push the vessel through the calms, all the time wondering what caused the peculiar change in the atmosphere. Maury questioned everything: the puzzling hot and cold currents that sometimes moved through the sea at five knots, the driving trade winds that never stopped blowing, the odd change in winds after passing through the calm belts, and the persistent howling gales of the southern "roaring forties" and the "furious fifties." After returning from the cruise he became engaged to his cousin, Ann Hull Herndon, and in 1834 they wed.

In 1839, while traveling through Ohio, Maury toppled off the top of a careening stagecoach and became crippled for life. The accident would have ended his career, but the secretary of the navy needed someone who understood astronomy and the calibration of navigation devices to supervise the dreary Depot of Charts and Instruments in Washington. Lieutenant Maury took the humdrum job because it did not involve sea duty, for which he had been declared unfit. Stuffed in crates and boxes he discovered thousands of old, musty sea logs turned in by skippers, some of which dated back to the Continental Navy. They were dead storage to the navy but provided a mother lode of information for Maury. Among the soiled pages he found records of weather and sea conditions for each month of the year from every part of the world. He began to wonder whether such loosely organized data could be compiled into constants of the sea and condensed into a form that could become as valuable to a navigator as his sextant and chronometer.

Maury and his small staff began assembling thousands of observations taken from ships' logs. Work progressed slowly because the data was unsystematic, so Maury persuaded the navy and commercial shipping firms to use a standardized form for reporting weather, winds, currents, and other hydrological and meteorological observations. As new data began pouring into the depot, Maury

ABOVE *Matthew Fontaine Maury photographed in 1853, when he announced to the maritime nations of Europe: "Every ship that navigates the high seas with these charts … on board, may henceforth be regarded as a floating observatory, a temple of science."*

applied it in different ways. He created charts and sailing directions from one distant port to another, and he continuously updated the information, taking into consideration climate change by month in different latitudes.

Over time, he eventually mapped the winds and currents of the world. Though his detailed charts and sailing directions became specific to an area, he developed a worldview (see map below) showing the prevailing winds that enabled the sailing master to exploit weather patterns, especially when choosing long-distance routes. One of the first skippers to try Maury's sailing directions shaved 37 days off his normal 110-day trip to Brazil and back.

By 1847 Maury had captured enough information—in fact, millions of observation——and produced the first track charts for the North Atlantic. As more observations became available, he revised and republished the charts. He then expanded into the South Atlantic, the Indian Ocean, and the Pacific. The track charts became an amalgamated collection of hundreds of voyages crossing through the same areas of the sea. The passage from New York to London was always faster because trade winds blew from the west. But there were pockets where a navigator might find more easterly winds than westerly winds, and vice versa, providing he knew where to go, and Maury's track charts revealed those secrets. Sometimes by going a hundred miles out of his way, a sailor could find more favorable winds and cut his voyage by weeks.

Sailors preferred winds off the stern quarter, and track charts, augmented by Maury's Sailing Directions, showed where the most favorable winds most often blew. To show wind direction, Maury designed a symbol that looked like a small dart with a tail of whiskers. The point of the dart represented a ship's position, but was actually the

BELOW *Maury's Winds and Routes map of the world from* The Physical Geography of the Sea *shows the prevailing trade winds of the oceans, the seasonal migration of the doldrums, and the recommended routes for finding the most favorable winds.*

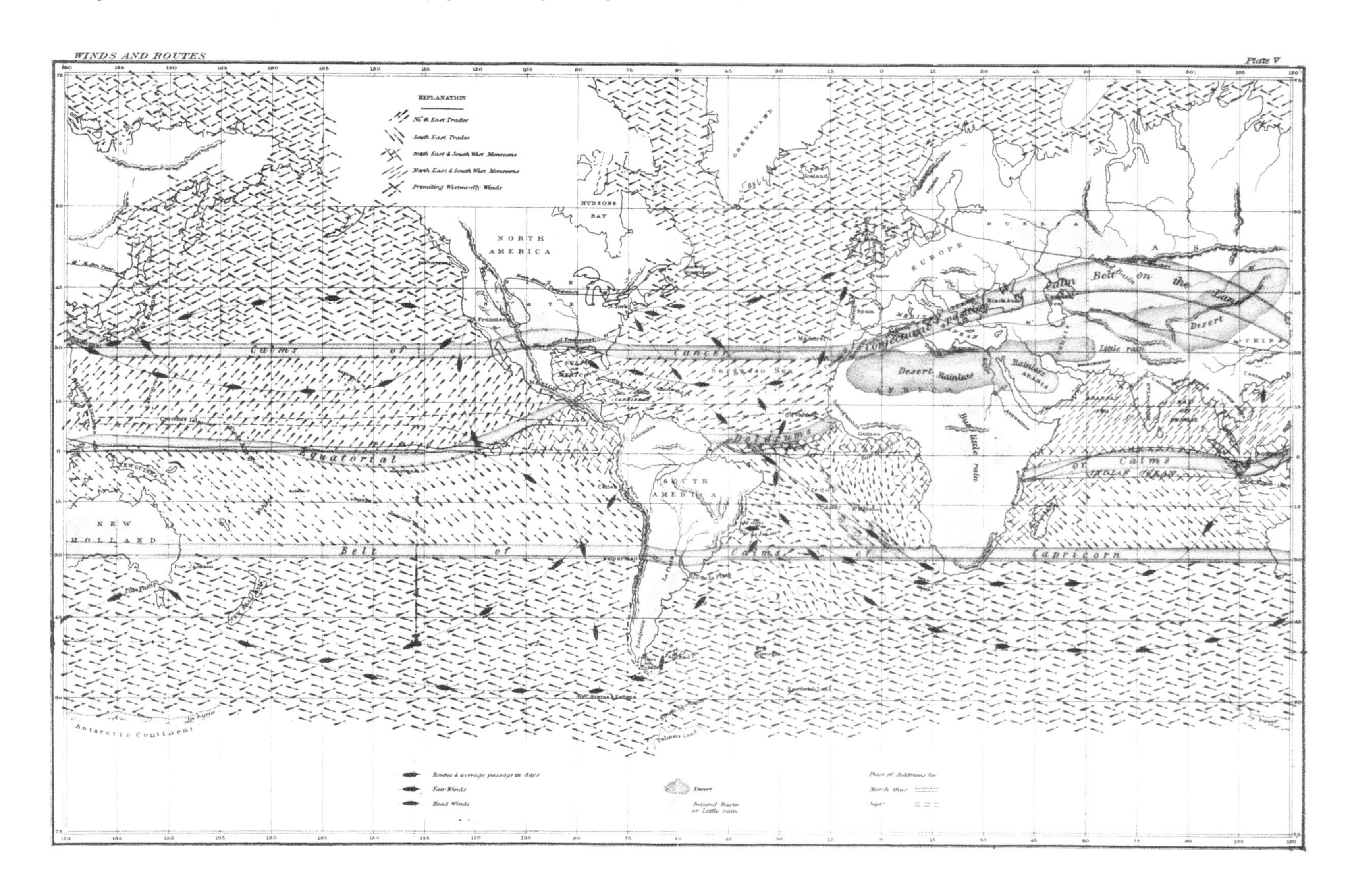

relative location of many ships taking wind direction readings near that spot. The wind blew toward the head of the dart, and each whisker represented the direction of the incoming wind when the navigator took the reading. Depending upon the time of year, a sailor could choose his route based upon the experiences of dozens of sailors who came before him.

Over time Maury compiled the accumulated knowledge acquired from the study of the sea and produced pilot charts, whale charts, the first orographic and bathymetrical maps of the North Atlantic, charts on thermal and tidal data, and discovered the effects of the oceans and atmospheric pressure on meteorology.

In between the years he plotted the undersea route for laying the Atlantic cable, and in 1855 gave the world the first authoritative work on the physical geography of the sea.

With the outbreak of the Civil War, the Virginian Maury resigned his commission as a U.S. Navy commander and joined the Confederacy. He spent the war in the South, as well as abroad in England, acquiring ships for the Confederacy. He was instrumental in the development of the tethered naval mine, or torpedo. Following the war, Maury accepted a teaching position at the Virginia Military Institute in Lexington. He died in 1872 during a lecture tour.

Maury never forgot the lowest point in his life. While recovering from the near fatal accident in 1839-40, he worried he would have to content himself "with cultivating a few little patches of knowledge." He never stopped thinking about winds and currents and desperately sought a scientific role that finally, in 1842, brought him to the navy's musty Depot of Charts and Instruments, where he became the 19th century's greatest scientist of the seas.

When faced with some oceanic phenomenon, he attempted to unravel the mystery, or use it as supporting evidence of his own theories. For example, in *Explanations and Sailing Directions* (1858) he writes: "Seamen tell us of 'red fogs' which they sometimes encounter, especially in the vicinity of the Cape de Verde Islands. In other parts of the sea, also, they meet showers of dust. What these showers precipitate into the Mediterranean is called 'Sirocco dust,' and in other parts 'African dust,' because the winds which accompany them are supposed to come from the Sirocco Desert … Were it possible to take a portion of this air, as it travels down the southeast trades, representing the general course of atmospherical circulation, and to put a tally on it by which we could always recognize it again, then we might hope actually to prove, by evidence the most positive, the channels through which the air of the trade-winds, after ascending at the equator, returns whence it came … As difficult as this seems to be, it has actually been done. Ehrenberg, with his microscope, has established, almost beyond a doubt, [by examining the Sirocco dust] that the air which the southeast trade-winds bring to the equator does rise up there and pass over into the northern hemisphere."

Maury continued to look, to measure, and to wonder to the very end of his life. His correspondence, letterbooks, diaries, journals, speeches, articles, notebooks, electrical experiment book, charts, and printed material in the Manuscript Division of the Library of Congress number nearly 15,000 items, a record of a lifetime's diligence and practical research.

Defining the Sea Roads

As the years passed, steamer traffic began replacing sailing ships, like the clipper ship **Three Brothers** pictured here in 1875, at 2,972 tons then the biggest sailing ship in the world. But both classes of vessels crossed through the same routes and collisions occurred. The New York Board of Underwriters came to Maury and asked if something could be designed to prevent collisions. By then the aging self-taught scientist had become Superintendent of the U.S. Naval Observatory and the world's foremost authority on navigation and oceanography. Maury went back to his charts, found the best routes for steamers and the most favorable winds for sailing ships, and created one-way highways in the North Atlantic. Collisions ended, except for the wayward sailor who followed his own lights instead of Maury's charts.

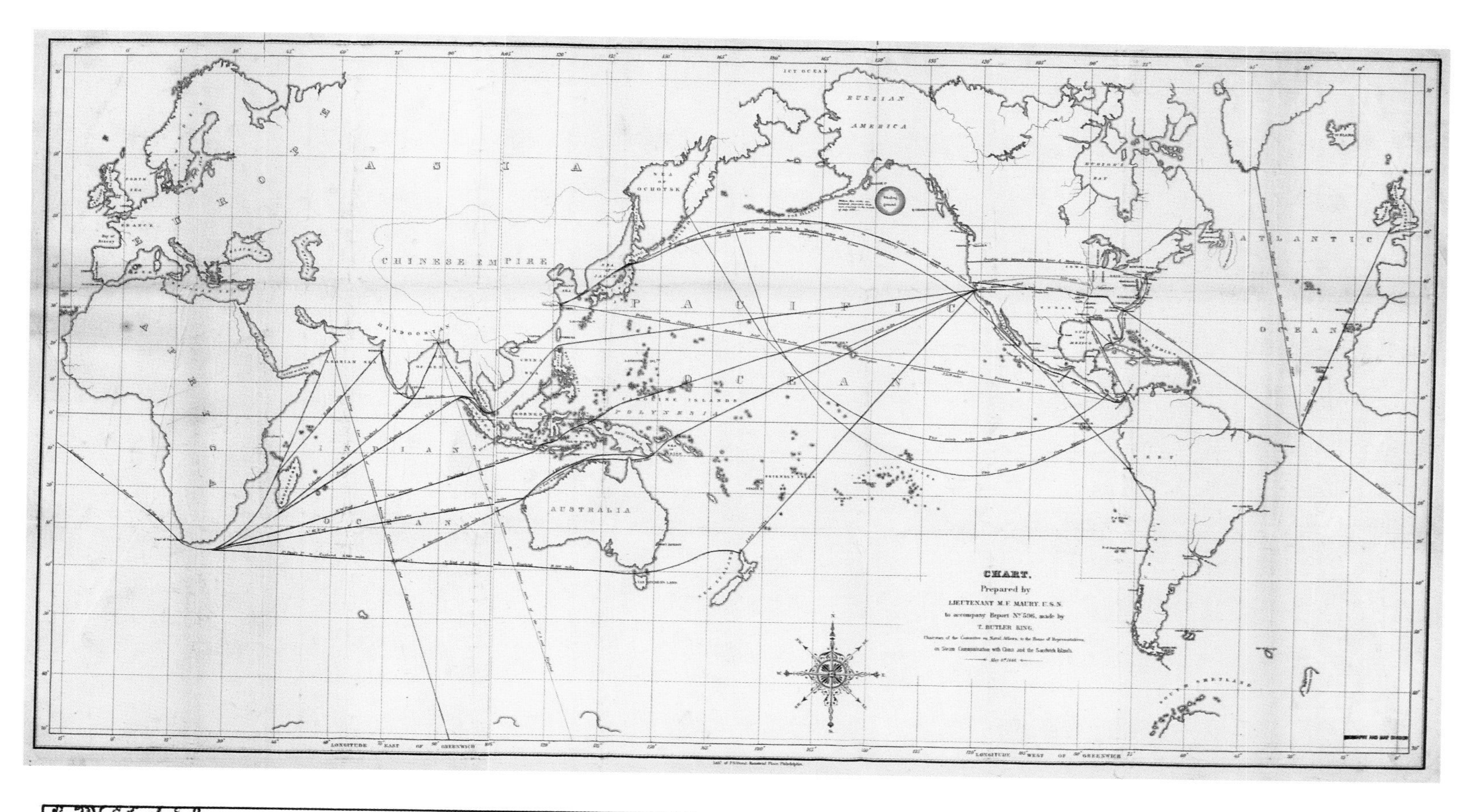

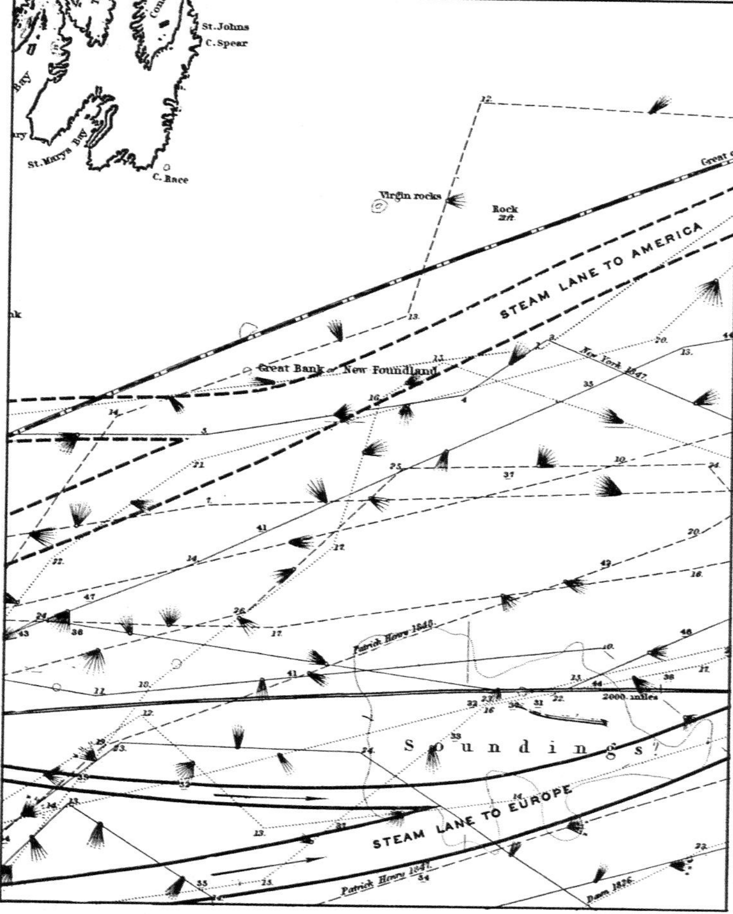

ABOVE *"Chart prepared by Lieutenant Matthew F. Maury, U.S.N. to accompany report No. 596 made by T. Butler King, Chairman of the Committee of Naval Affairs, to the House of Representatives on steam communication between Shanghai via the Sandwich (Hawaiian) Islands and Panama City." Maury's position as* the *authority on nautical matters became unassailable.*

LEFT *Section of a North Atlantic track chart Maury started in 1847 and continued to expand with data collected until 1855, when he inserted one-way steamer lanes to avoid collisions between transatlantic vessels. The small darts with feathery whiskers are the recorded accumulation of years of observations made by sailors reporting the direction of the incoming wind. (Wind and Current Chart, No. 6, series A, 3rd Edition, with steamer lanes published by the New York Board of Underwriters, 1855.)*

John Snow:

The Killer Cholera

In the mid-19th century, cholera was a killer disease. It particularly attacked poor people who lived in the unsanitary conditions of inner cities, just as today it gets most of its victims from refugee camps around the world. In 1849 alone, cholera accounted for 50,000 deaths in England. The puzzling question facing doctors at the time was: How is the infection spread?

In the summer of 1854 a cholera epidemic broke out in the Soho district of London. In the first week of September, 93 people died of the disease. At that time nobody knew what caused cholera or indeed any other infectious disease. Many people, doctors among them, blamed "miasmas" that were supposed to be poisons carried in the air (the word malaria comes from Italian meaning bad air). But cholera's chief symptom is copious diarrhoea and obviously affects the intestines, not the lungs.

Living in Soho during the 1854 epidemic was London's leading anaesthetist, Doctor John Snow—in 1853 he administered chloroform to Queen Victoria during the birth of Prince Leopold. Snow was born in York in 1813 and practiced as a surgeon in Newcastle upon Tyne. He witnessed an epidemic of cholera in nearby Sunderland in 1831. He moved to London in 1836 and soon began to suspect contaminated water supplies as possible vehicles for carrying infection. People living in Soho obtained their water from hand-operated pumps in the street, which connected to underground cisterns. These cisterns belonged to either of two water companies, both of which supplied water, sometimes both to the same street. The Lambeth Company got its water from the River Thames upstream, whereas the Southwark & Vauxhall Company drew water from the dirty part of the river within the city, not far from where the sewers discharged their detritus into the Thames.

In September 1854 Snow carried out a survey of Soho to establish where the drinking water came from. He went round knocking on doors asking which pump the people used. He also noted the addresses of all houses in which somebody had died of cholera. The results convinced Snow that most of the victims had drunk Southwark & Vauxhall water. Furthermore, when he plotted the "death houses" on a map, he immediately saw that they centered on a public pump in Broad Street. He petitioned the Board of Guardians of the parish to disconnect the water supply. They were desperate and reluctantly agreed to remove the handle of the pump. The cholera outbreak, which was already tailing off, ceased almost immediately. Doctor Snow was vindicated, knocking one more nail in the coffin of the miasma theorists. Urban myth claims that Snow personally marched down Broad Street and unscrewed the pump handle. But this is just myth and the only dominant presence Snow ever had in the street is the John Snow pub, which still stands there (although it is now named Broadwick Street).

Snow's results reached a wider audience in 1855 with the publication of his book "On the Mode of Communication of Cholera." He cited the case of a whole family in Yorkshire wiped out when they wore the unwashed clothes of a cholera victim. He claimed that cholera is caused by particles that can multiply, like "veritable animals." This was literally the germ of an idea: or the idea of a germ. He also believed that one organism is specific to one particular disease, unlike the miasmatists who thought that one miasma can account for several different disorders. Although the Italian physician Girolamo Fracastoro (c.1483-1553) had first put forward the idea that diseases are caused by germs in his book *De Contagione* of 1546, the idea was not taken seriously until German pathologist Jakob Henle (1809-85) proposed in 1840 that infection is caused by parasitic organisms. Then in 1883 German bacteriologist Robert Koch (1843-1910) discovered the causative organism of cholera, a type of bacterium called Vibrio. By that time Snow had been dead for 24 years; he was only 45 when he died.

RIGHT *There is some evidence that Snow "massaged" the mortality statistics a little to point the blame squarely at the Broad Street pump; though proved epidemiologically correct, it's the kind of thing that gives scientists a bad name!*

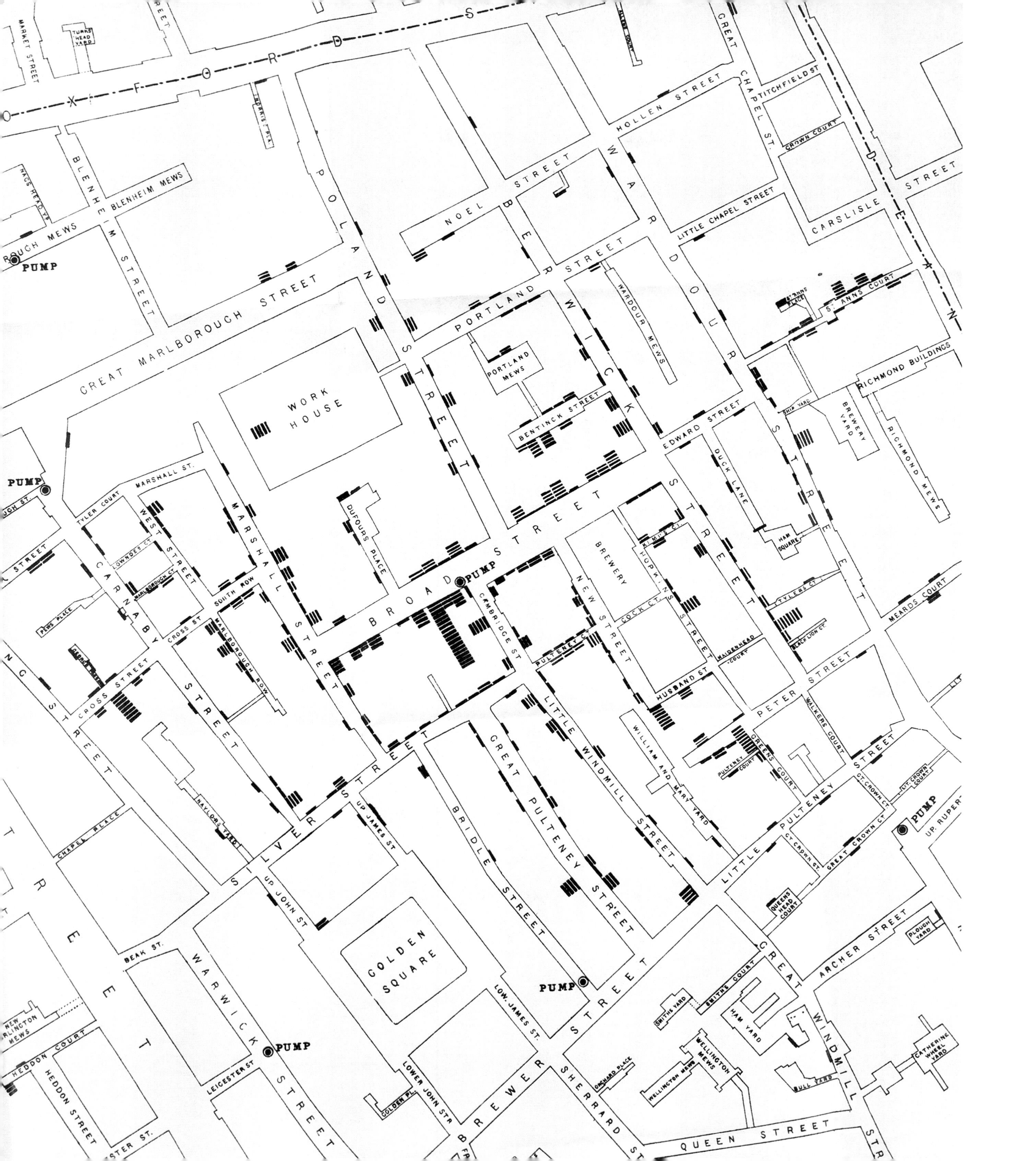

OXFORD
MARKET STREET
TURKS HEAD YARD
GREAT CHAPEL ST.
TITCHFIELD ST
HOLLEN STREET
CROWN COURT
BLENHEIM MEWS
BLENHEIM STREET
POLAND STREET
NOEL STREET
BERWICK STREET
WARDOUR STREET
LITTLE CHAPEL STREET
CARLSLISLE STREET
DEAN
PUMP
GREAT MARLBOROUGH STREET
PORTLAND STREET
WARDOUR MEWS
ST ANNS COURT
PORTLAND MEWS
RICHMOND BUILDINGS
BREWERY YARD
RICHMOND MEWS
SHIP YARD
WORK HOUSE
BENTINCK STREET
EDWARD STREET
DUCK LANE
MARSHALL ST.
TYLER COURT
WEST STREET
MARSHALL STREET
DUFOURS PLACE
BROAD STREET
PUMP
HAM SQUARE
BREWERY
HOPKINS STREET
NEW STREET
CARNABY STREET
SOUTH ROW
CAMBRIDGE ST
PULTENEY CT
COCK CT
TYLERS CT
MEARDS COURT
PEWS PLACE
CROSS ST
MARLBOROUGH ROW
MAIDENHEAD COURT
BLACK LION CT
CROSS STREET
HUSBAND ST
PETER STREET
LITTLE WINDMILL STREET
WILLIAM AND MARY YARD
PULTENEY COURT
GREENS COURT
WALKERS COURT
SILVER STREET
GREAT PULTENEY STREET
NAYLORS YARD
UP JAMES ST
LITTLE PULTENEY STREET
CT CROWN CT
LT CROWN COURT
GREAT CROWN CT
PUMP
CHAPEL PLACE
BRIDLE STREET
UP JOHN ST
QUEENS HEAD COURT
ARCHER STREET
PLOUGH YARD
BEAK ST.
GOLDEN SQUARE
WARWICK STREET
PUMP
LOW. JAMES ST.
PUMP
GREAT WINDMILL
SMITHS COURT
SMITHS YARD
HAM YARD
HEDDON COURT
HEDDON STREET
LEICESTER ST
LOWER JOHN STR
GOLDEN PL
BREWER STREET
SHERRARD STREET
ORCHARD PLACE
WELLINGTON MEWS
BULL YARD
CATHERINE WHEEL YARD
QUEEN STREET

Maps a Dollar Each:

The City Panorama Map

A year before America's first centennial, as a measure of national and civic pride, the city of St. Louis commissioned the largest panoramic map ever published. This was Camille N. Dry's massive **1875 Pictorial St Louis: the Great Metropolis of the Mississippi Valley.** The map was the grandest execution of a popular lithographic art form that flourished from 1870 to 1930. Dry's massive work might be described as the Sistine Chapel of panoramic maps.

The map consisted of 110 individual finely engraved color lithographic plates that could be assembled together into one massive panoramic map of St. Louis measuring 9 by 24 feet (2.9 by 7.3 meters)! Every building in the region was drawn; including businesses, schools, churches, homes, and institutions. Some 1,999 sites were individually drawn and specifically identified. The map was issued with a preface by Dry explaining its creation:

> "A careful perspective, which required a surface of three hundred square feet was erected from a correct survey of the city ... Every foot of the vast territory within these limits has been carefully examined and topographically drawn in perspective, and the faithfulness and accuracy with which this work has been done an examination of these pages will attest."

St. Louis was not the only American city to commission epic scale panoramic maps in color. An impressively detailed panoramic map of Baltimore in 1869 measured 5 by 11 feet (1.5 by 3.4 meters); while another beautiful map of Washington in 1884 measured 4 by 5 feet (1.2 by 1.5 meters). These maps were spectacular expansions of the more common commercially produced single sheet panoramic maps of the time. In North America, for the last thirty years of the 19th century and the first thirty years of the 20th century, single sheet panoramic city maps were printed in several colors. The quality and size of these lithographic maps made them popularly acceptable as wall mounted art. Other editions of maps were printed in two tones, and in a smaller format. All commercially produced types and sizes of panoramic city maps were intentionally modestly priced—at one to five dollars—to make them affordable to the average American family.

Perspective maps were not new inventions. Once the rules of perspective were established in the early Renaissance, artists and architects often drew and painted panoramic views of cities. Military engineers often drew them to plan an attack or defense of a fortification. The very first city atlas was the *Civitates Orbis Terrarum* in 1572. The first edition of the "Cities of the World" was published by Georg Braun and Frans Hogenberg in Cologne, Germany in a single volume. By 1617 this atlas consisted of six volumes and with over 363 panoramic maps of cities from Cologne to Paris and Rome, from Peking to Goa and Mombasa. In all, 46 editions were published in Latin, German, and French; many other city atlases were based on this great work.

Each *Civitates* map was a bird's eye perspective of the natural geography and street plan of each city viewed in vertical relief showing individual buildings, ships, canals, harbors, and other landmarks. The atlas also provided notes on each city's history, location, resources, trades, and industries. In an age of European expansion, these maps were extraordinarily valuable to merchants, missionaries, traders, and colonists. Many other nations competitively followed suit with atlases of their own.

It was not until about 1860 that advances in lithography, photolithography, and photoengraving allowed for the production of inexpensive, high-quality panoramic maps in color—and in a format large enough for framing. Although comparable works were produced

RIGHT *Number one of 110 equally detailed plates from* Pictorial St. Louis, the Great Metropolis of the Mississippi Valley *by Camille N. Dry. The numbered key runs from "1. Cathedral" to "117. Proctor, Trask & Co." People and carriages throng the streets. This is civic pride manifested cartographically.*

in some European cities, it was only in Post-Civil War America that the panoramic map proved such a popular, widespread, and endemic commercial art form. Elsewhere, only great metropolises were portrayed in such maps. It was only in North America that mapmakers found an extensive market for panoramic maps of thousands of small towns and communities. Their popularity resulted in a "Golden Age" for panoramic artists in America that lasted for six decades.

Uniquely in North America, realtors, or members of city councils, or local chambers of commerce, commissioned maps. City fathers used them to advertise the community's potential for commercial and residential growth. Realtors used them to promote sales. Panoramic maps were popular as wall hangings that enhanced civic pride. A citizen might proudly point out his own property, since for a fee the artist could include illustrations of private homes or businesses.

THE WORLD'S INDUSTRIAL AND COTTON CENTENNIAL EXPOSITION.
MAIN BUILDING. GOVERNMENT BLDG.
HORTICULTURAL BLDG. ART GALLERY.
FACTORY AND MILLS.
Bayou Metairie
Mississippi Valley R.R.
Bayou Poydras
St. Alphonsus Church.
Church, St. Mary's Assumption.
St. John's Ch.
Temple Sinai.
Annunciation Square.
Lee Monument.
West End.
(Lake Pontchartrain.)
1st.
St. Patric
COPY
THE CITY O
AND THE MISSISSIPPI RIVE

NEW ORLEANS,

LAKE PONTCHARTRAIN IN DISTANCE.

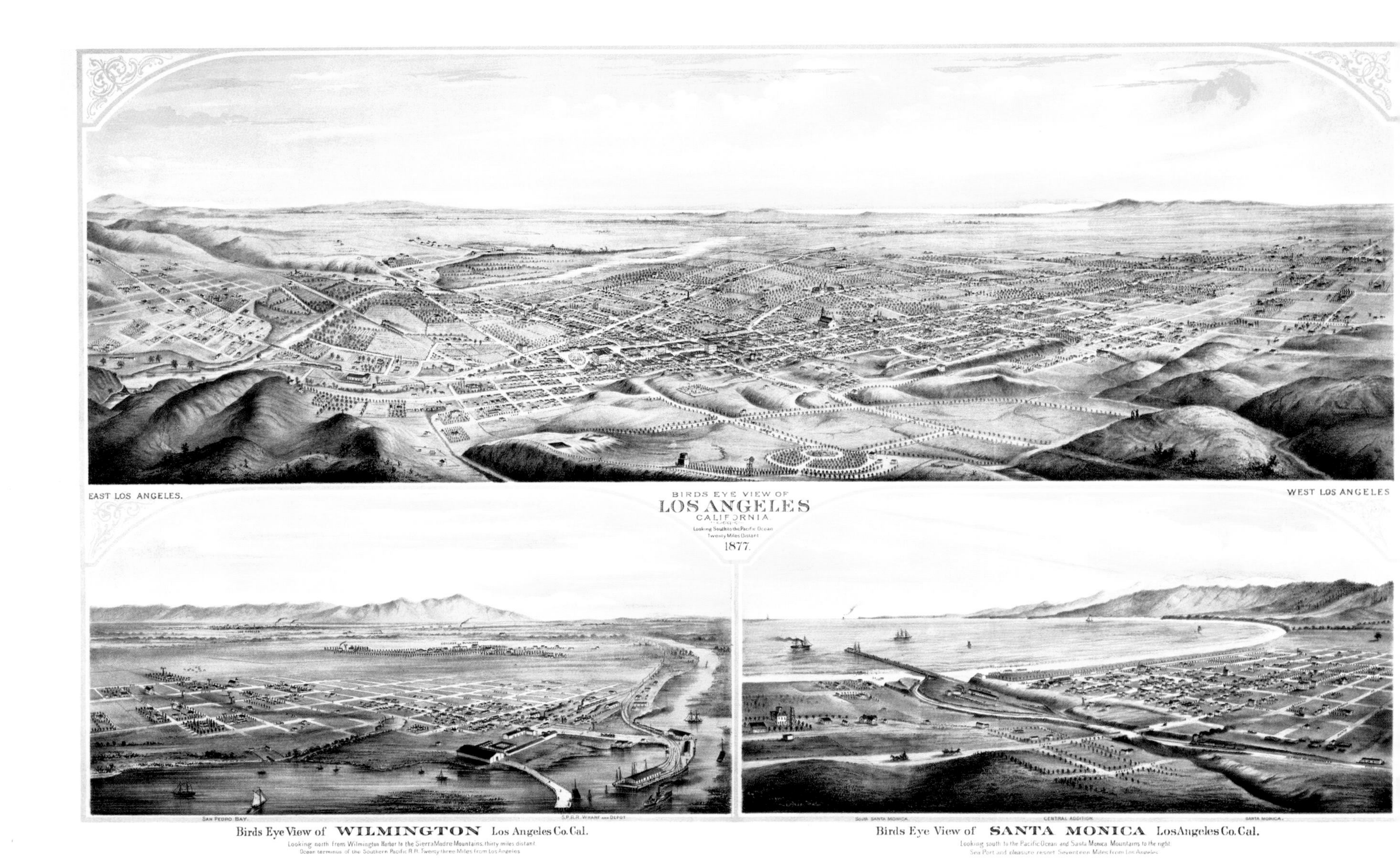

PREVIOUS PAGE *This map of New Orleans and the Mississippi was published by Currier and Ives of New York in about 1885. Prominent buildings and places, like the Temple Sinai and the Pickwick Club, are listed below; though in comparison with the St. Louis map on the previous page the listing is positively slapdash! The "cartographic breakthrough" in these panoramic maps is not in the use of perspective, which is actually archaic: it is in the lithography.*

Panoramic maps wonderfully illustrate the vitality of the cities of the era. Harbors and canals bustle with sailing and steam ships and barges. Steam locomotives hurtle along railway lines pulling passenger carriages, or cars loaded with cattle, or coal, or timber. Steel mills and factory chimneys belch out smoke. Streets are filled with people walking, shopping, and riding in carriages.

As panoramic maps were drawn from a high altitude viewpoint, it was often believed that artists used photographs or sketches made from air balloons, dirigibles, or later, from airplanes. Sometimes the artists and publishers encouraged this popular idea of high technology airship photography by calling their works "aerial panoramas" and so pandered to the public enthusiasm for balloons and flying machines. However, this was never how panoramic maps were produced. Each panoramic map began with a decision about the area to be framed within the city or town and the perspective on the street plan or landscape. The artist then walked the streets, sketching in buildings, factories, docks, rivers, and other aspects of the area as seen from an elevation of 2,000 to 3,000 feet (610 to 914 meters). All these sketches were then entered into the frame of the overall map.

Although thousands of panoramic maps were produced in the US and Canada, there were only a handful of publishers who specialized in this industry. Furthermore, there was only a very small community of artists who worked in this art form. Over half of all the maps collected

LEFT AND ABOVE *One of the joys of browsing through these panoramas (owing to the longevity of the particular art form) is of course watching the city grow, as here, with Los Angeles in 1877 (left) and 1894 (above). A second is seeing how different artists approached the same task. B.W. Pierce's lithograph (above) unusually and effectively provides two viewpoints. looking north from the "electric power house" on the corner of Wilder Street and Central Avenue; and then looking south from the corner of Tenth and Pearl Streets.*

by the American Library of Congress were produced by just five men: Albert Ruger, Thaddeus Mortimer Fowler, Henry Wellge, Oakley H. Bailey, and Lucien Burleigh. Just two of these—Bailey and Fowler—produced over 1,000 maps of cities and towns in over 20 American States and Canada. As late as the mid-1920s, panoramic maps were still in vogue, but by 1930, the business was no longer commercially viable.

Thaddeus Mortimer Fowler was a panoramic mapmaker for 55 years. In 1932, five years after his retirement, Fowler commented:

> "The business has been practically without competition as so few could give it the patience, care and skill essential to success. But now the airplane cameras are covering territory and can put more towns on paper in a day than was possible in months by hand work formerly."

Many of these panoramic maps give a vivid and extremely accurate pictorial record of urban America of the time. For many of these communities, these are the only maps that have survived. Consequently, beyond their artistic merit, these graphic portrayals of American life also prove to be documents of historic significance.

Charles Booth:

Lies or Damned Lies?

"There are three kinds of lies: lies, damned lies, and statistics." This condemnation of statistics is attributed variously to British Tory statesman Benjamin Disraeli (1804-81) and American writer Mark Twain (1835-1910), who in his 1924 **Autobiography** acknowledged Disraeli as the source. But not every 19th-century worthy was so cynical. Here is that rarity, a statistical analysis with a lasting and benevolent effect.

Statistics were a new analytical tool and in London there was even a Royal Statistical Society. Between 1892 and 1894 its president was the British social reformer Charles Booth. Booth was born in Liverpool, Merseyside, where he and his brother Alfred formed the Booth Steamship Company. They also founded Alfred Booth & Company, which made leather and leather goods. In his youth, Charles Booth was an ardent radical, but by 1875 he had settled down and went to live in London. Disturbed by the widely contrasting social conditions under which Londoners lived, he undertook a social survey of the city, which became the prototype for all similar surveys. Over the next 18 years he traveled the streets interviewing people and observing the conditions for himself. He finally published his results in 1902 in the monumental 17-volume *Life and Labour of the People in London.*

Cooincidentally his namesake William Booth (1829-1912), invariably known as General Booth, the founder of the Salvation Army, had in 1890 published a book called *In Darkest England and the Way Out.* General Booth also observed the poverty of unfortunates living in the slum areas of cities, and with like-minded people formed the volunteer Salvation Army to do something about it. Statistician William Booth also wanted to bring the plight of these people to the attention of the authorities. He did it through his book and, for the first time in detail, with maps. He color-coded the buildings in seven categories on the street plan, called *Descriptive Map of London Poverty*. These ranged from "Upper middle and Upper classes, Wealthy" at the top end to "Lower class, Vicious, Semi-criminal" at the lower end.

In Booth's mind, poverty and lawlessness were inextricably intertwined. Urban segregation correlated with social deprivation. The organizers of the 1841 Ireland census had come to similar conclusions when they stated that so-called third-class streets in Dublin were inhabited by "artisans, huxters, and low population". (A huxter or huckster was a pedlar or hawker.)

In trying to show that poverty can be a source of vice, Booth was implying that the only way to reduce crime was to reduce poverty. He also determined that poverty is related to unemployment (no surprise there) and is often age-related. People who were too old or infirm to work had to turn to crime to survive. From this analysis he became a champion of the idea of state pensions for the elderly and partly for pioneering this approach was made a Privy Councillor in 1904. Four years later his influence finally saw the passing of the Old Age Pensions Act. The effect was truly life-enhancing for millions over time: Booth's map can thus be seen as one of the most important ever made.

When Booth died in 1916 the lowering of some social barriers that followed World War I (1914-18) had not yet taken place in Britain. But even 100 years after Booth's time, demographic mapping is still being employed as an indicator of wealth or at least creditworthiness. An insurance company will give you an instant quotation for the premium to insure your car, but only after you tell it where you live and even the post code of your exact location. Inner cities are higher risk areas than country districts, or so says the insurance company. Applications for credit cards, loans, and mortgages may be judged in a similar way. A satisfactory result depends on what has come to be known as the post, or zip code lottery.

RIGHT *A seven-color code classifies the social classes of the occupants of the dwellings on Charles Booth's map of London. The wealthy upper class is shown in yellow, middle class in red, "fairly comfortable" in pink, mixed populations in purple, very poor in light blue, and the lowest class in black. This map shows Regent's Park at the center surrounded by mostly upper- and middle-class houses.*

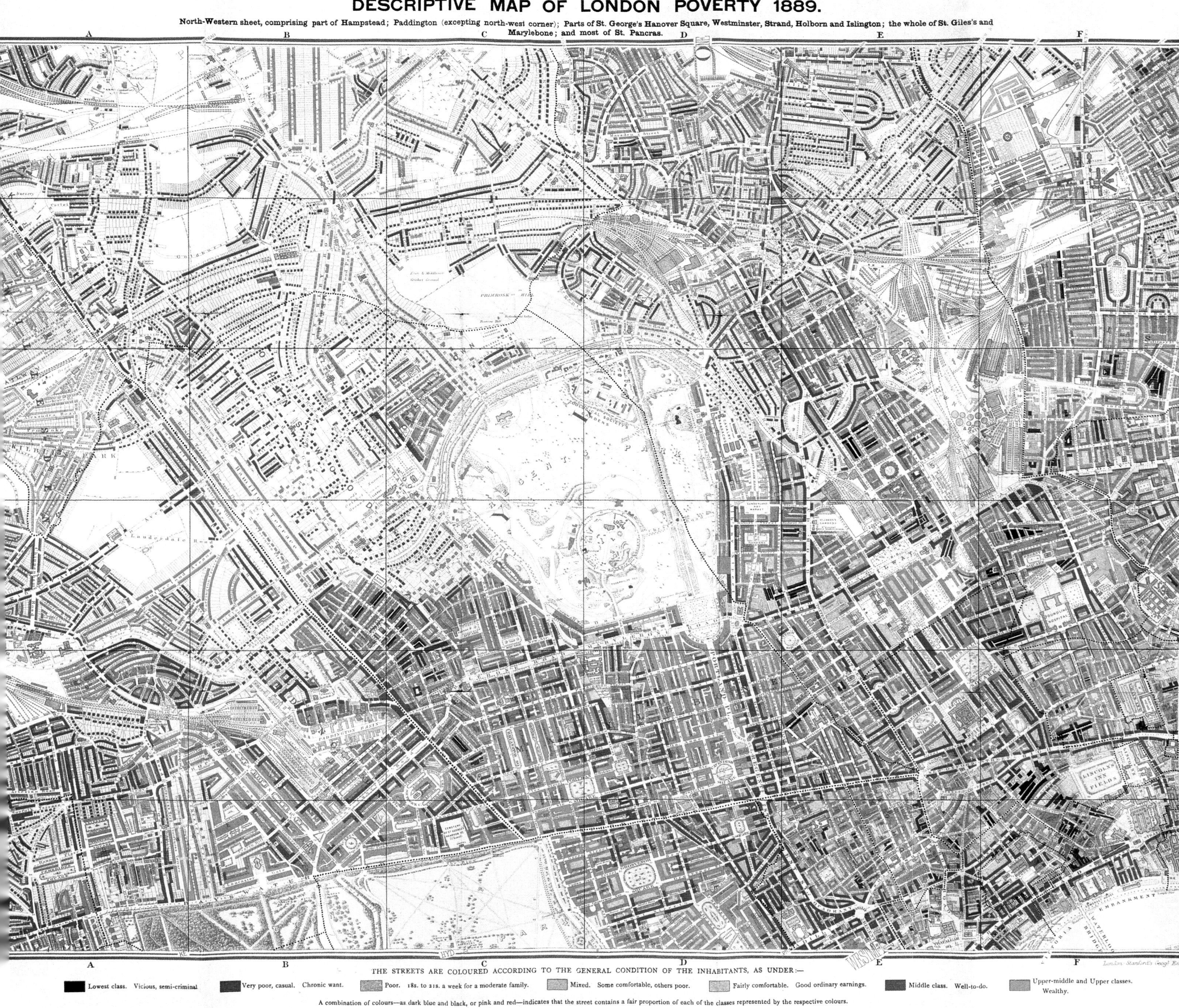
DESCRIPTIVE MAP OF LONDON POVERTY 1889.
North-Western sheet, comprising part of Hampstead; Paddington (excepting north-west corner); Parts of St. George's Hanover Square, Westminster, Strand, Holborn and Islington; the whole of St. Giles's and Marylebone; and most of St. Pancras.
THE STREETS ARE COLOURED ACCORDING TO THE GENERAL CONDITION OF THE INHABITANTS, AS UNDER:—
Lowest class. Vicious, semi-criminal
Very poor, casual. Chronic want.
Poor. 18s. to 21s. a week for a moderate family.
Mixed. Some comfortable, others poor.
Fairly comfortable. Good ordinary earnings.
Middle class. Well-to-do.
Upper-middle and Upper classes. Wealthy.
A combination of colours—as dark blue and black, or pink and red—indicates that the street contains a fair proportion of each of the classes represented by the respective colours.

Utilitarian and Beautiful:

Harry Beck's Idea

Harry Beck's 1933 London Underground map was a brilliantly simple solution to a spatial representation problem that would be copied throughout the world. A train map doesn't have to show the real distance, just where to get on and off. The Freudian art critic Anton Ehrenzweig (1908-1966) pointed out an added delight of Harry's map in his book The Hidden Order of Art: A Study on the Psychology of Artistic Imagination. This functional diagram assumes an extraordinary plasticity if we ignore its information content. In fact, it affects many viewers as abstract art, when it is viewed *upside down*.

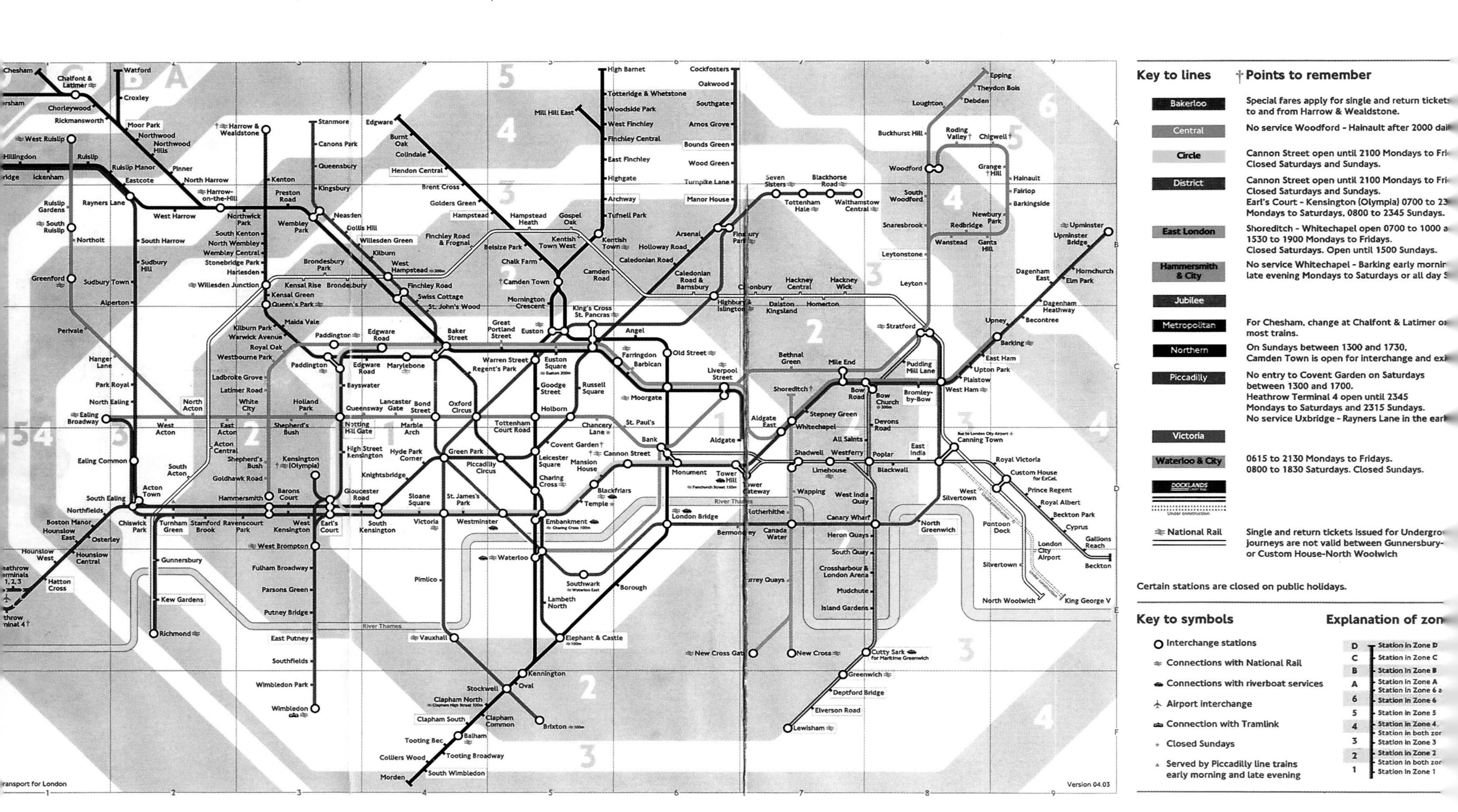

Harry Beck was an electrical draftsman, so was trained in the production of drawings of electrical circuits; circuits used in the running of the London Underground, the subway, itself. His training had given him knowledge of the symbols and techniques used to depict wiring, resistors, diodes, and junctions. He was versed in the color coding used and had the impressive drafting skills to produce concise drawings—this was after all his job with the Underground. He lived and worked in London, so had experience of the system.

The flow of electrcicity is yes-no-reverse (ignoring "volume" and "speed") decisions. All you need is an on-off switch, plus a little more stuff going on to reverse flow. Even less primarily complex is the binary system of the computer: yes-no, one or zero: and if we all only ever traveled along a line (as some philosophers argue we do all our lives) that is the model: you either go forward, or you don't. The model is perfect for a journey on a train. You either choose that train, or another. You can of course go a long way around the circuit to arrive where you want to be, but *at any one moment* when you are on the subway, there is only one correct decision—forward or stop, get off.

This is the first clever aperçcu of Beck's new schematic. Whatever is happening at street level is irrelevant, you don't need geographical coordinates to decide yes or no. The second is to limit the pattern of lines to 45-degree (orthogonal) turns only, again, ignoring the actual world. The third is to forget distance: if you're going to Shady Grove in Washington D.C., or Arnos Grove in London, then you're going. You can always ask how long it will take (as opposed to how far it is). In fact, the traveling times between most stations on the map, certainly in the central section, (which Beck expanded in "scale" for clarity) are similar. The third is color-coding the lines. Beck only had eight to deal with, there are now 14, but it still works. Not everyone adores Beck's design; accordng to Yale's Professor Edward Tufte:

> "The Underground Map and Minard's famous *Carte Figurative* [a thematic, "flow" map] of the French Army's disaster in Russia in the war of 1812 are alike in important respects: both are brilliant, and neither travels well. The Underground Map and Napoleon's March are perfectly attuned to their particular data, so focused on their data sets. They do not serve, then, as good practical generic architectures for design; indeed, revisions and knock-offs have uniformly been corruptions or parodies of the originals."

To which one might respond, Napoleon retreated from Moscow, but its subway has a Beck-inspired map (though without the 45-degree rule, and it's not as "clean"). Unlike Napoleon, Beck conquered the world—plus the heavens, in the form of cvil airline maps. Professor Tufte also says that the map nevertheless exemplifies "the deep principles of information design in operation, as well as the craft and passion behind great information displays." And he is right that this schematic approach is not always appropriate: London Transport has replaced some of its bus street plan maps with schematic maps—fine if you need only to transfer from one bus to another, hopeless if you want to work out where you are.

Beck took many years to perfect his map, without the use of a computer it was all drawn by hand. He was also not fully recognized, nor rewarded, by his employer. Even he would be amazed that his map is reproduced over 60 million times a year by companies *other than London Transport*, on postcards, plates, and so on; which brings us to Anton Ehrenzweig's observation. Turn the book around and see if the pattern of lines assumes a greater depth and "energy," as he suggested.

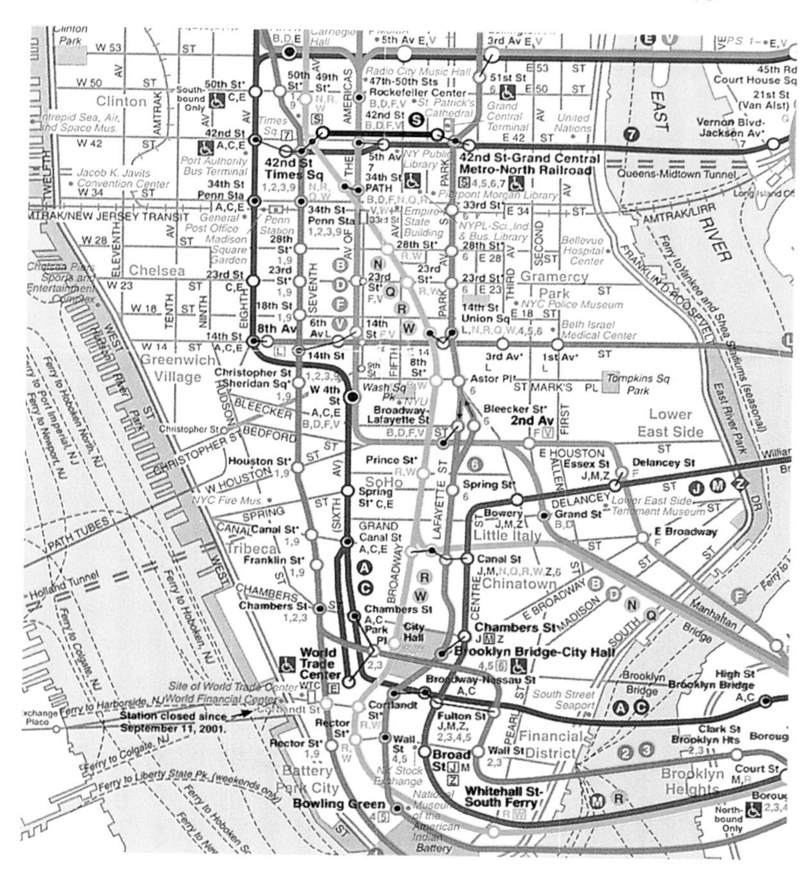

LEFT *Harry Beck's map shows the relationship between stations, rather than a geographical actuality. In 1947 Beck quit London Transport to become a design teacher. He continued to update his map, but LT gradually eased it from his grasp.*

ABOVE *This Manhattan subway map may look in need of more of the Beck touch; but its ambitions are greater, telling you not just how to get there but what you will find. We all interpret complex graphics better today than in the 1930s.*

Hermann Bollmann:

Distorting to Reveal the City

For charting the seas and mapping the land, cartographers produce a range of maps at various scales showing details of depths, heights, and topography. Street plans are meant to help people find their way within towns and cities; but the conventions of flat projections, whatever the scale and method of height description, are simply inimical to the human brain and eye at street level.

Additionally, all the conventional types of maps presuppose that the people using them know where they are in the first place. Users of road maps, for instance, have reference to signs showing place names and route numbers. But ask for directions within a city and you will be told: "Go that way for three blocks and then turn left at the Chrysler building." In 1962 German cartographer Hermann Bollmann used this tourist's approach to produce a pictorial map of midtown Manhattan, New York City. It showed the city from above, as it might appear to a soaring bird or a passenger in a helicopter. It depicted the individual buildings, block by block, as well as how tall they were compared to their neighbors. Though this was not merely a bird's-eye view, because perspective distortions were a deliberate part of the design. The idea of the axonometric city map had come to fruition.

Bollmann was a German graphic artist who made his first picture maps of German cities after World War II to illustrate how the buildings had been mangled by Allied bombing raids, thus completely changing the apparent perspective of the cities. He later applied his technique to illustrate the Alpine ski slopes that soon became a magnet for winter tourists. To make his maps, Bollmann and his coworkers used aerial photographs, as well as pictures taken on the ground. For example, his 1962 New York map used no fewer than 17,000 aerial pictures taken by a special camera designed according to his instructions, supplemented by a further 50,000 pictures taken at ground level. He drew each building to a scale of 1 to 4,800 (1 inch = 400 feet). Individual windows were shown, as well as details such as each lamppost and every statue. Famous landmarks, like the now-disappeared Pan Am building, the Empire State, and the Rockefeller Center, were given special treatment. Streets were presented overscale, the extra width preventing overcrowding on the drawing, and building heights were also increased slightly.

Bollmann expanded the exercise for the wider-ranging New York City map for the World's Fair in 1964. Ten years later Curt Anderson published a similar (isometric) map of a small section of the city, while Bollmann went on to produce his axonometric maps of several other major cities around the world. To celebrate the millennium in 2000, the Japanese cartographer Tadashi Ishihara repeated the exercise with a new map of Midtown New York. For accuracy he used more than 7,000 new aerial photographs and was careful to show the actual number of floors in each of the skyscraper buildings.

The unique feature of Bollmann's original maps is the axonometric projection—otherwise known as parallel perspective—he used. It has no vanishing points, rather like the third-angle approach of technical draftsmen and unlike the conventional perspective of traditional artists. At any point on a map drawn in this way, the perspective is the same. Distant objects have the same size as objects that are close up (that is, they do not get smaller as they get farther from the viewpoint). All parallel lines remain parallel, no matter their orientation, rather than tapering toward the vanishing point as in standard perspective. As a result, any distance (or height on a 3-D map) can be measured and translated into real terms on the constant scale of the map's axes. Hence the name axonometric. The whole idea has now graduated from street maps to inspire the creators of computer graphics—especially those employed in creating the high-speed action of computer games.

RIGHT *This map of downtown Chicago is drawn in the unmistakeable style of Hermann Bollman, who pioneered the technique with his maps of New York prepared in the 1960s. To give sufficient separation between the buildings, the mapmaker has drawn the streets slightly wider than in real life. The intense blue of the Chicago River is another visual clue to position, simple but effective.*

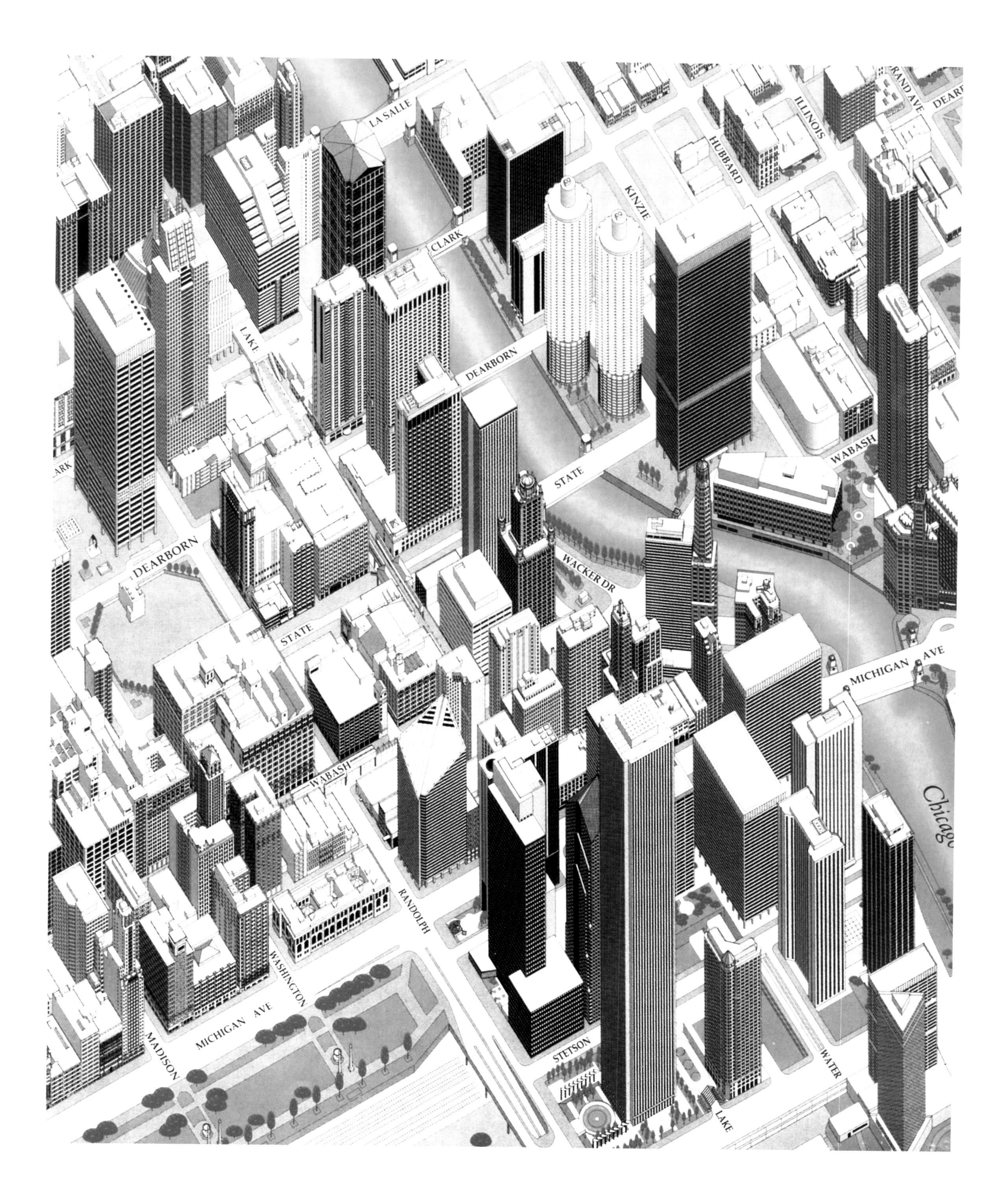
LA SALLE
ILLINOIS
HUBBARD
KINZIE
CLARK
LAKE
DEARBORN
STATE
WABASH
DEARBORN
WACKER DR
STATE
MICHIGAN AVE
WABASH
RANDOLPH
WASHINGTON
MICHIGAN AVE
MADISON
STETSON
WATER
LAKE

NASA:

Surveying Venus

By the fourth quarter of the 20th century, surveyors and geologists had worked out accurate ways of recording and mapping the permanent physical features of the Earth. But what about less permanent features, which change over time? And more challenging still, what about the surface features of worlds other than the Earth? The problems are just the same: the heavenly bodies—and the universe itself—stubbornly remain curved.

In 1972 the American National Aeronautics and Space Administration (NASA) launched into orbit the first of its so-called Earth Resources Satellites. Soon renamed Landsat, these satellites orbited the Earth at low altitude to map details of the surface such as soil, forests, crops, and so on. They used two types of sensors. Passive sensors picked up light and infrared radiation reflected from or emitted by the Earth, whereas active sensors transmitted microwave radio waves and detected signals that bounced back to the satellite. A computer assembled the reflected signals into a detailed map.

This technique of using microwave "echoes" is of course radar, and if it works for land, why not the sea? The answer to this question was the satellite Seasat, which from out in orbit mapped the hills, valleys and mountains of the Earth's ocean beds. If it works for the Earth, then why not for another planet? Venus, for example. Scientists had already launched space probes that flew past and even orbited Venus, but all they could see was an unbroken layer of dense yellowish cloud 60 miles (100 kilometers) thick. They knew the atmosphere consists mainly of carbon dioxide, liberally laced with droplets of sulfuric acid. But what was the surface like underneath? Was the terrain similar to Earth's, as some scientists had predicted?

Fleeting images from Soviet *Venera* space probes, which landed on the surface of Venus in the 1980s, showed a landscape of barren stony desert. But scientists needed a more global view, and this was finally provided beginning in 1990 by NASA's space probe *Magellan*. The probe was launched in May 1989 from the Space Shuttle *Atlantis*. It reached Venus in August the following year and went into an elliptical orbit that brought it to within 190 miles (300 kilometers) of the surface. It takes Venus 243 Earth days to rotate once on its axis, and between September 1990 and September 1992 it did this three times. Throughout this period *Magellan* made radar maps of the surface, so that at the end of the three cycles it had mapped 98 percent of the surface. The resulting images captured details as small as 330 feet (100 meters) across.

The maps revealed that more than three-quarters of the planet is covered by flows from ancient volcanoes. Rolling plains have wind-blown dunes as well as sinuous valleys and channels. There are also belts of highly deformed mountains. Magellan picked out more than a thousand impact craters, resulting from meteorites that bombarded the planet earlier in its history. The ground is littered with what astronomers call ejecta, which is material that splashes out of meteor craters at the time of impact. Some of it may also have been hurled out of erupting volcanoes. The craters are up to 16 miles (25 kilometers) across, and the tallest volcanoes tower 2.5 miles (4 kilometers) high into the acrid sky. Some of the volcanoes may still be active.

After three years *Magellan* descended to a lower, more circular orbit to map the gravity field around Venus. But it could not detect a magnetic field. Like the Earth, the planet has a large nickel-iron core. There is no magnetic field, however, because Venus rotates so slowly. (The Earth's magnetic field results from a dynamo effect created by the fairly rapid rotation of the core.) NASA finally lost radio contact with *Magellan* in October 1994. In the four years it functioned, it revealed the Ancient Greek's evening star to be an inhospitable, barren furnace of a world. And if they had had the smallest inkling of what it is like, the Romans would surely not have named it after their goddess of love.

RIGHT *NASA scientists mapped mosaics produced by Magellan's synthetic aperture radar onto a computer-simulated globe to produce this total global view of Venus. It is centered on latitude 180 degrees east, and slight gaps in the scans were filled with data from the earlier* Pioneer Venus Orbiter *missions.*

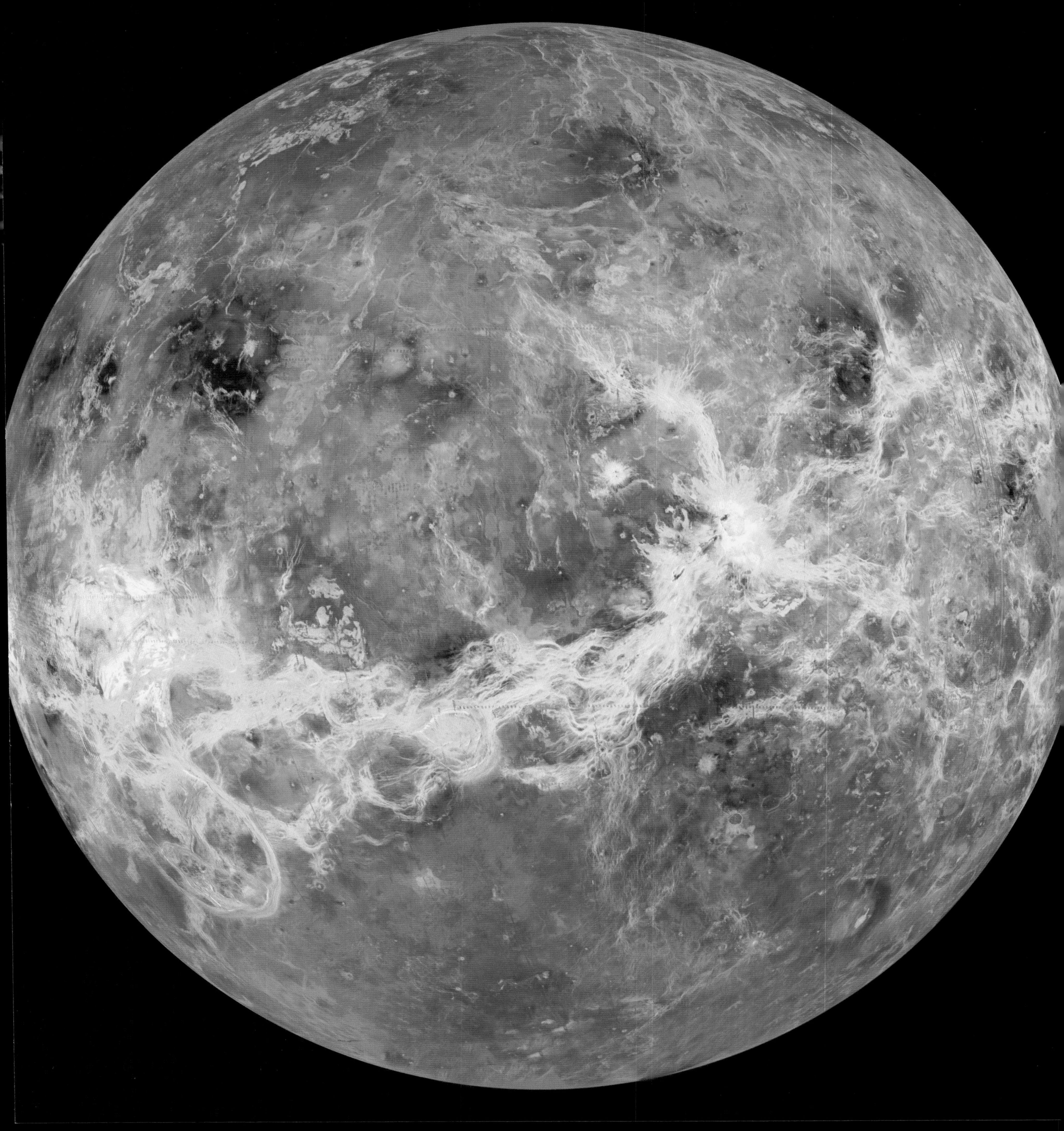

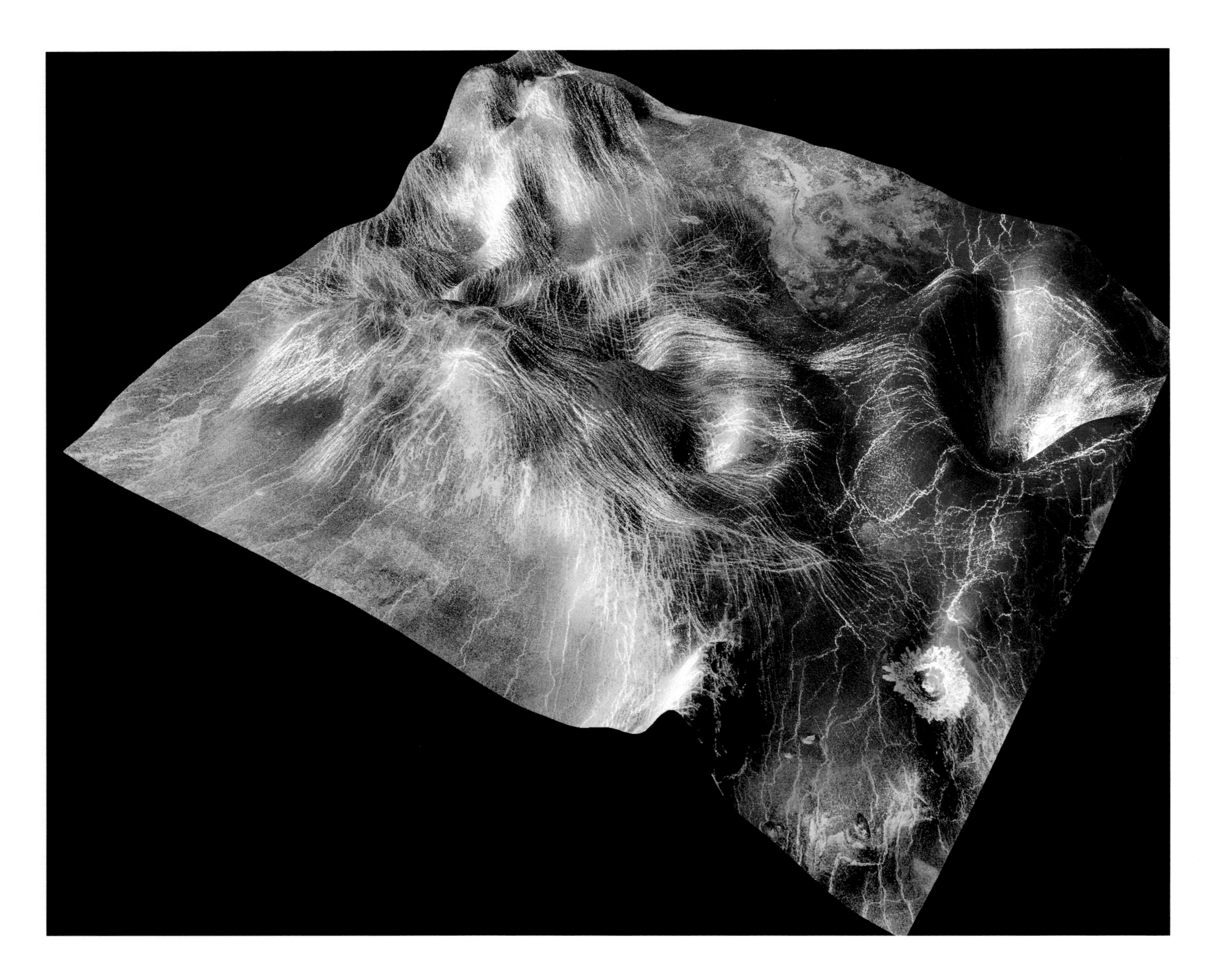

ABOVE *On the lowland plains of Venus, molten magma from below the crust sometimes rises to produce a localized "hot spot." It first pushes up the surface into a dome, after which cooling and contraction take place as the dome collapses to create a depression called a corona. The purple corona on the right is about 62 miles (100 kilometers) across and 0.6 mile (1 kilometer) deep. The vertical scale here is 100 times larger than the horizontal scale. It is a type known as an arachnoid because the radial fractures resemble spider's legs. An early stage of corona formation can been seen as the fractured ridge on the left. The computer-generated colors represent the emissivity of (amount of incident radiation reflected by) the surface, from high (purple) to relatively low (red).*

RIGHT *The mountain on the right horizon of this 3-D computer-generated view of the surface of Venus is the huge volcano Gula Mons, which towers to a height of 9,850 feet (3,000 meters) above the surrounding plane. The smaller volcano to the left, Sif Mons, is 6,560 feet (2,000 meters) tall and 1,860 miles (3,000 kilometers) across. The viewpoint is 3,940 feet (1,200 meters) above the surface. The small-scale structure of ridges and valleys in the foreground has added color simulation based on previous photographs taken by landers deployed by the Soviet* Venera 13 *and* 14 *space probes. The overall image was produced by the Multimission Image Processing Laboratory of NASA's Jet Propulsion Laboratory*

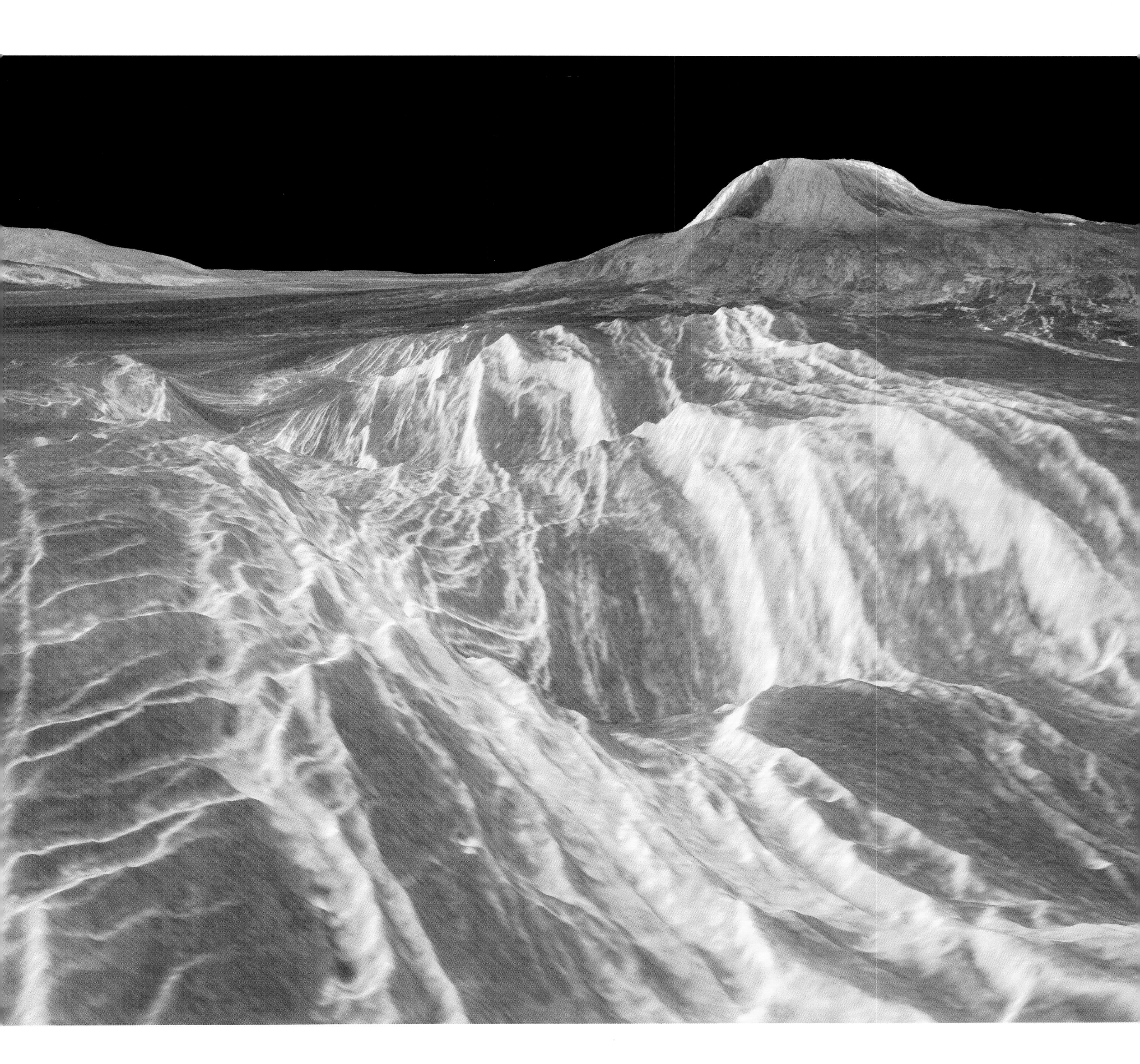

Bathymetry and Predictive Computer Models:

Mapping the Tsunami

The Sumatran-Andaman megathrust earthquake occured on Sunday, December 26th, 2004 at 00:58:53 GMT (7:58:53 am local time at epicenter) with Mw=9.0 NEIC, (later estimated 9.1 or even 9.3). Epicenter Latitude 3.32 North, Longitude 95.85 East, or 3.09N, 94.26E southwest of Banda Aceh in Northern Sumatra. What a weight of human suffering lies behind the "technical" description.

The earthquake occurred on the interface of the India and Burma plates and was caused by the release of stresses that develop as the India plate subducts beneath the Burma plate. The India plate begins its descent into the mantle at the Sunda Trench, to the west of the earthquake's epicenter. The trench is the surface evidence of the plate interface between the India and Australia plates, situated to the southwest of the trench, and the Burma and Sunda plates, situated to the northeast.

In May 2005 scientists reported in the journal *Science* the earthquake that unleashed the fatal waves had lasted for about ten minutes, in comparison with ten seconds for the most powerful earthquakes normally experienced. It produced the largest rupture in the ocean's surface ever observed (800 miles/12,900 kilometers). Charles Ammon, Associate Professor of Geosciences, Pennsylvania State University, stated:

> "Globally, this earthquake was large enough to basically vibrate the whole planet as much as half an inch, or a centimeter. Everywhere we had instruments, we could see motions."

The tsunami, which traveled at about 500 mph (800 km/hr), even reached the eastern coast of Africa, almost 3,000 miles away from the epicenter. The death toll has been estimated at well over 230,000 and will never be known with any accuracy, with bodies washed out to sea. Two months later, 500 bodies a day were washing up on the shores of Indonesia. The tsunami devastated the coastlines of Indonesia, Sri Lanka, South India, Thailand, and other countries with waves of up to 100 feet (30 meters). Their terrific speed and distinctive wavelength (a tsunami is effectively a series of waves, not just one) allow tsunamis to be identified by buoys moored in the ocean. Although seismic networks recorded the massive earthquake, there were no wave sensors in the Indian Ocean because tsunamis are rare in the region. Thus, no warnings were issued. A single wave station south of the earthquake's epicenter registered tsunami activity less than two feet high heading south toward Australia. The Pacific Ocean international tsunami warning system was started in 1965, the year after tsunamis from a 9.2 earthquake struck Alaska in 1964. It is administered by the US National Oceanic and Atmospheric Administration (NOAA). Member states include all the major Pacific Rim nations in North America, Asia, and South America, plus the Pacific islands, Australia, and New Zealand. If India and Sri Lanka had been members, both would have had several hours try to evacuate coastal areas. Even the best and most expensive early warning system would have been next to useless for Indonesia, however, since the epicenter was so close.

The magnitude of the disaster led not only to charitable donations from around the world on an unprecedented scale, but also a huge scientific examination, not least those under the auspices of the Intergovernmental Oceanographic Commission of UNESCO. It included, for example, the height of water marks on the walls of the port of Sibolga, Sumatra, recorded by representatives of the Institute of Computational Mathematics and Mathematical Geophysics, Siberian Division, Russian Academy of Sciences. HMS *Scott* quickly took members of the British Geological Survey to acquire high resolution "swath" bathymetric data over that part of the area of the main fault rupture in Indonesian waters—to map the newly formed seabed.

Much of the research is encapsulated in computer-designed mapping, in the hope of designing better early warning systems, now being put in place. Though the events of December 26th 2004 remind us that even if we may have mapped the plates of the earth's surface with great accuracy, yet we do not control them, and never will.

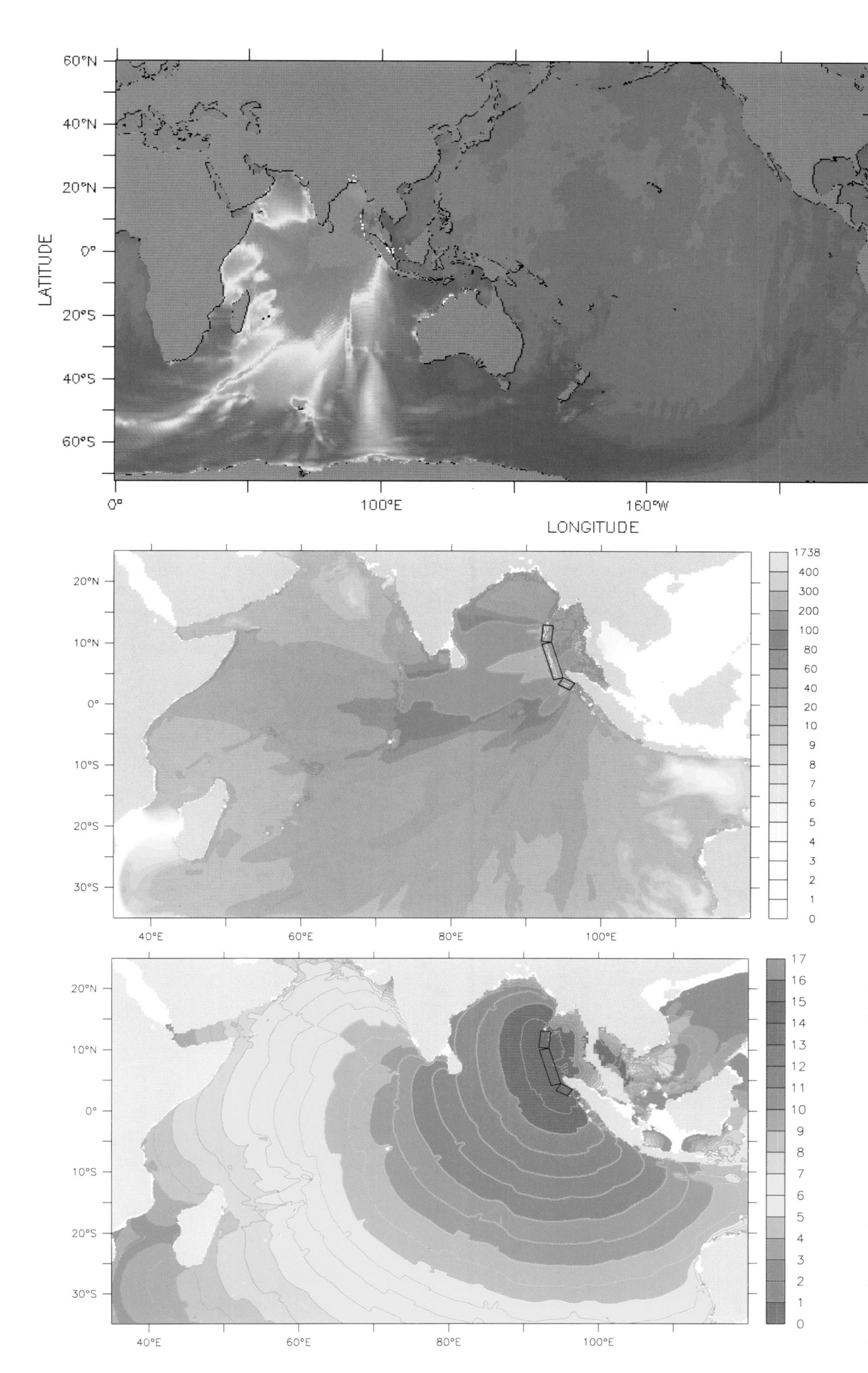

ABOVE AND LEFT *The wave amplitude in centimeters (above) and millimeters (right). The differences lies in the time elapsed and of course scale. As is common on many thematic maps, red spells danger (above), or the extremity. Although the maps are produced through computer software, we choose the color symbology. Tsunamis only get massive when right inland in the shallow water. Several days after the Indian Ocean tsunami, scientists from the NOAA Laboratory for Satellite Altimetry in Silver Spring, Md., examined data from four satellites, two operated jointly by NASA and the French space agency, CNES, the European Space Agency's* Envisat, *and the U.S. Navy's* Geosat Follow-On. *By chance they were all in the right position to measure sea level variations on the fateful day. This data can be used to test tsunami computer models.*

LEFT *Arrival time of the first wave, in hours. By studying the arrival time of of the* p *and* s *waves from an earthquake to seismographic stations, it is possible to predict the arrival time of a tsunami; but the variables are complex, not least the shape of the seafloor.*

Human Geography:

Invisible Values, Invisible Lives

Through the development of satellite mapping systems, it is now possible to accurately map the physical geography of the planet in astonishing detail. Satellite maps are used to track hurricanes and cyclones. They are also a means of monitoring the slower geographic changes brought about by desert expansion, coastal erosion, and the retreat of glaciers. We have also greatly enhanced our understanding of our world and ourselves through maps created by combining geography and statistics.

One reason for combining these disciplines has been the development of human geography. And an obvious use of such a geographic map system is to reveal the relative health and wealth of humanity throughout the world. The United Nations has been the foremost producer and user of maps based on the principles of human geography. They have created maps that present abstract ideas and statistical data in a visible form. Many of the UN's maps are attempts to reveal what is invisible to the eye of the satellite camera, but essential to our existence: the quality of human life. Over the years, the United Nations has had to prioritize its distribution of financial aid to nations and regions throughout the world. Mapping systems have been used, for example, to indicate regions most in need of medical aid to prevent an outbreak of cholera. Other mapping systems have been used to determine geographic boundaries of crop failures and so evaluate the volume of food aid required to alleviate famines.

For the UN, the difficulties involved in estimating levels or measurable degrees of poverty in the Third World have proved to be extremely difficult. With so much poverty in the world, the UN found that it needed some system, or Richter Scale (as used in measuring earthquakes) to compare the poverty of one nation against the poverty of another. It was obvious enough that Third World poverty could not be meaningfully translated into terms that were comprehensible in terms of European and North American monetary systems. Annual wages or financial estimates in the developing world meant very little when comparing one nation with another. Some nations with large populations of landed peasants managed to live and prosper through systems of barter and trade. In monetary terms they seemed to survive without any income whatever. Other Third World poor forced to live in urban slums require money for everything: water, food, rent, electricity, fuel, and clothes. In terms of annual wages, these urban poor may appear to be financially better off than their rural compatriots, but in terms of health, housing, and sanitation they are far poorer.

The United Nations began to look for a mapping system that would truly reflect the human condition—and not just a duplication of the World Bank statistics based on annual per capita incomes, or the gross national product. In 1993, the United Nations adopted the Human Development Index This was devised by Mahbub ul-Haq (1934-1998), at various times Director of the World Bank, Pakistan's Minister of Finance, Planning, and Commerce, and UNDP Special Advisor. It was considered to be the most accurate system to evaluate the relative "wellbeing" of nations.

The Human Development Index did not just look at the financial status of each nation, but attempted to find a statistical means of evaluating the quality of life. This system is based on a combination of statistical figures related to wealth, health, and education. Financial rating is based on the gross production per capita but is adjusted in terms of local purchasing power. The health rating is based on life expectancy at birth. The educational rating is based on student enrollment figures and adult literacy percentages.

Each nation's score is based on a scale rated from zero to one (to three decimal points) and is ranked in order from one to 162: the number of countries acknowledged by the UN. In the 2001 UN report based on figures from 1999, Norway was at the top of the ranking with a score of 0.939; while Australia, Canada, and Sweden shared second place with a score of 0.936. These were closely followed by Belgium, United States, Iceland, Netherlands, Japan, and Finland as the top ten ranked nations, with scores ranging from 0.935 to 0.925. Statistics vary from time to time, in one year France was rated joint first with Canada.

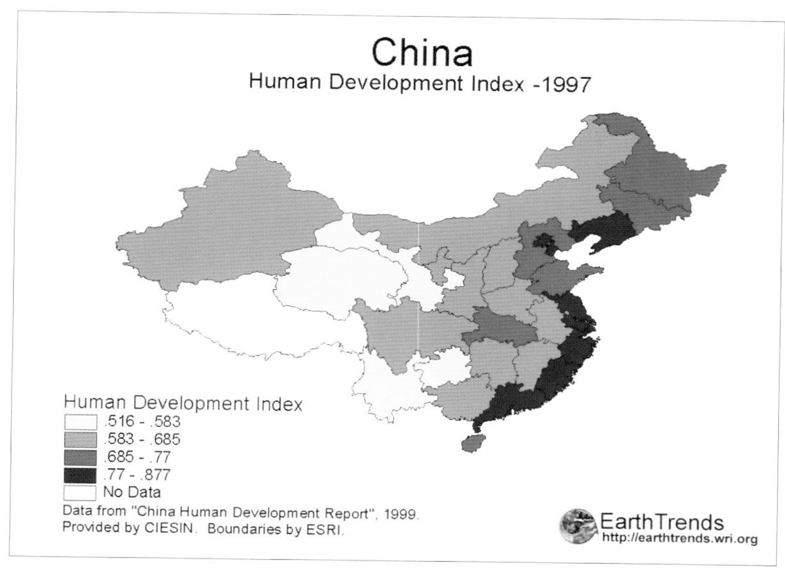

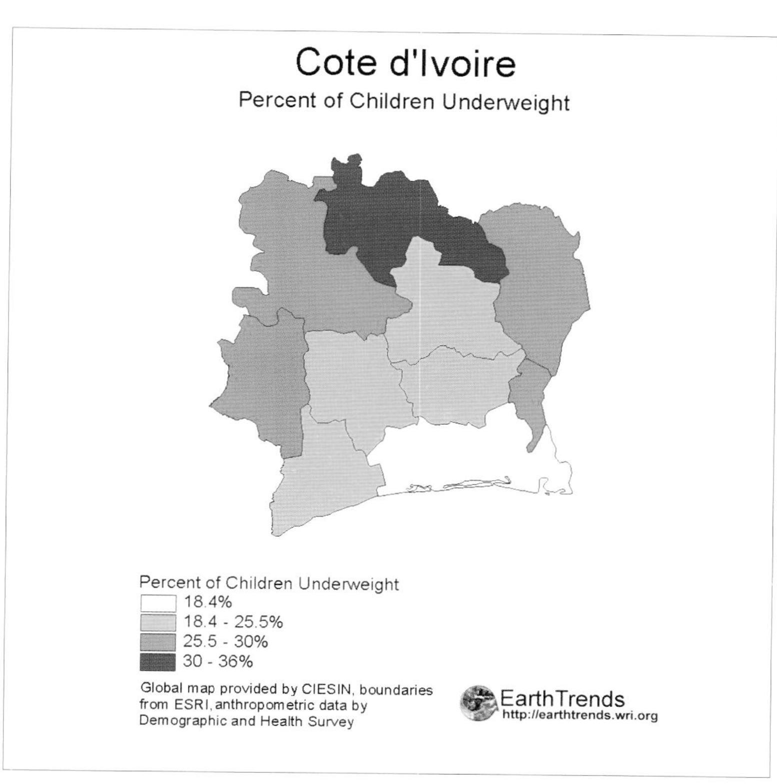

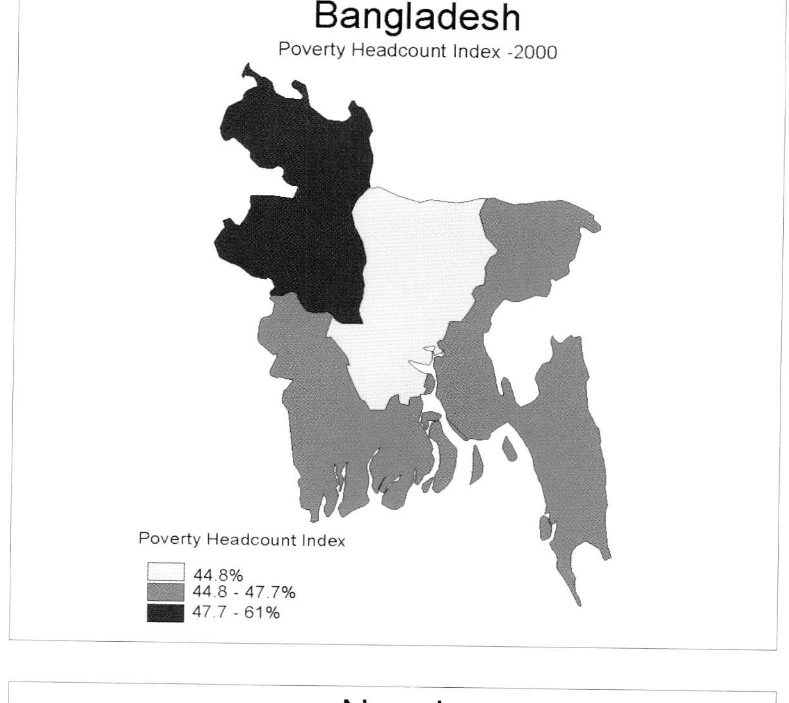

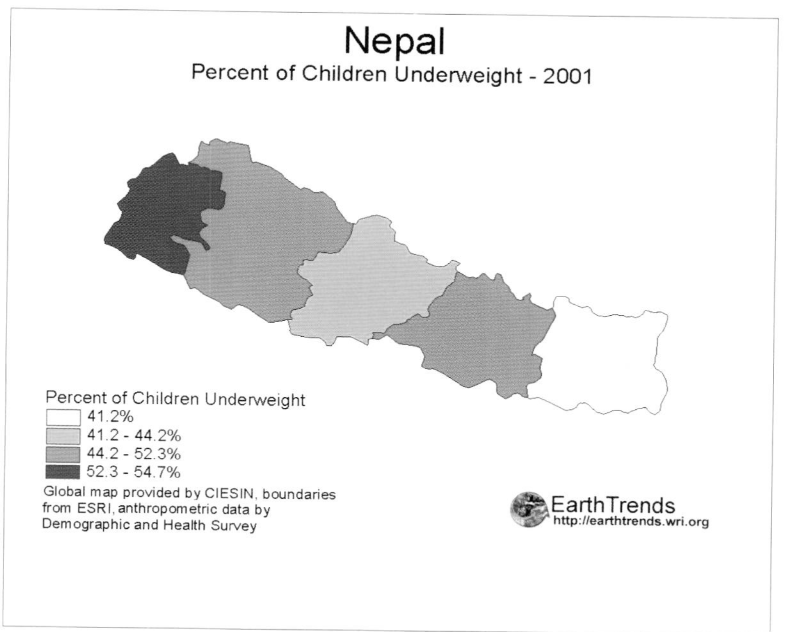

However, the pattern is consistent and the difference in life quality for most top industrialized nations is hardly noticeable from year to year.

The poorest ranking in the world in 2001—with a score of 0.258—was Sierra Leone with a ranking of 162. Second from the bottom was Niger with a score of 0.274. Burundi, Burkina Faso, Ethiopia, Mozambique, Guinea-Bissau, Chad, Central African Republic, and Mali complete the list of the ten lowest ranked nations, with scores ranging from 0.309 to 0.378. Again, the ten poorest countries form a predictable and consistent pattern. All are impoverished African ex-colonies torn

ABOVE *Very little can be garnered from the China HDI map (top left) in isolation. It must be compared with other countries and used in conjunction with other information: for example, with 1989 as base "100," Chinese per capita food production rose from 58 in 1960 to 150 in 2000. Similarly, what does the Bangladesh poverty map tell us without similar maps? To even begin to understand, we need to know the population density is 954 per square kilometer (2,470 per square mile), in comparison with 136 for Asia (excluding the Middle East) and 45 for the world. A more immediate comparison can be made betwen the two underweight children maps, though the legends must be carefully read.*

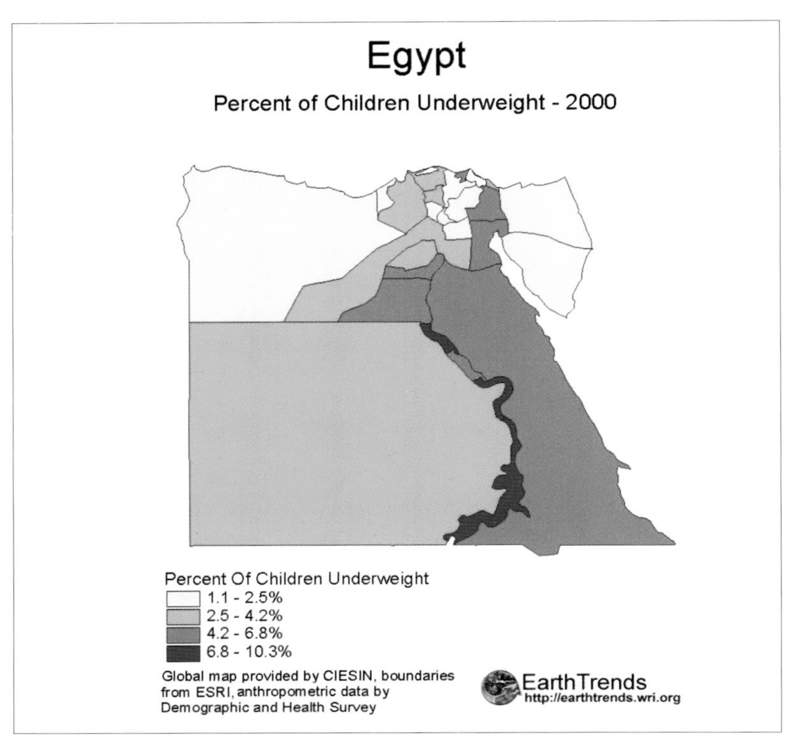

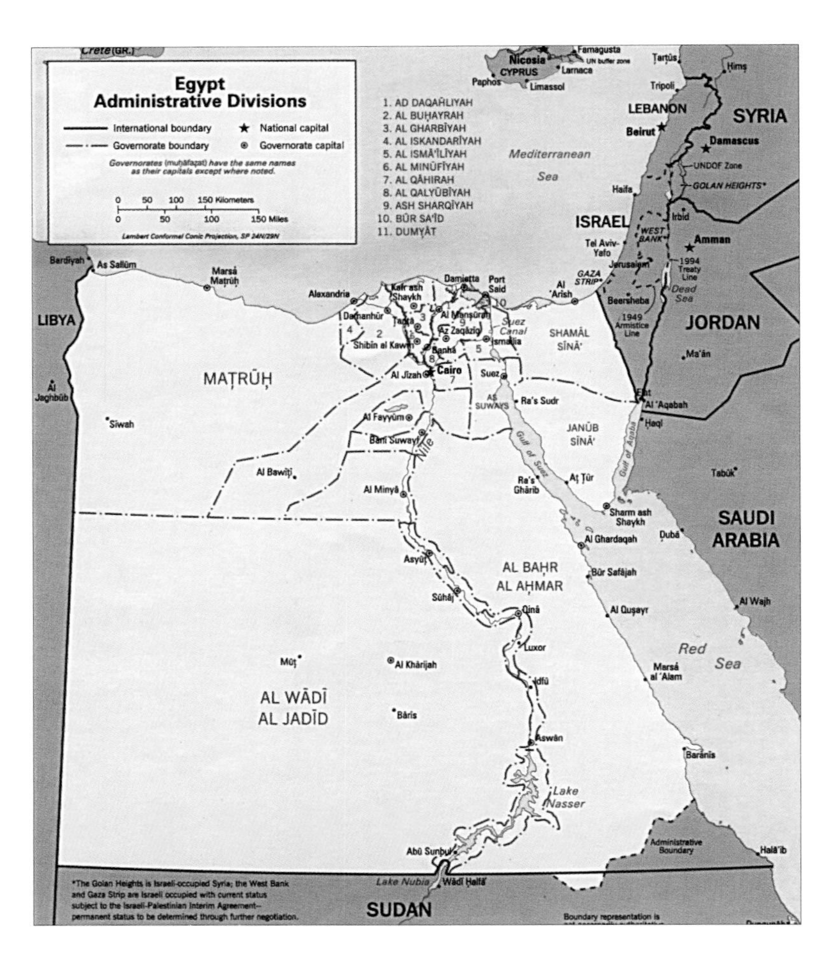

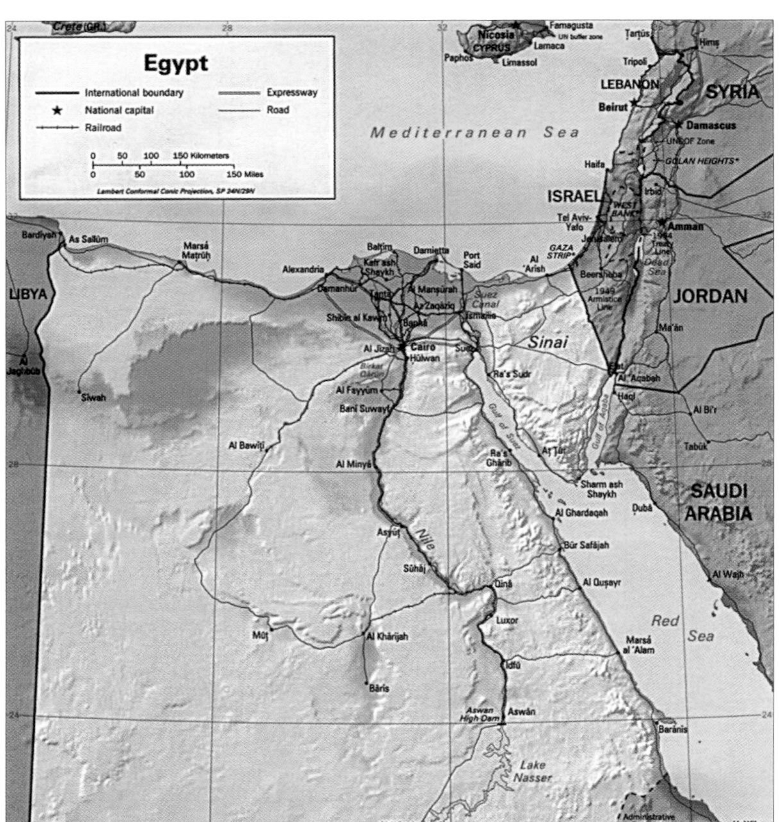

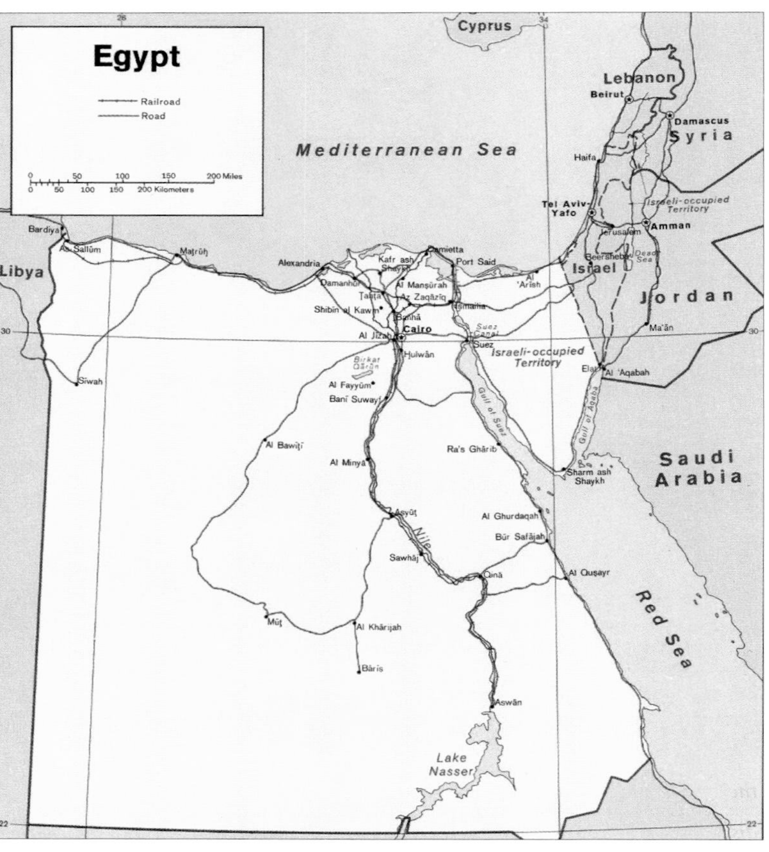

THIS PAGE *Four maps of Egypt, from top left, clockwise, underweight children, administrative (1997), political (1972), and topographical. In 1972 the Arab Republic is preparing to attack, with Syria, Israelis in the Sinai. These maps illuminate one another over time. All maps, even the relief map, are "thematic."*

apart by civil wars. All have very little in the way of natural resources and all are undergoing terrible environmental disasters that will further impoverish them. The United Nation's Human Development Index (HDI) has become the most popularly acknowledged means by which the relative wealth of nations is measured and mapped over the last decade. And yet, even with the acceptance of this system, many questions often arise. In this Index, the US has never ranked among the first five. It has been consistently ranked from 6th to 12th. Is there something wrong with this ranking of America?

How does this rating system of nations relate to the real world? In one sense this index system is quite ludicrous. The US is without doubt the most phenomenally wealthy and powerful nation on earth. Its military and financial power is in any real terms many times greater than the rest of the world combined. One might reasonably expect it to have a better ranking. After all, if America's wealth were evenly distributed, every citizen in America would be a multi-millionaire. However, most of the nation's wealth is in the hands of less than 2% of its population. There is real poverty in many of America's urban slums, and over 30 million Americans do not even have basic medical coverage.

This is the difficulty with systems using national averages: poor minority groups can be ignored. Also, the enormous wealth of America's financial elite tends to inflate estimates of a citizen's "average income." The most outrageous example of an impoverished minority is that of the Native Americans. Many live in rural ghettoes in substandard housing. The infant mortality rate among Native Americans is four times that of the national average. The unemployment level is over fifty percent. The average Native American annual income is one-third that of national average. Average life expectancy is 44 years.

The United Nation's maps based on the HDI have many valid uses when viewed on a global scale. However, they are of limited relevance to poor individuals or impoverished ethnic minorities (though they do show revealing regional variations within countries). Nor is the US unique. Similar examples can be found in every nation. Despite the best efforts of the UN and the Index there still is no single unit or set of statistics for accurately measuring human poverty.

There is, however, a set of Human Geography maps devised by EarthTrends, a project at the World Resources Institute in Washington DC, that employs a convincing and provocative unit. These maps were created to reveal the geographical distribution and the percentage of underweight children in the populations of the world's nations. Economists and politicians may argue endlessly about the relative nature of poverty, but a starving child is—after all—simply a starving child. In a shocking way, using malnourished children as basic units is a sadly appropriate scale for measuring poverty. Cartographic information, of course, like all information, is not always power. Knowing more precisely where the suffering lies does not relieve it.

RIGHT *Children at work in the Nile Delta. Why are children more underweight in the fertile Nile Valley, as shown on the map opposite? Perhaps it is the debilitating disease bilharzia, endemic to the region. Perhaps it is sheer hard labor.*

VIRGIN
POWHATAN
Held this state & fashion when Capt. Smith was deliuered to him prisoner 1607
MONACANS
MANNAHOACKS
POWHATAN
MANGOAGS
CHAWONS
KUSKARAWAOKS
TOCKWOGHS
HONI SOIT QVI MAL Y PENSE
Jamestowne
Powhatan
The Fales
Cape Henry
Cape Charles
Poynt comfort
Smyths Iles
Russels Iles
Winstons Iles
Brookes Forest
Rickards cliffes
Powhatan flu
Chesapeack
PEACK BAY
Scale of Leagues
and halfe
Leagues
THE VIRGINIAN SEA
Discovered and Discribed by Captain John Smith 1606
Graven by William Hole

The Ages of Exploration

"Virginia: discovered and discribed by Captayn John Smith, 1606 ; graven by William Hole." This example of John Smith's map was first published in London in 1624; north is to the right. The myth-making has begun: the famous story of his capture by Powhatan is included top left. The map is not even based on Smith's own explorations, it owes much to an earlier map, and an earlier explorer (see page 124). By the time this map was published, Smith had been back in England for 15 years, never to return to the New World.

Mysterious Origins:

Portolan Sea-Charts

In 1270 St Louis, King of France, led the eighth Crusade. His fleet sailed from Aigues Mort in the south of France. They were bound for Tunis but a sudden storm near Sardinia forced them to take refuge in Cagliari Bay. The king demanded to know where they were and the sailors brought him a map to show him their position. This is the earliest reference to the use of a chart on a ship.

One of the most remarkable things about medieval sea-charts is that even the earliest examples are highly detailed and surprisingly well surveyed. They were used in conjunction with the mariner's compass, which was not invented in Europe until the late 12th century.

These charts are known as "portolans" after the written sailing directions, *portolani*, which they supplemented and superseded. Clear conventions governed the way they were made. They were drawn in colored ink on vellum, usually of sheep or goat skin and while some skins were cut to a rectangular shape, in many cases the neck area of the animal was left on. The basis of the chart was created by drawing a network of direction lines or "rhumbs." The chart maker drew a "secret circle" around a central point and, by drawing a vertical and horizontal line through the center, then by repeatedly bisecting the spaces between, he marked sixteen equidistant points on the circle's circumference. Each of these subsidiary circles was then subdivided in the same way. The rhumb lines were used by the chartmaker for plotting the coastlines and by sailors for planning their courses. Different colored inks were used to distinguish different directions.

Because headlands, islands, and estuaries were important features for navigation, they were enlarged out of proportion on the charts. To avoid obscuring the coastline, place names were written on the land, perpendicular to the coast. They conventionally run clockwise around the Mediterranean.

Scales were provided and though some were simply a functional line of dots or bars, many provided the opportunity for elaborate decoration. Compass roses could also be exquisitely drawn and brightly colored, heightened with gold. Kingdoms were often identified by shields or depictions of the monarch, and cities were represented in a stylized form. The most ornamental charts also included religious symbols and drawings of animals and ships. It is questionable whether such charts were ever intended for use at sea or whether they were made for merchants, statesmen, and ecclesiastical officials to consult (and display) ashore.

The oldest surviving sea-chart is the Carte Pisane, drawn in the late 13th century. Although it was preserved in Pisa, it was not necessarily made there, and is now in the Bibliotheque Nationale in Paris.

The chart is drawn on a framework of two circles of rhumbs, which touch in the sea off southern Italy. Eight wind directions are labeled around the circles' rims. Rectilinear grids occupy some of the chart area not covered by the rhumb circles. On the neck of the chart and near its mid-point there are circles containing a scale with divisions corresponding to the dimensions of the grid squares.

The strong geometric basis to this chart has led to speculation that it may be based on surveys carried out by the Roman fleet and army, with distances measured by counting marching paces and the oar-beats of galleys, and with directions determined by astronomical observations. Julius Caesar is known to have ordered such survey work, continued by Caesar Augustus, and corrupted reflections of classical mapping may be preserved in world maps drawn in the Middle Ages, before the advent of portolan charts. Perhaps once the compass was invented, the utility of the ancient mapping was realized. Or perhaps a medieval trading state such as Venice had sufficient organization, resources, and motivation to begin the project from scratch.

RIGHT *On Mateo Prunes' 1559 chart of the Mediterranean world both real and mythical islands appear in the northwest. The real ones include* Fixlanda *(Iceland) and* Isola Verde *(probably Greenland). Among the mythical islands are* Isola de Brazil *and* Isola de Maydi, *the latter probably a name of Arabic origin.*

AFRICA
ASIA

On the Carte Pisane, the Mediterranean coastlines and islands of Europe and the Levant, and the north African coast to the west of Tripoli, are drawn in detail and are lined with closely-packed place names. By contrast, the Atlantic coast of Europe is known only in the sketchiest terms. Britain (*Isula Engreterra*) features prominently but in a barely recognizable form. The river to London is marked halfway between Cornwall and Dover, probably as a result of confusion with Southampton, a port which had strong trading connections with Venice at this time.

The coverage of medieval portolan charts was restricted to the coasts of Europe and north Africa but later ones like the atlas of portolan charts by Battista Agnese, 1544, record the progress of European exploration. Agnese was a prolific Genoese chart maker. The chart shown on the opposite page indicates how rapidly the coasts of the Americas had been surveyed in the half century since Columbus crossed the Atlantic. The Caribbean is drawn in detail, though Yucatan is shown as an island rather than as a peninsula. Further north, a patch of green suggests that mariners had encountered the phenomenon of the Gulf Stream and the weed of the Sargasso Sea. Inland in South America, the Andes mountains are prominent and the magnitude of the Amazon rain forest is impressive.

It was less than fifty years since da Gama had opened the sea route to India by sailing round the Cape of Good Hope and not much more than twenty years since Magellan had commanded the first circumnavigation of the globe. However, while new geographic knowledge was being acquired, old falsities had not yet been dispelled. The north-east Atlantic ocean is still peppered with mythical islands

BELOW *This portolan chart of the Mediterranean and Black Seas on vellum, probably drawn in Genoa, is the oldest cartographic artifact in the enormous archives of the Library of Congress. Did they know when they acquired the chart in Munich in 1914 that its estimated date would be 1320-1350?*

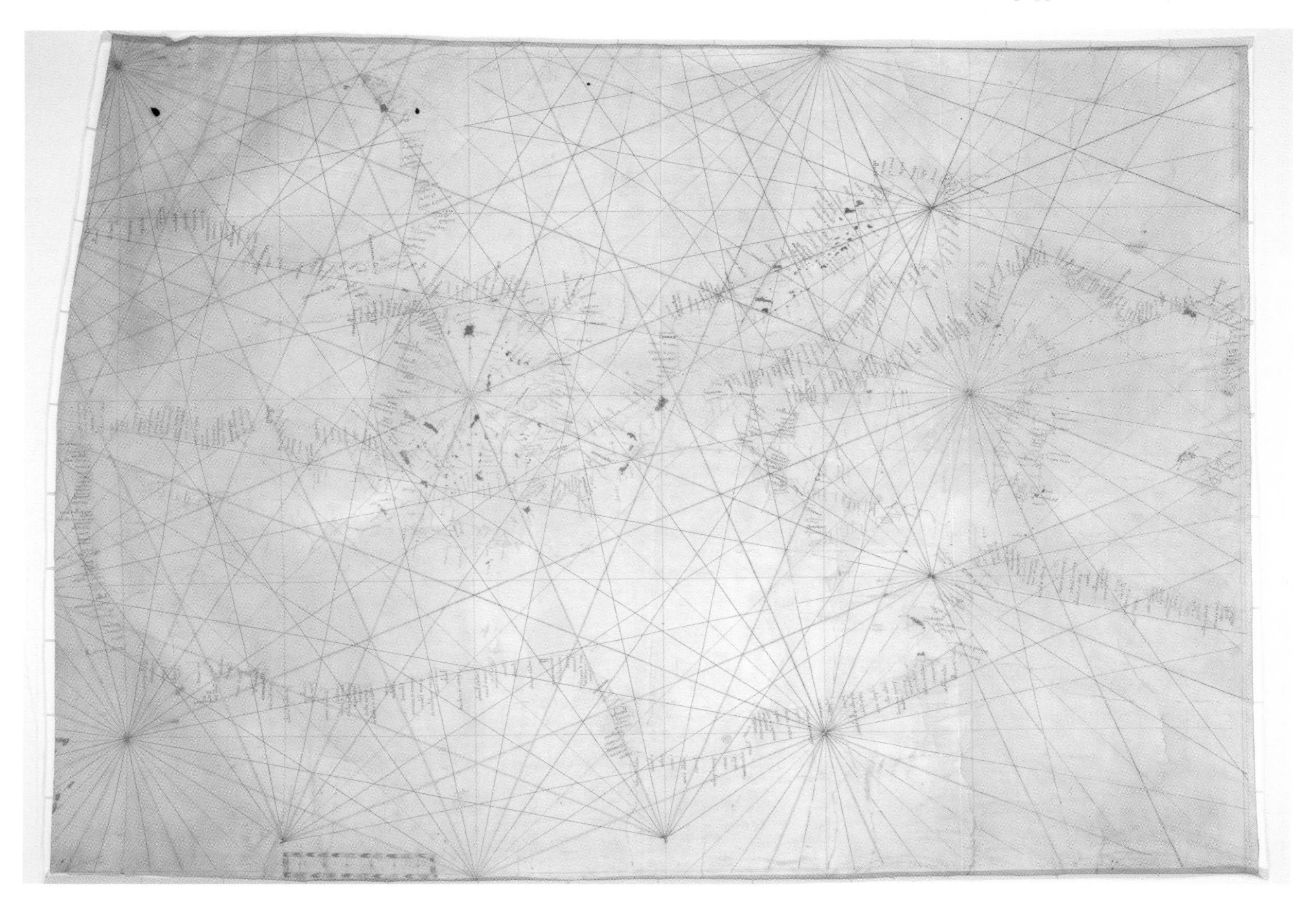

such as those believed to have been discovered by St Brendan. From the 15th century, latitude scales were introduced to portolan charts for determining positions north and south of the Equator. Navigators found their latitude by observing the height of the stars above the horizon. This 16th century chart also marks degrees of longitude, starting from Ptolemy's prime meridian, which ran through the Canary Islands. It was to be more than another two hundred years before navigators were able to find their longitude at sea.

The chart shown on page 93 of the Mediterranean Sea and the coasts of Atlantic Europe and northern Africa was made in Majorca by Mateo Prunes in 1559. This chart was made nearly three centuries later than the Carte Pisane, but its basic cartographic style is very similar. The principal difference is that exuberant decoration has been applied to the simple functionality of the early portolan. Prunes' chart is oriented with west at the top and has three religious figures at its neck. Above them is one of eight windheads: cherubs' faces with bulging cheeks representing the principal wind directions. The chart is drawn with its network of rhumbs centered on the Tyrrhenian Sea and with compass roses on five of the sixteen circumference nodes. The roses divide the compass into 32 points. Scales are drawn on both long sides of the chart, elaborated to look like ribbons. The density of place names is greatest along the northern Mediterranean shoreline and still rather sparse in northern Europe. Banners identify the great cities, and rulers are shown enthroned in tents. Mount Sinai is depicted and alongside it flows the Red Sea, colored red, naturally. Africa is enlivened with indigenous and mythical wildlife.

BELOW *Battista Agnese's 1544 portolan atlas was the most accurate portrayal of nautical distances then available; and at the same time a luxury item for the wealthy merchant class to buy and show off with pride. It consisted of nine charts and a world map. The Gulf of California had only been discovered by Spaniard Francisco de Ulloa five years before, in 1539, but features in the atlas.*

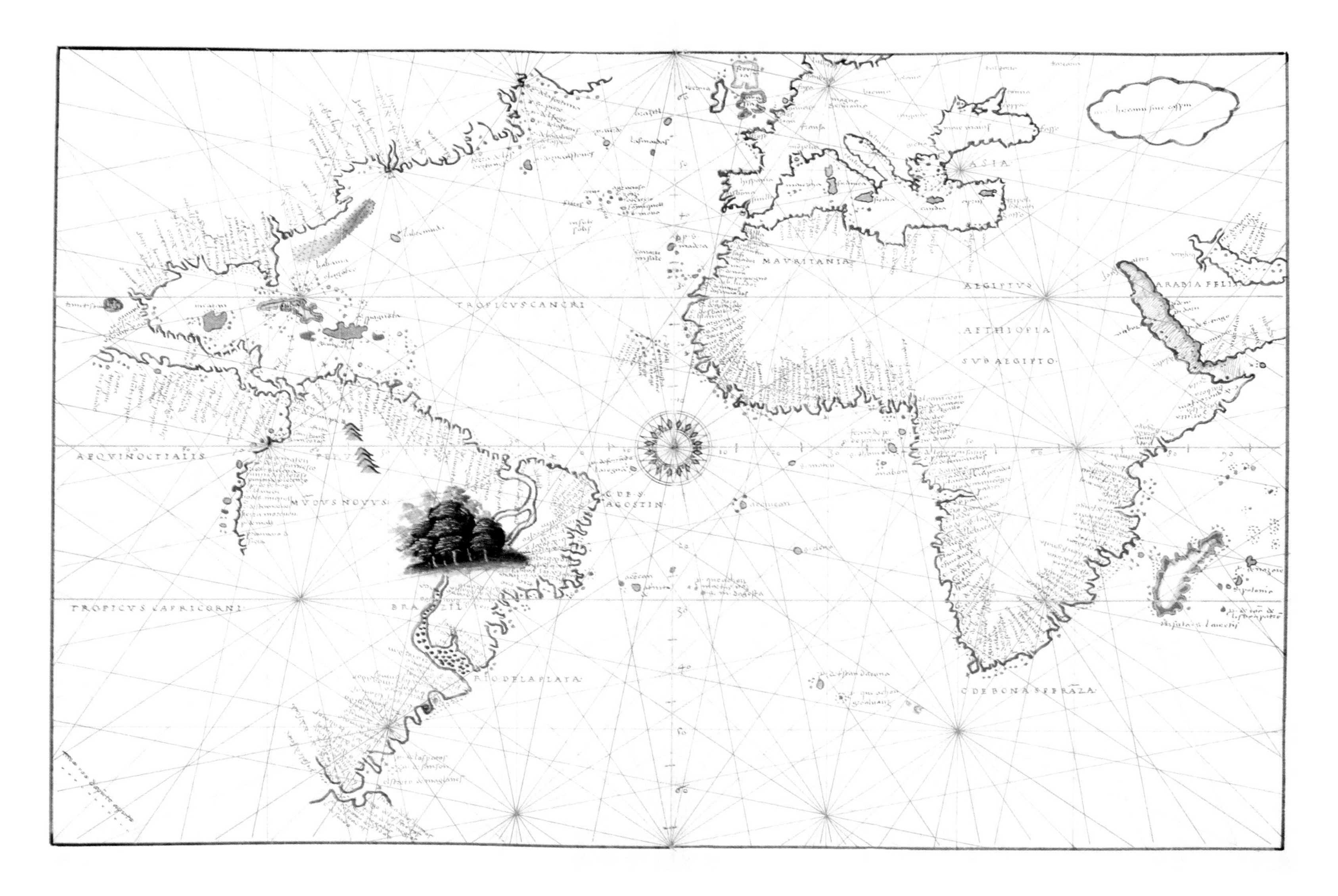

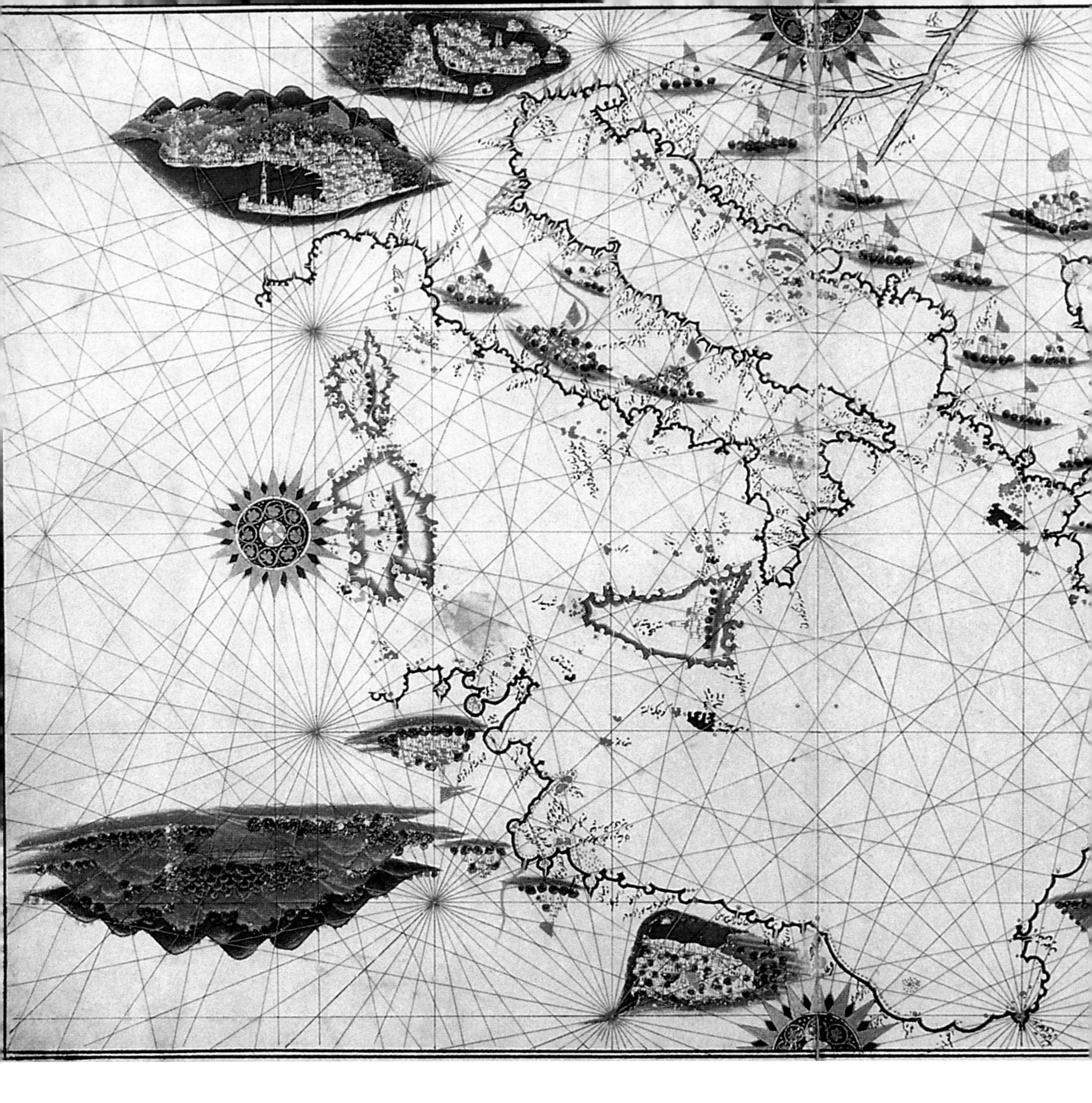

LEFT *Arabic portolan chart of Italy and Greece from the 1500s. One feature of the early portolans is their relative consistency of scale: they cover the entire Mediterranean or a large segment of it, such as its western or eastern part, or the Aegean, or the Adriatic. There are no known large-scale portolan charts of a small area, of a bay perhaps, or the potentially treacherous entrance to a harbor, as you might expect. Though there were "Island Books,"* Isolarii, *which showed the individual islands of the Aegean Sea. These lacked the distinctive network of rhumb-lines. The portolans came into being after the introduction of the compass: the first mention of a magnetic needle in the Mediterranean is in 1187. Pilots could plot their courses according to the points of the compass and their knowledge was captured over time in the charts. The limitations of the portolan, however, are manifest. There is no allowance for magnetic variation or the curvature of the earth; this does not matter within the confines of the Mediterranean, but is a potentially fatal flaw on the oceans. The portolan would gradually give way to the chart based upon celestial navigation. It took a while; portolans were still in use throughout the 16th century, when explorers were going beyond the Atlantic to explore all the world's oceans.*

Apocryphal Vision:

The Greatest Navigational Error in History

"The eyetie will kill us all. Cipango!" He wanted to spit to reinforce the point but daren't for fear of losing more teeth. His body was as one with the ship, his bones seemingly grinding to dust as the scurvy took hold, just as the sea and worms were whittling the hull. They could actually see pieces of timber breaking away now, floating on the dead calm.

As the terrible mix of fear and fury that was mutiny began to infect the crew, and his ship rotted 50 miles off an unknown coastline, the greatest dead reckoning navigator in history was in his tiny cabin writing to his patrons, their Majesties King Ferdinand and Queen Isabella of Spain. "We are very close. From here to the Ganges River there are ten days."

The Italian is of course Christopher Columbus: but the date? And the location? This is not 1492, but September 1502, and Columbus' fourth and last voyage. The rotting caravel lies off the coast of Honduras. Drill a hole through the middle of the Earth from there and you will emerge almost exactly at the Ganges. A decade's exploration of the New World, including several years as Viceroy of the West Indies, had done nothing to shake the delusions of *Cristobal* (he preferred the Spanish form) about a short western route to the riches of the Orient. An extraordinary narrative of triumph and disaster flows from that scribbled map of the New World, a story of self-belief and hubris bordering on insanity that would see the Admiral of the Ocean Seas returned home in chains.

The premise was simple. The overland route to the riches of the east was long and arduous; it also entailed the payment of taxes to a string of middlemen. The western edge of Europe was actually far closer to Cathay than was the Levant. Brave the westward ocean crossing and you owned the world; and there would surely be new lands to claim, as a kind of bonus. Looking for funding, first from the King of Portugal and then in Spain, Columbus sold this vision with a messianic fervor for ten years. What made him so sure?

The first reason is probably the only one that looks, in hindsight, entirely rational. By 1492, with his three ships the *Santa Maria*, *Pinta* and *Nina* ready to sail, the ex-pirate Columbus was an experienced mariner and brilliant navigator. He had sailed to the southernmost point of the known world, the Portuguese trading post of Mina on the Gold Coast, 10 years before. There he saw the first hints of land to the west with his own eyes. Big canes, pine tree trunks, and pieces of apparently wrought wood had floated towards him in the south Atlantic from the western horizon, with all the allure of the sirens. The second reason was a piece of misinformation 13 centuries old. Columbus studied Ptolemy's ancient map of the world, that extended Asia to cover the unknown Americas and the Pacific.

The most important confirmation of his own theories, however, came from an extraordinary source. We know it is the most important because he tells us, in a letter to the Spanish monarchs written on that last desperate voyage of 1502:

> "In the carrying out of this enterprise of the Indies, neither reason nor mathematics nor maps were any use to me. Fully accomplished were the words of Isaiah."

The bedrock of Columbus' belief was the apocryphal Biblical Second Book of Esdras! Here was all the confirmation he needed: it told him the earth was round, the distance west from Europe to Asia, or "India" was very short; and to compound the error, the distance of one degree was $56^{2}/_{3}$ miles. An Italian mile that is, not an Arabic one, which made the proposed route even shorter (see box, right). The man who discovered another world, who had the uncanny skill to navigate thousands of miles across an unknown ocean using nothing but a log and a compass, born in the heart of the Renaissance, was actually *looking back*. His superstitious reliance upon the ancient prophecies is redolent of the our crude idea of the Dark Ages.

Fate was about to play a bizarre trick on the autocratic Admiral. His

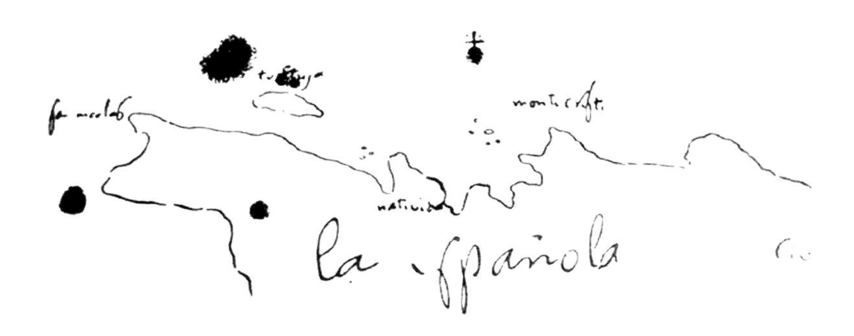

ABOVE This sketch of the coastline of what is now the island of Haiti and the Dominican Republic—the earliest extant map of the New World indisputably by a European—is very probably by Columbus himself through handwriting analysis. It was made on the first voyage, at the end of November. Columbus found the island so beautiful he gave it the most honored name possible: Hispaniola, the Spanish island.

erroneous calculations led him to believe that Asia lay just 78° west of Spain, the equivalent of 3,900 miles (6,280 kilometers) from the Canaries using his "short" degrees. For 36 days and nights the three ships sailed west from the Canaries, out of sight of land after the first three days. 3,900 miles across an unknown ocean.

There was a perfect example on that terrifying voyage of the strange mixture of overweening confidence and tactical cunning that characterized Columbus. He gave an instruction to the crew that after 700 leagues (2,400 miles) they must not navigate at night because the land would surely be near. How many of the desperadoes on board the *Santa Maria* would have fallen for that bluff? In the dead of night, with rations and water all but exhausted and the hold full of nothing but cheap trinkets with which to beguile the "Indians"—most of them, probably. The lovely Bahamian island of Guanahani was finally sighted on October 12th; and here is the fateful trick. The distance from the Canaries to that first landfall in the New World was 3,900 miles. Columbus had been proved triumphantly right in all his calculations, here was Asia; and he lost no time in sailing off to find "Cipango," Japan. The great island the Bahamian natives called Cuba must surely be it, and the *Caniba*, or cannibals, who occasionally came to capture and eat them must be the subjects of the Great Khan himself. No evidence, no setback, would ever shake the great explorer's certainty.

Martin Waldseemüller's World map, published in 1507 (probably in Saint-Dié in the Rhineland) shows North and South America as separate and extremely "thin" longitudinally; understandable errors. And at least he accepted the fourth continent, unlike its discoverer.

Columbus' Miscalculation

Columbus' view of the World, (below right), based on theories from the Ancient World, an arithmetical oversight, and Biblical apocrypha, was a huge miscalculation. Asia is too big; the world is too small. This simple polar projection summarizes the "scale" of the misjudgement. On Christmas Day of the first voyage, exactly at midnight, the **Santa Maria** ran aground on a sandbar near what is now Cape Haitien, Haiti, and became a total loss. Surely another sign from God: Columbus decided that he should leave behind a settlement on the blessed island, which he did. He left 35 men with stores enough for a year (and ammunition, so that the natives would obey "with love and fear") and sailed for Spain. He returned with his second and much larger expedition, (16 vessels and 1,500 men), in November 1493, to find they had all been massacred.

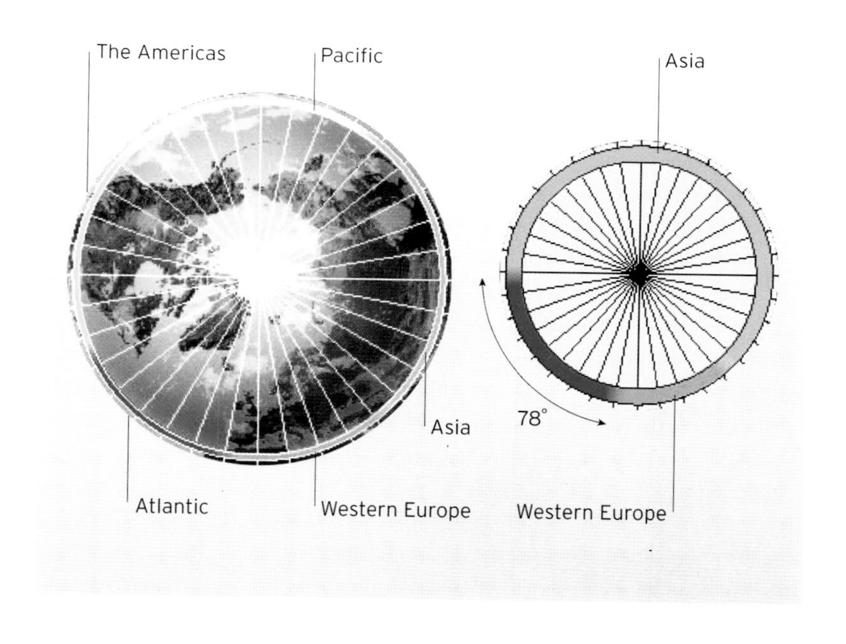

OVERLEAF *Martin Waldseemüller's 1507 world map is the first to name America (in South America) in honor of Amerigo Vespucci. Unlike Columbus, Waldseemüller was a man prepared to adjust his thinking. On this map he gives too much credit to Vespucci for the discovery of the New World; at the top the two hemispheres portray Ptolemy as the cartographer of the Old World, Vespucci of the New. On a 1513 chart he tried to play down the role of Vespucci, but it was too late: the name stuck. Only one copy of the large woodcut map shown exists, in Wüttenberg, Germany.*

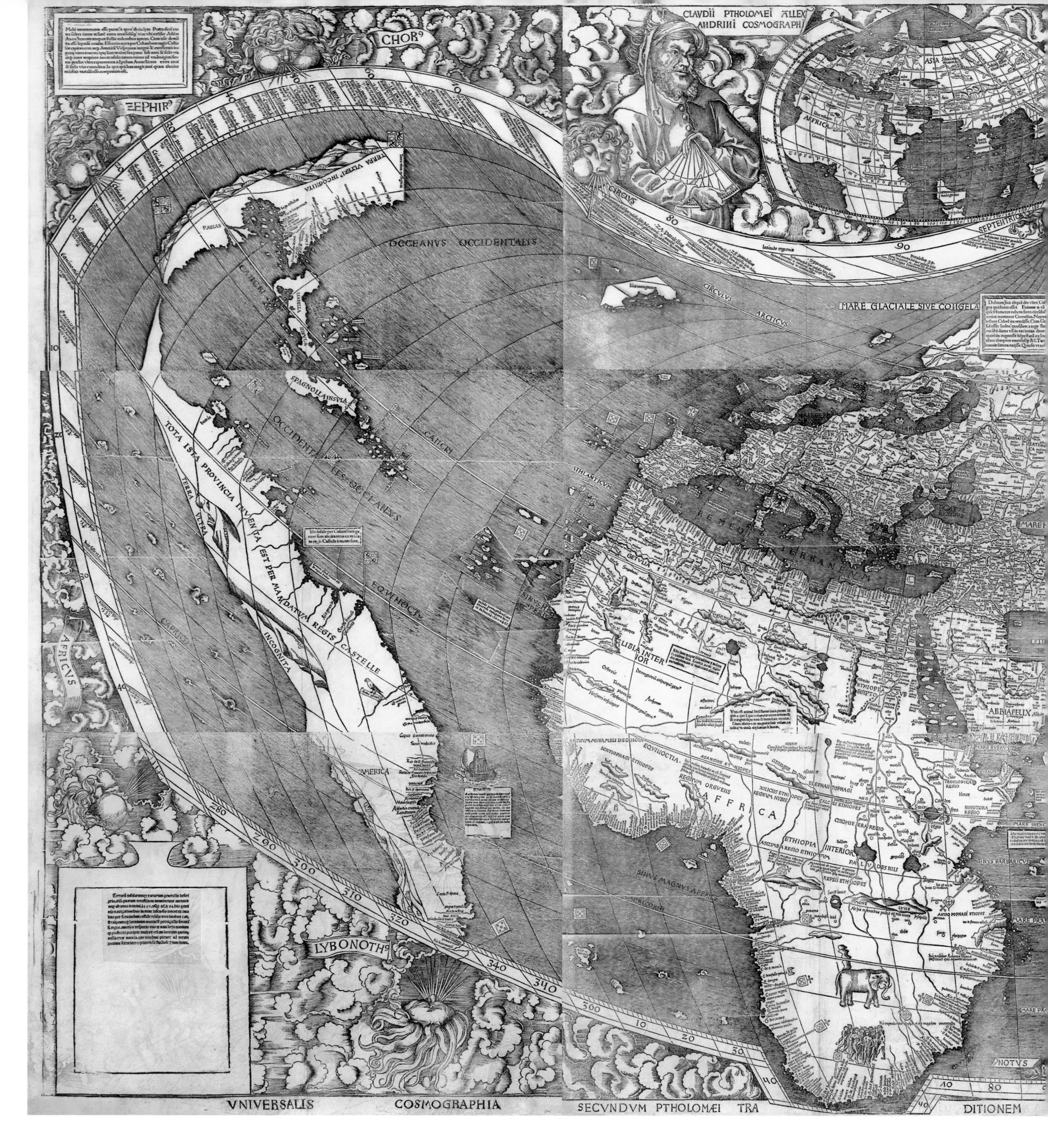
CLAVDII PTHOLOMEI ALEX ANDRINI COSMOGRAPHI
CHOR°
ZEPHIR°
OCCEANVS OCCIDENTALIS
SPAGNOLLA INSVLA
TOTA ISTA PROVINCIA INVENTA EST PER MANDATVM REGIS CASTELLE
TERRA VLTRA INCOGNITA
OCCIDENTALES OCCEANVS
AFRICVS
AMERICA
LYBONOTH°
CIRCVLVS ARCTICVS
MARE GLACIALE SIVE CONGELA
SEPTENTRIO
LIBIA INTERIOR
AFFRICA
ETHIOPIA INTERIOR
SINVS MAGNVS AFFRICE
SINVS BARBARICVS
NOTVS
UNIVERSALIS COSMOGRAPHIA SECUNDUM PTHOLOMAEI TRA DITIONEM

AMERICI VESPVCII
CECIAS
AQVILO
TIPVM ORBIS GENERALEM DESCRIBENDO: VE
TERVM INVENTA PONERE, ET EA QVE A NEO
TERICIS INTERIM REPERTA SVNT (SICVT EST
CATAIA REGIO) CONIVNGERE PLACVIT. VT
TALIVM RERVM STVDIOSI DVM VARIA COGNO
SCERE CVPIVNT VOTORVM COMPOTES LABO
RI NOSTRO SINT CRATI PLAERAQVE OMNIA
TAM PASSIM COGNITA QVAM NOVITER LVS
TRATA DILIGENTER AC DISTINCTE SVB VNO
ASPECTV COLLOCATA PROSPICIENTES.
SVBSOLANVS
GLACIALE
ARCTIC
CIRCVS
CHA TAY
SERICA REGIO
SCITHIA EXTRA IMAV
SCITHIA INTRA IMAVM
ASIA
INDIA MERIDIONALIS
CYAMBA PROVINCIA MAGHA
INDIA EXTRA GANGE
INDIA INTRA GANGEM
SINVS GANGETICVS
TAPROBANA INSVLA
EQVINOCTIALIS
OCCEANVS ORIENTALIS INDICVS
OCCEANVS INDICVS MERIDIONALIS
TROPICO CAPRICORNII
CAPRICORNII
IAVA MAIOR
IAVA MINOR
SEYLAM
PETAM
ANGAMA
NECVRA
VLTVRN' EVRVS
EVRONOTVS
LICET PLAERIQVE VETERVM DESCRIBENDI TERRA-
RVM ORBIS STVDIOSISSIMI FVERINT: NON TAMEN
PARVM IPSIS EISDEM INCOGNITA MANSERVNT, SI-
CVT EST IN OCCASV AMERICAE: AB EIVS NOMINIS
INVENTORE DICTA, QVE ORBIS QVARTA PARS PV
TANDA EST. SICVT ET VERSVS MERIDIEM APHRICE
PARS: QVAE SEPTEM PENE GRADIBVS CITRA CAPRI
CORNVM INCIPIENS, VLTRA TORRIDAM ZONAM ET
EGOCERI TROPICVM AD AVSTRVM LATISSIME PRO
TENDITVR. SICVT QVOQVE IN TRACTV ORIENTA
LI REGIO CATAIAE, ET QVICQVID INDIAE MERIDIO
NALIS VLTRA CENTESIMVM ET OCTOGESIMVM LON
GITVDINIS GRADVM EST SITVM. QVAE NOS PRIO-
RIBVS OMNIA ADIVNXIMVS: VT ISTIVSCEMODI RE
RVM AMATORES, QVAECVNQVE SVB HANC DIEM
NOBIS PATENT OCVLIS INTVENTES, DILIGENTIAM
NOSTRAM POBENT. ID AVTEM VNVM ROGAMVS
VT RVDES ET COSMOGRAPHIAE IGNARI HAEC NON
STATIM DAMNENT ANTEAQVAM DIDICERINT CHA
RIORA IPSIS HAVD DVBIE POST CVM INTELLEXE-
RINT FVTVRA.
O MADAGASCAR
VSTER
ET AMERICI VESPVCII ALIORV QVE LVSTRATIONES

Peter Bienewitz:

Apian's Cosmographia

From a modern cosmological standpoint, dominated by the highly mathematical model of the Universe that is the Big Bang theory and supported by advanced observation technology like the Hubble space telescope, we might be tempted to regard the understanding of former times as little more than crude superstition. But earlier models of the Universe, such as that presented by the German astronomer Peter Apian (1495–1552) were also constructed with great precision, and based on a long-established body of thought.

Peter Apian (a Latinized form of the name Peter Bienewitz), who studied at Leipzig and Vienna and was appointed professor of mathematics at the University of Ingolstadt in 1527, was one of the most celebrated scholars of his age. The work that first brought him fame was his *Cosmographia seu descriptio totius orbis* (Cosmography, or a Description of the Whole World) of 1524. The multifaceted science of cosmography aimed to provide a mathematical explanation for the configuration of the universe, and embraced a wide range of theoretical and applied disciplines: astronomy, geography, cartography, surveying, and navigation, as well as the design and construction of mathematical instruments. (Until the discovery of Waldseemüller's map of 1507 in 1901, Apian's map of 1520 was thought to have been the earliest to apply the name America to any portion of the newly discovered lands.)

In addition to the many illustrations of celestial phenomena in the *Cosmographia*, such as lunar and solar eclipses, and its geographical maps, Apian's work included an important diagram visualizing the geocentric structure of the universe. This model had been proposed by the ancient Greek philosopher and scientist Aristotle (384–322 B.C.) in his books *On the Heavens* and *Physics*, refined by the Egyptian Ptolemy in the 2nd century A.D., and still held sway at the beginning of the 16th century. The model also attempted to account for the existence of various forms of motion, and hence for the very different laws that were held to govern celestial and terrestrial bodies. In contrast to the Earth (the sublunary world), which was made up of matter (comprising the four elements fire, water, earth, and air) that moved in straight lines, heavenly bodies were composed of an entirely different, fifth element called aether, which moved in a circular way. This idea forms the basis for the concentric, spherical shells of aether, each fitting tightly around the next, that are shown—in two dimensional cross-section—on Apian's diagram. At the center lies the Earth with its four elements. Surrounding it, in turn, are the spheres labeled Luna (moon), Mercurii (Mercury), Veneris (Venus), Solis (Sun), Martis (Mars), Iovis (Jupiter), Saturni (Saturn), the fixed stars or firmament, and the crystalline heaven. Beyond this was the prime mover (primum mobile), the principle that causes all movement from the outside inward but which itself remains unmoved. All motions in the cosmos came ultimately from this prime mover. Finally, outside the ten spheres of the universe is a region Apian describes as "The Empire and Habitation of God and All the Elect." Aristotle conceived of this region, the Empyrean, as the abode of the Unmoved Mover, and Christianity later interpreted it as God in His Heaven. A fundamental principle of Aristotle's cosmology was that the Earth is mutable and subject to corruption and decay, while the heavens are regular, unchanging, and perfect. It is easy to see how this notion was readily adopted by the medieval schoolmen who formulated Christian theology.

Studies such as the *Cosmographia* and Apian's later *Astronomicum Caesareum* (1540) were firmly grounded in the canon of contemporary scientific knowledge. Both volumes stress measurement as the fundament of accurate calculation and expound important new ideas, such as the use of solar eclipses to determine longitude. Apian's meticulous scholarship made him much feted in his lifetime; his patron, the Holy Roman Emperor Charles V, appointed him court mathematician. But for all his painstaking calculation, the face of heaven changed in 1543, with the publication of Nicolaus Copernicus's heliocentric *On the Revolution of the Celestial Spheres.*

RIGHT *A colored woodcut volvelle, (a calculating device with movable concentric circles) showing the zodiac from Apian's* Astronomicum Caesareum.

CANCER
GEMINI
LEO
TAVRVS
VIRGO
ARIES
LIBRA
PISCES
SCORPIVS
AQVARIVS
SAGITTARIVS
CAPRICORNUS
LATITVDO LVNAE MERIDIOLIS
LATITVDO LVNAE SEPTENTRIO

Diego Gutiérrez:

The Fourth Part of the World

In the second half of the 16th century, three European nations laid claim to what they saw as the rich pickings of the newly discovered America. France, Portugal, and Spain all had what they perceived as legitimate claims. In 1562 Spanish cartographer Diego Gutiérrez produced a map that was designed to settle the argument. In 1554 Gutiérrez took up the position of Pilot Major as principal cosmographer—what we would now call a cartographer or, simply, a mapmaker—in the Spanish *Casa de la Contratación*, a post he inherited from his father.

His family owned a business that supplied maps and navigational instruments to pilots and navigators, including those who ventured across the largely uncharted waters of the Atlantic Ocean. Diego Junior had learned well from his father and was paid a princely salary of 6,000 maravedis by King Philip II. Diego's brother Sancho also worked at the *Casa*. Diego was recognized as "an official who makes sea charts."

Anxious to consolidate its position as a world power, the Spanish crown commissioned the *Casa* to produce a large-scale map of America, or of the Fourth Part of the World as they called it. The result, the map produced by Diego Gutiérrez in 1562, includes the whole of Central and South America, with parts of the North American east coast and the western coasts of Europe and Africa. It was engraved in Antwerp by the famous Belgian engraver Hieronymus Cock (1510-70), and printed in sufficient numbers for the Spanish to distribute it—and establish their claims—among their European neighbors. Up to then, it was the largest printed map of America.

Although there are no lines of latitude on the map, the Equator and the Tropics of Cancer and Capricorn are boldly marked. The choice of these lines is significant because everywhere south of the Tropic of Cancer was claimed by Spain, while France and Portugal shared the remainder. These international divisions are confirmed by the coats of arms that decorate the corners of the map. East of Argentina on the Atlantic coast is the Portuguese coat of arms, the southwestern edge of North America carries the Spanish Habsburg Empire's coat of arms, while the southeastern side holds the arms of the French crown. In 1559, after the Thirty Years War, the Peace of Cateau Cambrésis allocated the area south of the Tropic of Cancer and west of the prime meridian to Spain. North of the tropic, France held sway while northern South America was acknowledged as Portuguese territory. Many other images decorate Gutiérrez's map. There are mythical sea monsters and mermaids in the oceans and armed giants in Patagonia. A volcano erupts in central Mexico, and a sinuous Amazon River meanders snake-like right across South America. Lake Titicaca, Mexico City and Florida are also located. Many capes and mountains are named, surrounded by images of monkeys and parrots. At the southern tip of the Baja California, "C. California" is inscribed, one of the first printed references to this name. America itself is said to be named on behalf of the King of Spain after Americus Vespucius, who is acknowledged to have discovered this "fourth part of the world" in 1497. Only two known copies of the map remain, one in the British Library in London and the other in Washington's Library of Congress.

The purely decorative additions are down to the map's engraver, Hyeronimus Cock, who was one of the 16th century's most important printmakers. Like his brother Mathias, he was a well-known painter and studied in Rome before opening his shop *Aux Quatre Vents* ("To the Four Winds") in Antwerp in 1548. In the early 1550s Cock produced engravings of the ruins of Rome and in 1559 published his famous *Pompa funebris* depicting the funeral in Brussels of King Philip II's father the previous year. In 1555 he engraved portraits of Philip II and of King Maximillian II of Austria. A series of 12 engravings from 1563, made a year after Gutiérrez's map, pictured the triumphs of Holy Roman Emperor and King of Spain Charles V. Cock also engraved more than a dozen other maps from 1550 to 1562.

RIGHT *On the truncated scale of the map of Central and South America drawn by Diego Gutiérrez in 1562 the coast of West Africa appears to be only a few days' sailing from Brazil. Just off the southern tip of Cape Horn, "Magellan's Land" represents the fringes of the unknown continent of Antarctica.*

AMERICAE SIVE QVARTAE ORBIS PARTIS NOVA ET EXACTISSIMA DESCRIPTIO.
AVCTORE DIEGO GVTIERO PHILIPPI REGIS HISP. ETC. COSMOGRAPHO. HIERO COCK EXCVDE. 1562
VICTORIA
SEPTENRIO
PARTE DI GRONLANDA
OCEANVS SEPTENTRIONALIS
TIERRA DE LABRADOR
CANADA PROV
TIERRA DE BACCALAOS
TIERRA FRANCISCA
TIERRA DI NORIMBERGA
AVACAL
APALCHEN
MOCOSA
CALICVAZ
OTAGIL
CAPAS CHI
TIERRA FLORIDA
LA NVEVA GALITIA
TERLI CHICHI
MECHI
ISLAS DELOS AÇORES
SPAGNA
ISLAS DE CANARIA
GOLFO DE LA NVEVA ESPANA
TROPICVS CANCRI
OCEANVS OCCIDENTALIS
AFRICA
ISLAS DE CAPOVERDE
NVEVA ESPANA
AEQVINOCTIALIS
NOVA ANDALVSIA
CARTAGENA
MAR DEL SVR
REGIO DEL PERV
REGIO DE BRASIL
LA FLOTA DE PORTVGAL QVE VA PAR CALICVTE
TROPICVS CAPRICORNI
MARE MAGELLANICVM SIVE PACIFICVM
TIERRA DE PATAGONES
CHICA
OCEANVS AVSTRALIS
Estrecho de Magallanes
REGIO PATALIS
TIERRA DE MAGALLANES
MERIDIES · zuyd
SERENISS. PRINCIPI MARGARITÆ AB AVSTRIA DVCISSÆ PARMÆ ET PLACENTIÆ ETC. GVBERNATRICI PROVINCIARVM INFERIORIS GERMANIÆ HIERONYMVS COCK PICTOR DEVOT. DEDICABAT
Hieronymus cock excude. cum gratia et priuilegio

Theatrum Orbis Terrarum:

Ortelius Brings the World to Book

When in 1590 Flemish cartographer Gerardus Mercator published his book of maps, he wanted a suitable image to illustrate the front cover. He chose an engraving of a kneeling Greek Titan holding up the world on his shoulders. The Titan's name was Atlas, and ever since nearly all books of maps have been named after him. But Mercator's was not the first atlas, even if it was the first to have that name.

Twenty years previously Flemish engraver and cartographer Abraham Oertel, or Ortelius as he came to be known, published a book of world maps. He was born in Antwerp in 1527 of German parents from Augsberg, hence his description of himself as Belgio-Germanus. He studied Greek and Latin as well as mathematics, and by the age of 20 was earning his living by painting maps; he joined the appropriate guild and was qualified to color in the black-and-white engravings produced by some of the many cartographers who thrived in the city.

The maps were needed by local merchants and traders, some of whom often traveled as far as the Spice Islands (Moluccas) in south-east Asia, many of them Dutch possessions at that time. In 1554 Ortelius set up in business on his own, dealing in rare antiquities, books, and coins. About ten years later, influenced by his contemporary Mercator, he started engraving his own maps. He produced a map of the world using a heart-shaped projection (1564), a map of Egypt (1565), a complete map of Asia (1567) and a new, large, eight-page world map

LEFT *Following classical Greek theories, using the Arctic and Antarctic Circles and the Tropics of Cancer and Capricorn, Ortelius divides the world into climatic zones in this "Roman" (i.e. historical) map. The Temperate Zones are habitable, the tropical Torrid Zone is just habitable, north and south Frigid Zones uninhabitable.*

BELOW *Ortelius's world map in the first edition of the* Theatrum Orbis Terrarum *of 1570 includes the Americas, although their size and shape are obviously distorted. The great unknown southern continent, which embraces Australia, was a remarkably persisting fiction.*

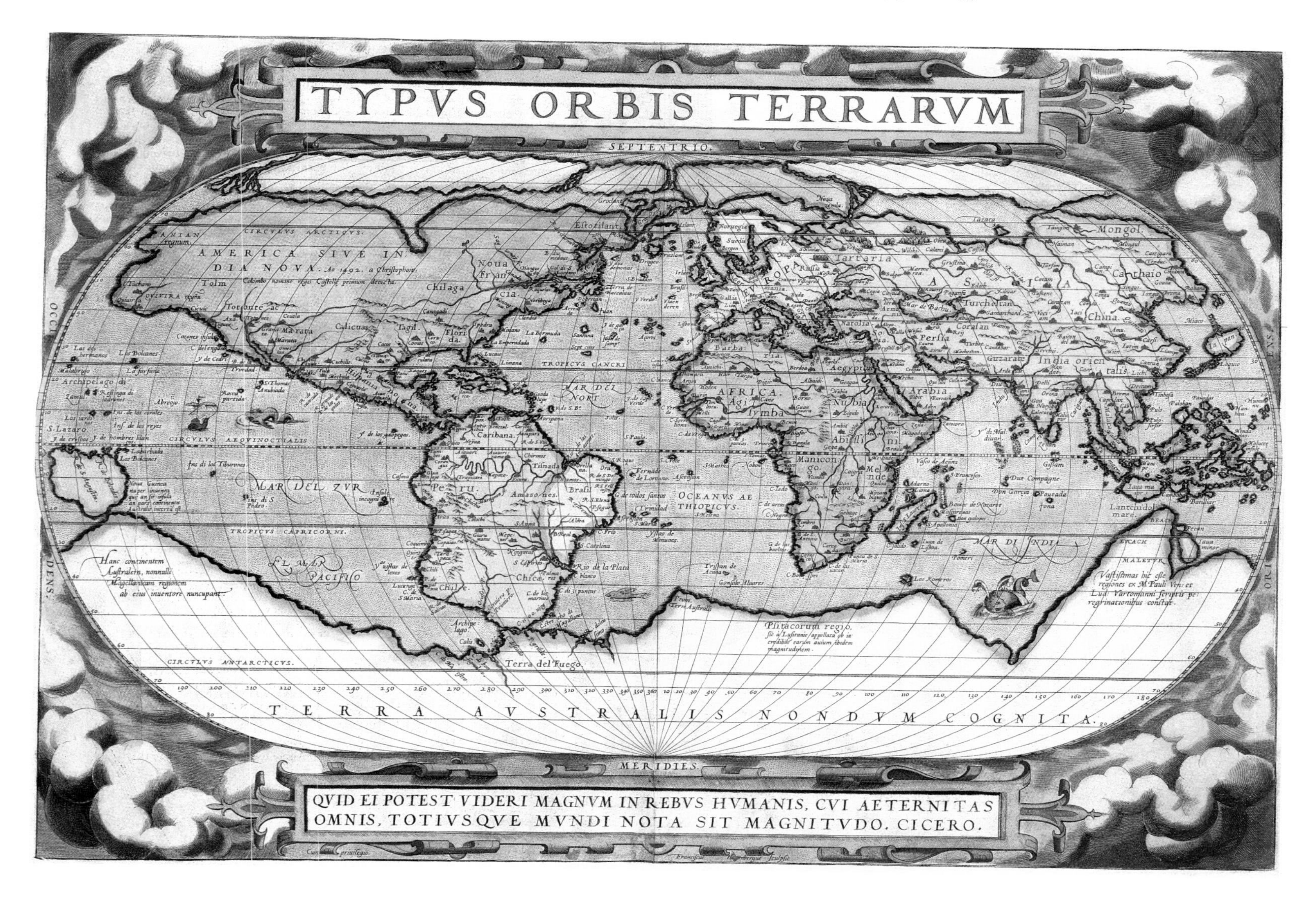

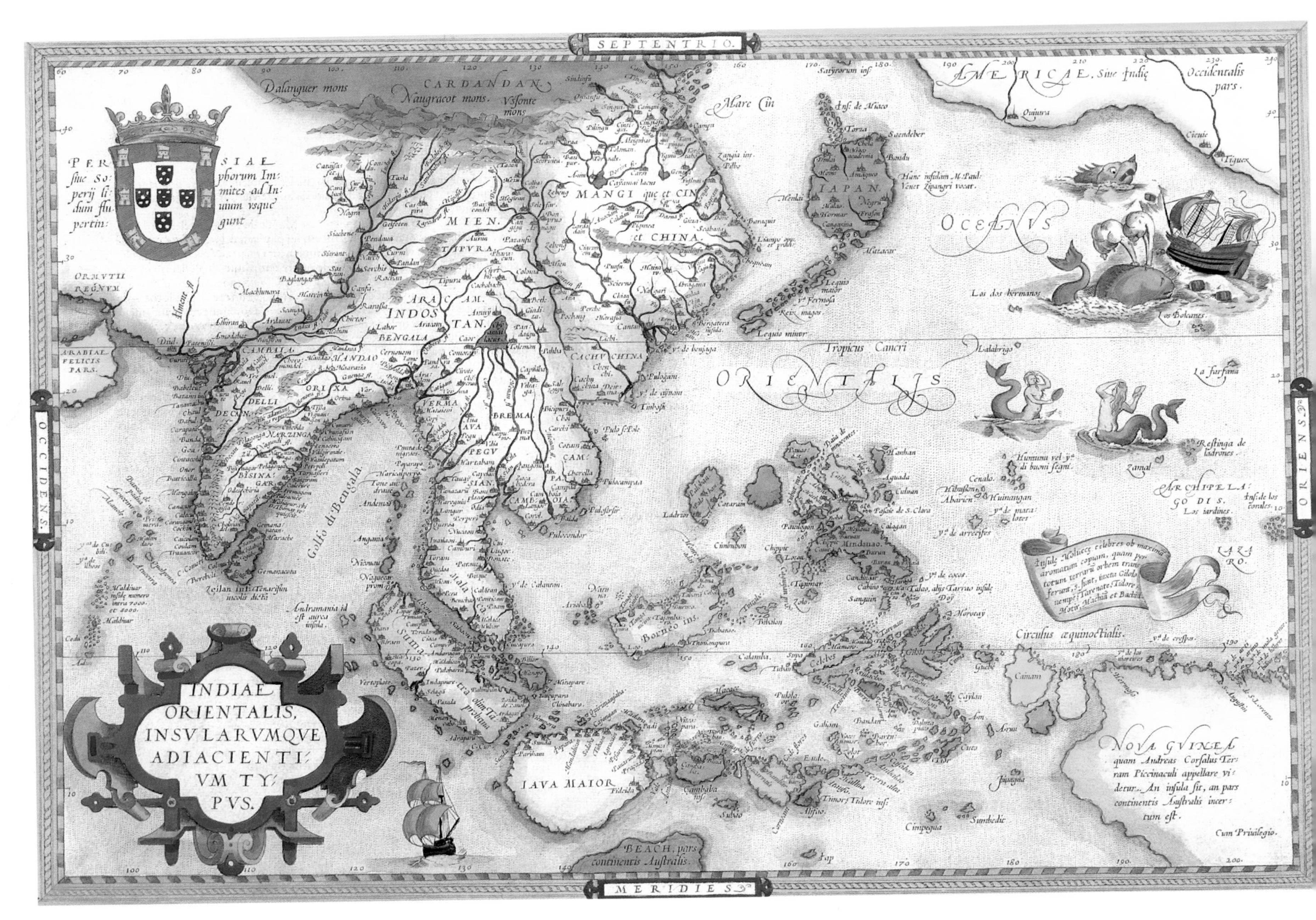

ABOVE *The map of India and "Adjacent Islands" reveals the importance of this region to 16th-century European traders. Featured prominently—and much larger than in reality—are the Spice Islands (Moluccas), Sumatra, Java, and Borneo (together they are much bigger than the subcontient of India). America is much too close to Japan, while New Guinea fades off into the unknown Australia.*

(1569). At first these maps took the form of individual parchment scrolls decorated with ornate borders and embellished with elaborate shields and escutcheons. They were pretty to look at but awkward to use, having to be unrolled and rerolled every time.

To get round this problem, in 1566 Ortelius began work on a book of maps—an atlas—which he called *Theatrum Orbis Terrarum (Theater of the World).*

To be strictly precise, Portuguese discoveries as early as the fourteenth century had been documented by manuscript charts bound together in volumes; and in the 1560s, many "I.A.T.O." (Italian atlas assembled to order) or "Lafreri" volumes had been printed. But these did not truly integrate text and images in the way we would expect from an atlas. Rather than appearing as a single edition, Italian atlases were assembled to suit the needs, desires, or whims of the individual customer.

Ortelius employed and generously acknowledged other map-makers for much of the work. The first edition of 1570 was engraved by Frans Hogenberg; it contained 70 hand-colored maps and refers to 87 different cartographers; this "Catalogus Auctorum" is itself a priceless rccord, many of the cartographers mentioned would have

been lost to history without it. It was reprinted four times in its first year. By the much larger 1603 edition there were acknowledgments to the work of 182 others, 44 people working on the world map alone. Originally in Latin, the *Theatrum* was soon translated into Dutch, French, German, Italian, Spanish, and eventually English. By 1598 Ortelius had sold 2,200 copies, which is an impressive number for an expensive illustrated book today—and represents a phenomenal success in the 17th century. The atlas increased in extent from 53 double-folio pages in 1570 to 129 pages in the Italian edition of 1612, containing 166 maps. With the 1724 printing (126 years after Ortelius had died) the total number of editions reached 89, with the numbers of sales topping 7,200. More than a quarter of these copies still exist in various editions with collectors and libraries throughout the world.

ABOVE *The 1595 map of Korea (shown as an island) and Japan was to remain the definitive Eiropean view for more than half a century. The text (on the back of the map) describes how the country is full of snow all year long, that the people make an artificial wine out of rice, and thay they are especially delighted to drink almost scalding hot water containing a herb called "Chia" (tea).*

After Ortelius's death in 1598 the atlas was published by Jan Baptiste Vrients, who purchased the rights from Ortelius's heirs.

The *Theatrum* begins with a frontispiece illustrated with allegorical females representing the continents. First is the authoritarian figure of Europe (naturally), holding a scepter and steering the world's affairs with a rudder in her other hand. An oriental princess represents Asia, holding a container of smoking incense. Africa's black maiden shines

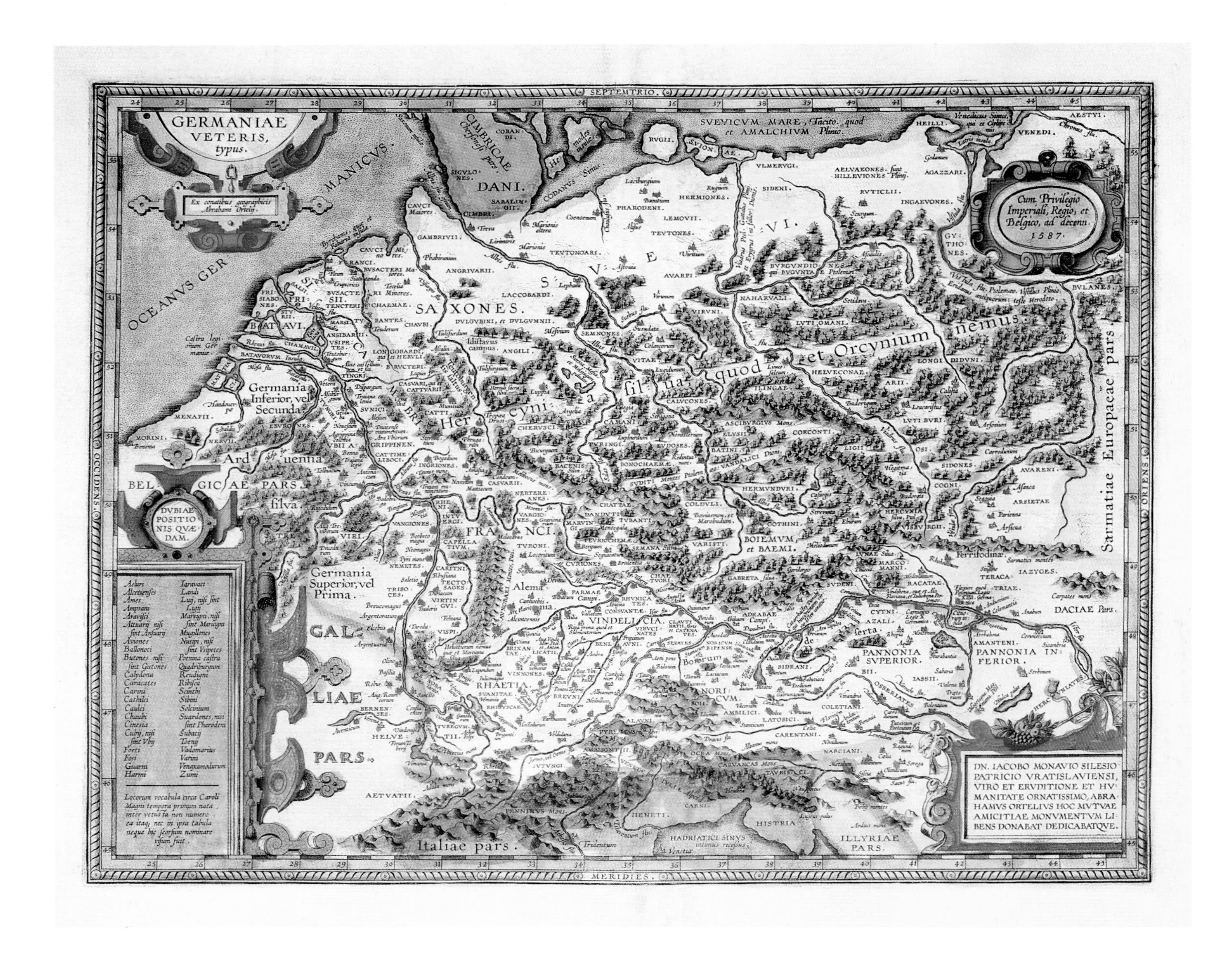

with the tropical heat, while an Amazonian warrior representing America completes the quartet. Inside the atlas, after a dedication to King Philip II of Spain, a recommendation from Mercator, and a portrait of Ortelius there is a list of sources. The atlas proper starts with a world map, followed by maps of the New World, Asia, Africa, Europe, and their regions. Later editions include separate maps of major islands, especially those in the Mediterranean Sea. The atlas concludes with a geographical dictionary of regions and places–which would later become known as a gazetteer–with ancient place names alongside the modern ones.

Some of the maps have a graticule representing lines of longitude and latitude, whereas others have coordinates around the borders and along the equator (the "Equinoctal Circle"). The variation presumably arises from the different approaches of the contributing cartographers. On maps of large areas the central borders of the original Latin version are labeled *Septentrio* (North), *Oriens* (East), *Meridies* (South) and *Occidens* (West). Most oceans seem to be home for whales and mermaids, and there are many caravel-type sailing ships; later maps illustrate galleons. Many of these devices seem to be merely to fill in some of the blank spaces. On land areas there are symbols for

ABOVE *European countries got more detailed treatment in later editions of Ortelius's atlas. This map of ancient Germany dates from 1587 from an edition of about 3,300 hand-colored copies, of which 200 to 300 still survive in libraries and collections throughout the world. It is noticeable that nearly all of the towns and cities lie along the courses of rivers, even in the non-mountainous areas*

mountains, forests, and towns–every town has a church with a spire, even in India and Sumatra. This "town" symbol was to persist on maps large and small for several centuries. Sometimes there are animals, with bears in northern Russia and camels to the south of the Caspian Sea. Major rivers and lakes are represented, although there are no roads marked on the maps. Text on the maps is sparse so that, as Ortelius himself said, they speak for themselves: "The mappe being layed before our eyes, we may behold things done or places where they are done."

When we compare the maps in the *Theatrum* with their modern equivalents, we notice that some areas are drawn disproportionately large. This distortion may represent the importance attached to some places—Java, Sumatra, and Japan, for instance—or it may result from ignorance of the true nature of the region.

Western North America, an unexplored region, is much too large and to the south of Cape Horn and the Cape of Good Hope there is a vast southern continent, which embraces Australia although reflecting the shape of its northern shore.

In 1575 King Philip II of Spain appointed Ortelius His Majesty's Royal Cosmographer and is said to have always kept a copy of the *Theatrum* to hand; perhaps he studied it to look for new places to conquer. The area of the Americas between the Tropic of Cancer and the equator was prominently labeled New Spain, and Philip may have had his eye on New France to the north.

The appointment swelled Ortelius's already great reputation—he was known as the Ptolemy of his century—which was due also in part to his shrewd approach to business. For each edition of the atlas, he bought the paper and paid the engravers for the copper engravings (the printer, Christophe Plantin, got paid merely for the actual printing), and he sold the printed books to publishers and distributors and direct to the public. He continually updated the work, and no doubt charged more for each edition as it included more maps than the previous one.

On the basis of this single publication, Ortelius became comfortably rich, and lived a wealthy bachelor lifestyle among the burghers of Antwerp. He also traveled widely, assembled a collection of archeological specimens, and published accounts of his travels as well as studies of ancient geography, such as the 1587 *Thesaurus Geographicus*.

The double-folio paper size of Ortelius's *Theatrum* was about about 23 by 34 inches (58 by 86 centimeters). Before metrication took over the paper trade, the size of a standard sheet of drawing paper measuring 26 by 34 inches was designated as *atlas*—another legacy from the work of Ortelius and Mercator.

Teamwork

Spectandum dedit Ortelius mortalib. orbem,
Orbi ſpectandum Galleus Ortelium.

It should not be forgotten that the marvelous cartographic creations of the Dutch Golden Age were the products of a sophisticated and hard-headed business, and were effectively the products of teamwork. The French bookbinder, typographer, printer, and publisher Christophe Plantin (or Cristóbal Plantino) commissioned this portrait of Ortelius for the **Theatrum**. His knowledge of typefaces, engraving, and the printing process was no doubt an important factor in the huge success of the atlas. Between 1569 and 1573 Plantin prepared for publication the **Biblia Polyglotta**, the Polyglot Bible, intended to fix the text of the Old and New Testaments on a scientific basis. Philip II of Spain supported him in this work against clerical opposition, rewarding him with the title **Prototypographus Regius** and the right to print all liturgical books in the king's dominions. Perhaps the king's enthusiasm for Plantin's work would have paled had he known that the printer was until his death a follower of an obscure band of mystical heretics!

The Apprentice Refugee:
Kaerius

Pieter van den Keere (1571-ca 1646) or Petrus Kaerius—as he is often referred to in the Latinized form—was a gifted Dutch engraver and cartographer. His life story embodies two common themes in cartographic history of this period: he was affected by the political and social upheaval of his age, and his family connections epitomize the increasingly close-knit nature of the Dutch mapmaking profession.

It is thought that Kaerius was born in Ghent around 1571 into a Protestant family. At the height of the bloody Dutch rebellion against Spanish rule in 1584—the year the rebels' leader Prince William of Orange was assassinated—Kaerius fled with his sister Colette to Elizabeth I's England to escape religious persecution. There, they met their fellow countryman and refugee Jodocus Hondius, eight years Kaerius' senior, who taught the younger man the skills of cartography and engraving. Hondius married Colette in 1587; in due course the daughters of this union would marry the celebrated mapmakers Jan and Henry Jansson.

Kaerius first distinguished himself by engraving a number of maps for the foremost cartographers of the day. For example, in 1592, he engraved a set of maps of Ireland, entitled *Hibernia novissima descriptio* for the Italian artist Baptista Boazio, then resident in London.

By the following year, conditions had improved sufficiently in the Low Countries to allow Kaerius to travel home. Settling in Amsterdam, he continued his cartographic trade, working both alone and in conjunction with his brother-in-law Hondius (who had also returned to the Netherlands); his collaboration with Hondius include large wall maps of Europe and the World. In around 1599 Kaerius engraved a series of plates for 44 maps of the English and Welsh counties, the regions of Scotland, and the Irish provinces. The English maps in this collection were based on the renowned 1570s atlas of Christopher Saxton, the Scottish on Ortelius and the Irish on a 1599 map by Boazio.

Despite being born into a time of great unrest, Kaerius thereafter enjoyed a long and fruitful working life. One of his most well-known publications was an Atlas of the Netherlands (*Germania Inferior*, 1617–22, with text by Pieter Montanus, another brother-in-law of Hondius). The frontispiece to this collection shows the Seventeen Provinces of the Low Countries in the form of a rampant lion; this motif, which was known as "Leo Belgicus" and was already popular when Kaerius re-used it, was much favored by Dutch cartographers of the "Golden Age" keen to make a nationalistic point about the region's great resilience and pride.

Another work in which Kaerius was involved, which underlines the dynastic nature of Dutch mapmaking, was Jan Jansson's *Atlas Minor* of 1628. Originally published in 1608, the map opposite is a copy of Willem Blaeu's map of two years earlier and was acquired by Jan Jansson in the 1620s. Hondius and Jansson issued it sporadically. An amended edition was issued in 1680 by Moses Pitt (c. 1654-96) in his doomed attempt to produce a twelve-volume *English Atlas,* which foundered due to lack of finance. (Sir Christopher Wren and scientist Robert Hooke were both foolish enough to back the enterprise.) It is a wonderfully elaborate piece of work and although a copy, it actually outshines the original engraving by Joshua van den Ende.

Kaerius' engraving and publishing skills were not, however, confined to cartography; in 1620, one of the first newspapers to appear in English was produced by Kaerius and exported from Amsterdam. He also engraved portraits of notable figures of the age; that of Elizabeth, Queen of Bohemia (and daughter of King James I of England) is in London's National Portrait Gallery.

Kaerius is also associated with a famous edition of John Speed's Atlas of the British Isles (*The Theatre of the Empire of Great Britaine*). In 1627, Speed's publisher George Humble was seeking to issue a pocket edition of the famous atlas to coincide with a new edition of the full resource. To do this, he combined plates from Kaerius' 1599 British maps on the recto with Speed's descriptive texts of the relevant area on the verso. This series of successful hybrids became popularly known as

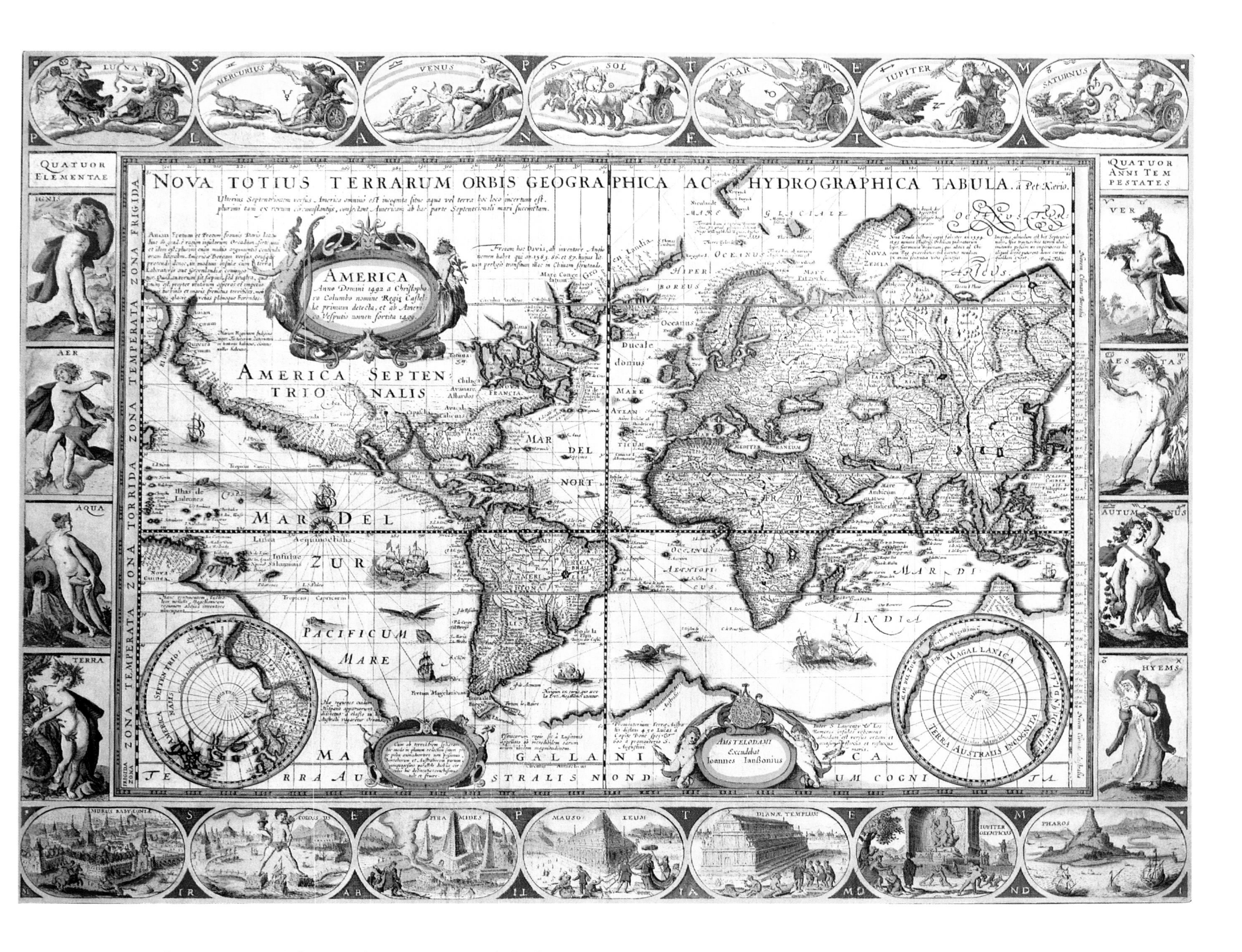

"miniature Speeds." Kaerius' maps constituted the large majority in this edition, testament to their quality. Kaerius died at the age of 75 in 1646. His maps are also sometimes signed "Keer" or "Coerius."

In 2003 Kaerius gained a particular honor in Romania. His copper map engraving, *Vetus Descriptio Daciarum Nec non Moesiarium*, was published about 1625 in Amsterdam. It shows the region of Dacia, north of the Danube River, and Moesia, south of the Danube. A stamp was issued in 2003 reproducing the map to celebrate and advertise the National Map and Book Museum in Bucharest.

ABOVE Nove Totius terrarum Orbis Geographica Ac Hydrographica Tabula, *from the "Atlantis Maioris Appendix," engraved by Kaerius and published by Hondius and Jansson in Amsterdam at various times. This example was a hand-colored map of about 21 x 16 inches (530 x 400 mm). The borders are identical to those of the Blaeu map, of which this is a copy; but the van den Keere map includes descriptive text, at the top left and top right sides, relating to the North-east and the Northwest Passages. There is a long note near the cartouche dealing with the question of whether or not the straits of Anian and Davis link up to create the tantalizing Northwest Passage.*

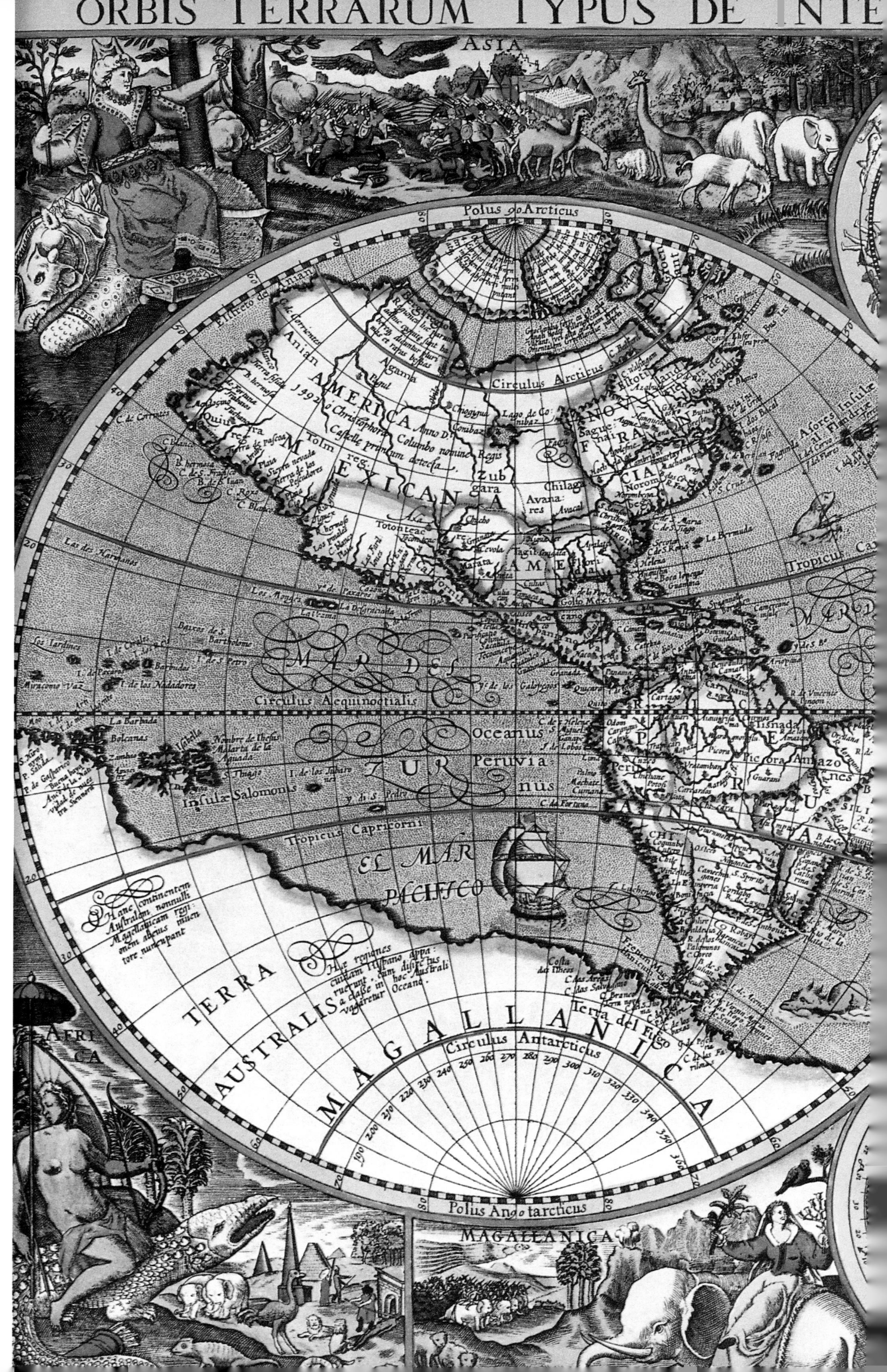

RIGHT *Whose map is this? It is a 1607 version of a world map by Kaerius. But if we were truly looking for the father of this map the answer is not so straightforward. Excluding for a moment Ptolemy, despite the fact that this a Ptolemaic view, we can trace it back via Jodocus Hondius to Gerard Mercator. Hondius obtained Mercator's plates after the Master's death in 1594. We can then trace the image forward: Jan Jansson produces the* Atlas Minor *based upon the same maps in 1628, some plates engraved by Kaerius. This version incorporates the improvement Ortelius made in his 1578 map, removing the "bulge" in South America. To confound matters still further, in 1630, cartographer to the Dutch East India Company Petrus Plancius produced a map with exactly the same name:* Orbis Terrarum Typus de Integro Multis in Locis Emendatus. *His map showed Japan more accurately as one small and three larger islands. This image—and most of the maps of the Durch Golden Age—are palimpsests, accretions of knowledge. On occasion, even the decorative motifs are copies. This particular map is nevertheless famous for the richness of its allegorical decoration.*

O MULTIS IN LOCIS EMENDATUS auctore Petro Kærio anno 1607
EUROPA
Polus 90 Arcticus
OCEANUS TARTARICUS
AFRICA
ASIA
Aequinoctialis
Aequator
MARE LANTCHI DOL
OCEANUS AETHIOPICUS
NOVA GUINEA
De Iaua minori
MAGALLANICA
TERRA AUSTRALIS
Circulus Antarcticus
Polus Antarcticus
PERUANA
MEXICANA

The Age of Atlases:

The Blaeu Family

The late 16th and early 17th centuries became known as the "Age of Atlases," as skilled cartographers produced ever more lavish collections of maps. The Dutch were the undisputed masters of this highly lucrative enterprise, which was pioneered by Ortelius and Mercator, and reached its zenith with the Blaeu family. The maps they published quickly became renowned for their great beauty and high standard of workmanship, and their printing house grew into the largest press in Europe.

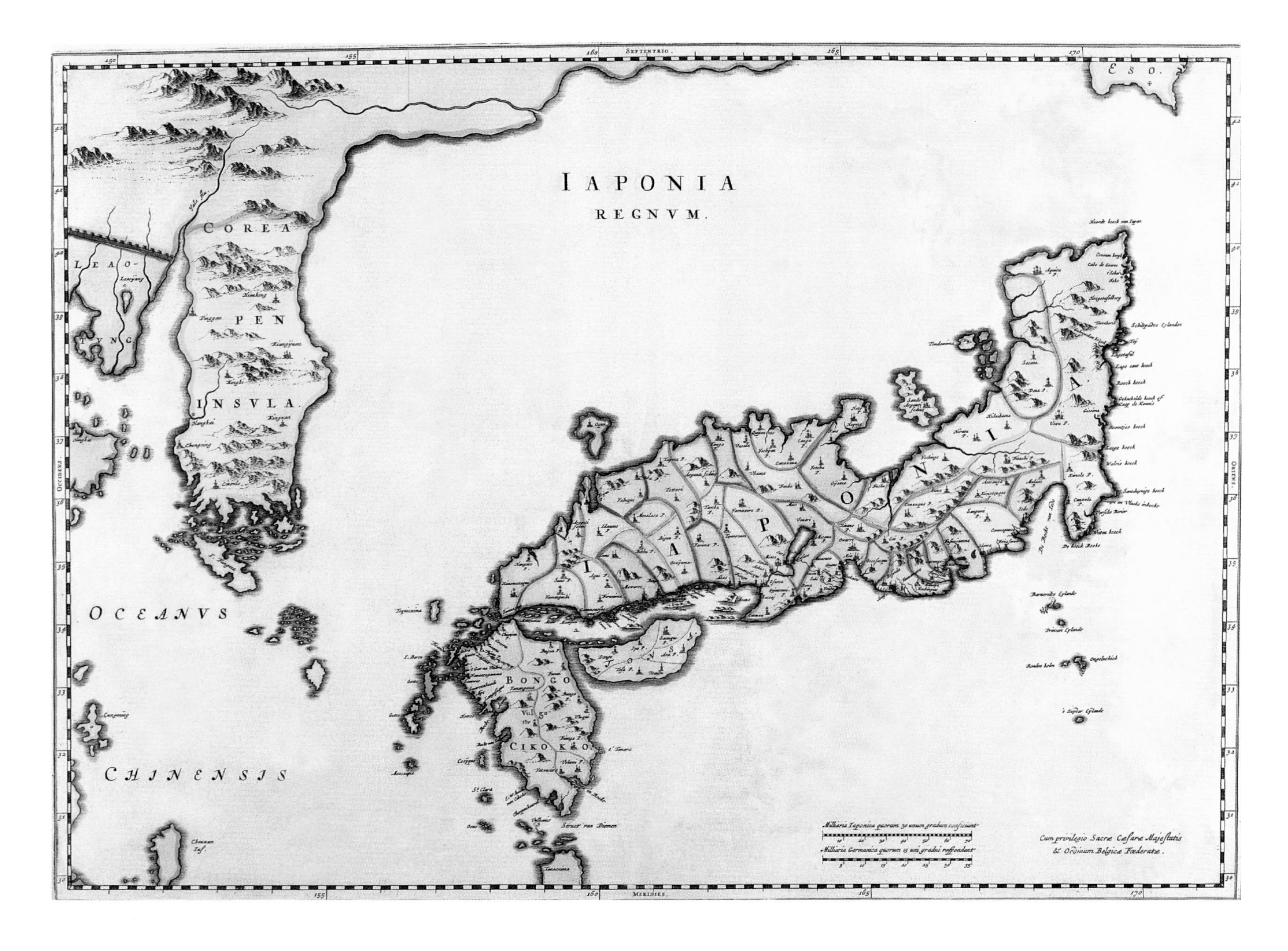

The Blaeu mapmaking dynasty, which flourished for around 100 years, was founded by Willem Janszoon Blaeu (1571-1638) of Alkmaar. After studying under the renowned Danish astronomer Tycho Brahe, he established himself in Amsterdam in 1596 as a manufacturer of globes and scientific instruments. By the early 17th century, Willem had acquired a printing press and diversified into producing maritime charts, wall maps, and books. The Blaeu printing works proceeded to build its reputation on its sea atlases, beginning in 1606 with *Licht der*

BELOW LEFT *This map of Japan, first published in Joan Blaeu's* Atlas Sinensis *of 1655, was based on information supplied by the Jesuit Martino Martini. At top right is the tip of the island of Hokkaido (Ezo). At this time, neither its Japanese masters (who annexed it c.1604), nor the Chinese or Europeans, knew whether it was an island or attached to the Asian mainland.*

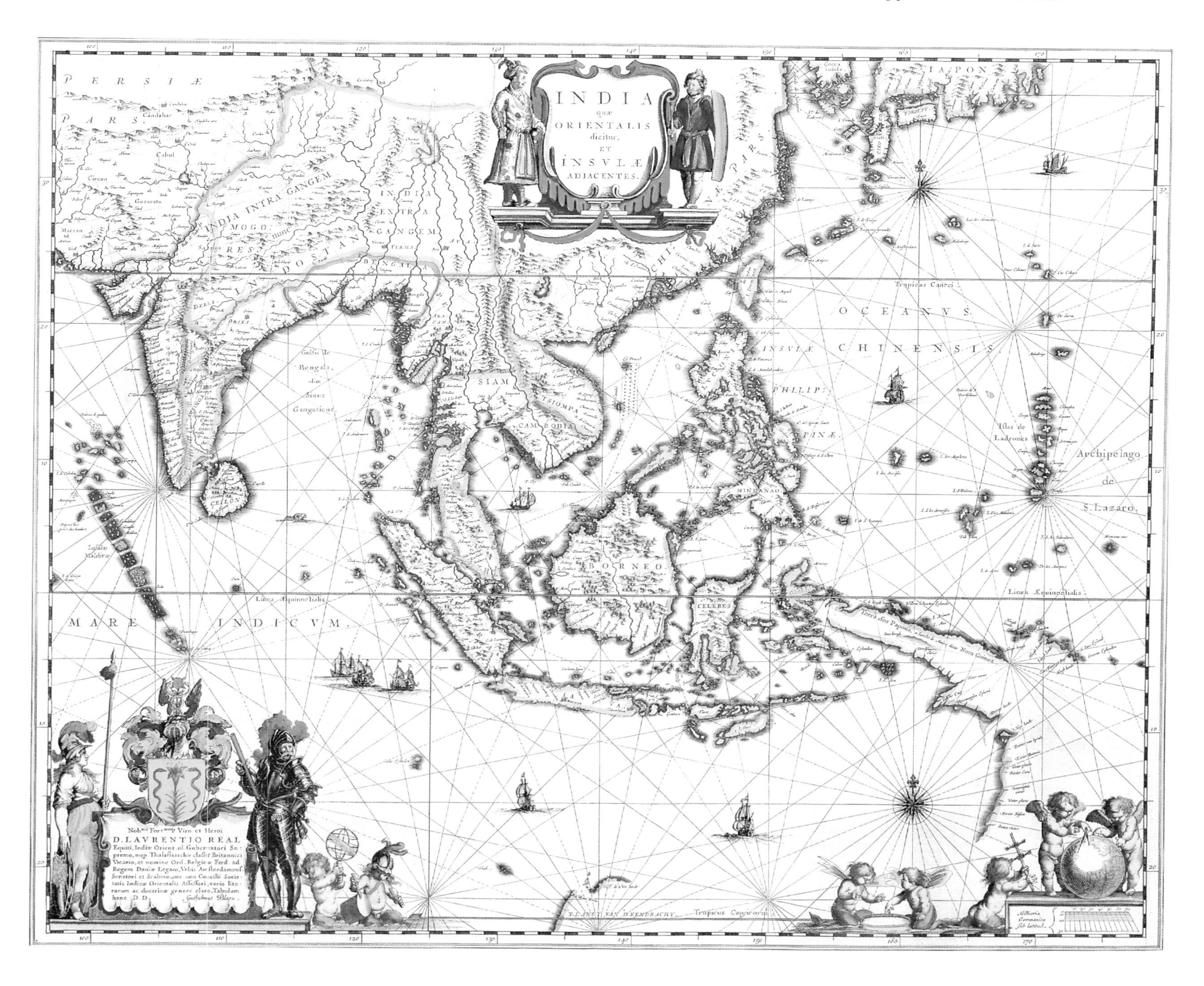

BELOW *Willem Blaeu was made Chief Cartographer to the Dutch East India Company (VOC) in 1633. This 1640 French edition of his Southeast Asian map—two years after his death—shows its sphere of operation. It is dedicated to Laurens Real, governor-general of the VOC's headquarters at Batavia (Jakarta), who introduced slave labor to the nutmeg plantations on Amboina.*

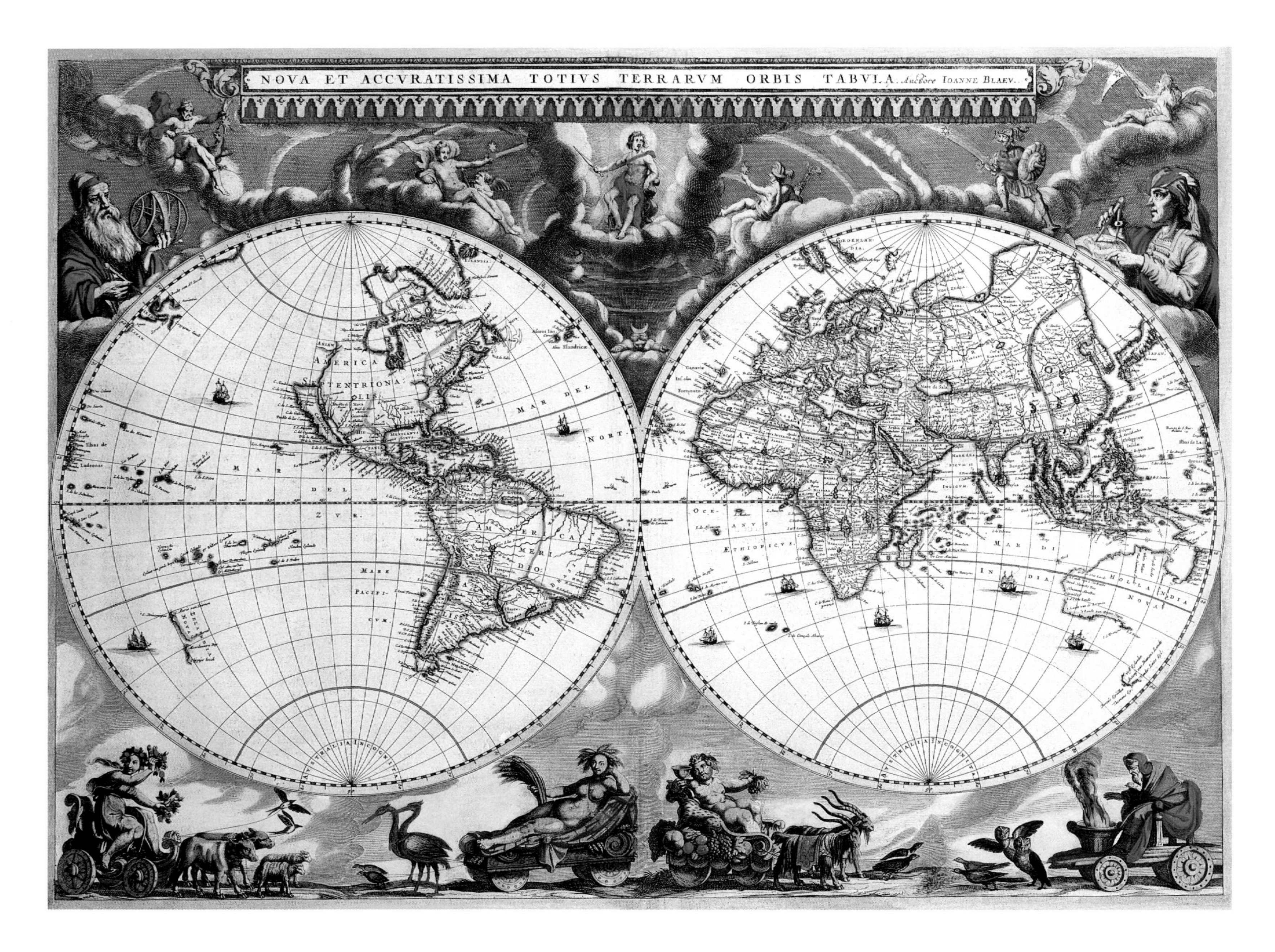

ABOVE *Joan Blaeu's double hemisphere map of the world, which introduced his 1662* Atlas Maior, *reflects the extent of contemporary geographical knowledge. California still appears as an island, and there is no trace of the Great Lakes. The figures below the map represent the four seasons, while above are the astronomers Galileo (top left) and Tycho Brahe, who taught Joan's father (top right).*

Zeevaert ("The Light of Navigation"), which contained charts of the North Sea around the Low Countries and southern England. This was followed by *Eeste deel der Zeespiegel* ("The Sea-Mirror") in 1623. From around 1630, Willem was joined by his son Johannes (generally known as Joan), who was born in the same year that his father set up the firm, and had studied law at the University of Leiden. Their names begin to appear together on the maps from 1631 onwards.

The publishing house's growing prestige and turnover saw the workshop move in 1637 to much larger premises on Bloemgracht. Here, nine presses were in operation, and at the height of the firm's prosperity it is thought that around 80 men were employed there full-time.

Most significantly, the father-and-son team began to publish a series of remarkable atlases, which surpassed in beauty and scale anything of the kind that had gone before. The first of these was the Atlantis Appendix of 1630, which was based upon copperplates that Willem had acquired from the stock of the older mapmaker Jodocus Hondius Jr. However, the Blaeus' first true masterpiece was to appear in 1635: an atlas of 208 maps, grandly entitled *Theatrum Orbis*

Terrarum, Sive Atlas Novus in quo Tabulae et Descriptiones Omnium Regionum ("The Theatre of the World, or New Atlas, including Tables and Descriptions of All Regions") with different editions in Latin, German, Dutch, and French.

This handsome two-volume work enjoyed instant success. Its popularity is attested by the fact that it was frequently reissued, with new editions greatly enlarging the original resource: by 1645, it had expanded to four volumes, and by 1662 to twelve.

The 1662 work contained 593 double-page maps and 3,000 pages of text. Fittingly, this later edition was known as the Atlas Maior, or "Greater Atlas." In this form, it was the largest, most expensive, and most sumptuously illustrated book to be produced in the 17th century. In common with other atlases from the Blaeu stable, it was renowned

ABOVE *A bird's eye plan of the southeastern French city of Avignon. This plan appeared in the atlas of Italy (1663), one of the series of renowned Blaeu townbooks. Since 1348 Avignon had belonged to the Papacy and only became part of France again in 1791. The famous broken Pont d'Avignon (top left), which today has only four spans, was more complete at the time this plan was made.*

for its high degree of craftsmanship, especially its vibrant, hand-colored maps printed on the finest paper, its meticulous typography, and its luxury binding. The world map from the Grand Atlas was decorated with border panels depicting the four elements, the four seasons, the seven planets of the Solar System, and the Seven Wonders of the World. The Atlas was treasured as an exquisite artifact by the prosperous Dutch burghers who made up its intended market (it cost

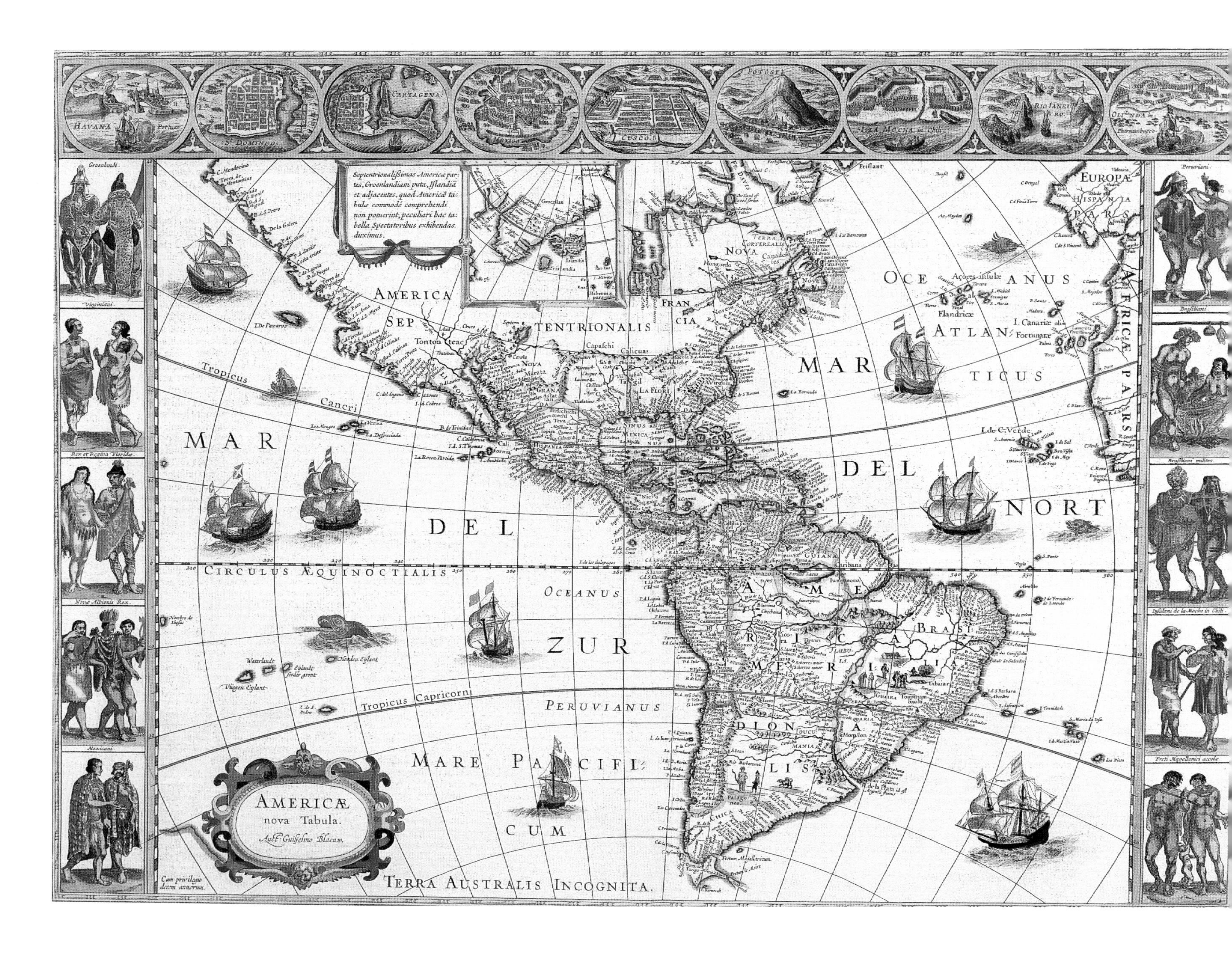

45,000 florins, a small fortune at that time). Its massive format meant that special atlas display cabinets were often constructed to house it, further enhancing its status as a prized family heirloom. Several heads of state also owned the work, which was often donated as a lavish gift by Dutch trade delegations; for example, it is known that a copy was presented to the Ottoman Sultan Suleyman II in 1688.

ABOVE *On this map from the Blaeus' 1635* Theatrum Orbis Terrarum, *it is clear that South and Central America, after a century of Spanish conquest, are extensively charted, but the interior of North America was still unknown; Europeans had no idea of the extent of the Mississippi River, whose upper reaches were first navigated by the French explorers Jolliet and Marquette in 1673.*

The 17th century witnessed a great increase in maritime exploration by the European powers, foremost among which were the United Provinces of the Netherlands, a dynamic young nation that had been engaged since 1568 in a protracted struggle to win its freedom from Spanish domination. The economy began to thrive as never before, as Dutch seafarers wrested control of key trading routes and stations from the moribund Portuguese empire; most notably, the fabled Spice Islands, the Moluccas in the Indonesian Archipelago, with their sought-after nutmeg, cloves, and mace, which fell to the Dutch in 1623 (though ostensibly shared with the English East India Company).

The center of European commerce, which had been at Antwerp in the Spanish Netherlands (Belgium) shifted quite quickly to Amsterdam. In this economic boom based on seaborne trade, maps were of paramount importance, both practically and as symbols of world trade dominance. The Blaeu's preeminence in their profession was formally recognized in 1633, when Willem was appointed Chief Cartographer to the Dutch East India Company (*Vereenigde Oostindische Compagnie*, or VOC). When Willem Blaeu died in 1638, Joan took over his position at the VOC.

After Willem's death, the business continued to be run by Joan and his younger brother Cornelis (this brief partnership ended in 1644, when Cornelis died at the young age of 34). Alongside the magnificent atlas editions, the Blaeus produced a series of town atlases during this period. In particular, their *Novum ac Magnum Theatrum Urbium Belgicae Regiae* of the 1640s, containing 220 maps, had a decidedly nationalistic flavor, celebrating the thriving and industrious cities of the Dutch Republic, which finally gained Spanish recognition of its independence at the Peace of Westphalia in 1648.

We may gauge just how prestigious the products of the Blaeu printing house were from the fact that King Charles I of England owned a copy of this work. A town atlas of Italy was also produced. Scholars have estimated that, at the peak of its success, the Blaeu press created more than 1 million impressions from 1,000 copper plates over a period of just four years.

Joan's official esteem continued to grow, with his appointment as Captain of the Civic Guard in 1650, election to Amsterdam's city council (1651-72), and several other responsible positions. He clearly enjoyed the favor of the country's young *Stadholder*, Willem II. In 1667 another printing premises was opened in Gravenstraat, containing fifteen presses.

The year 1672 was truly an "annus horribilis" for the Blaeu family. On 23 February, the Gravenstraat works were completely destroyed by fire. Both printed maps and copperplates were irrevocably lost. Later that year, in September, political intriguing saw Joan lose his seat on the council, when a new *Stadholder*, Willem III, was appointed. Joan's health duly suffered and he died at the end of 1673, aged 75.

Although the family business was carried on by his sons Joan II and Pieter, it never again reached the heights it had attained under Willem and his son, and the firm was finally wound up early in the 18th century. With the death of Joan Blaeu, the Golden Age of Dutch mapmaking is generally acknowledged to have come to an end.

Tools of the Trade

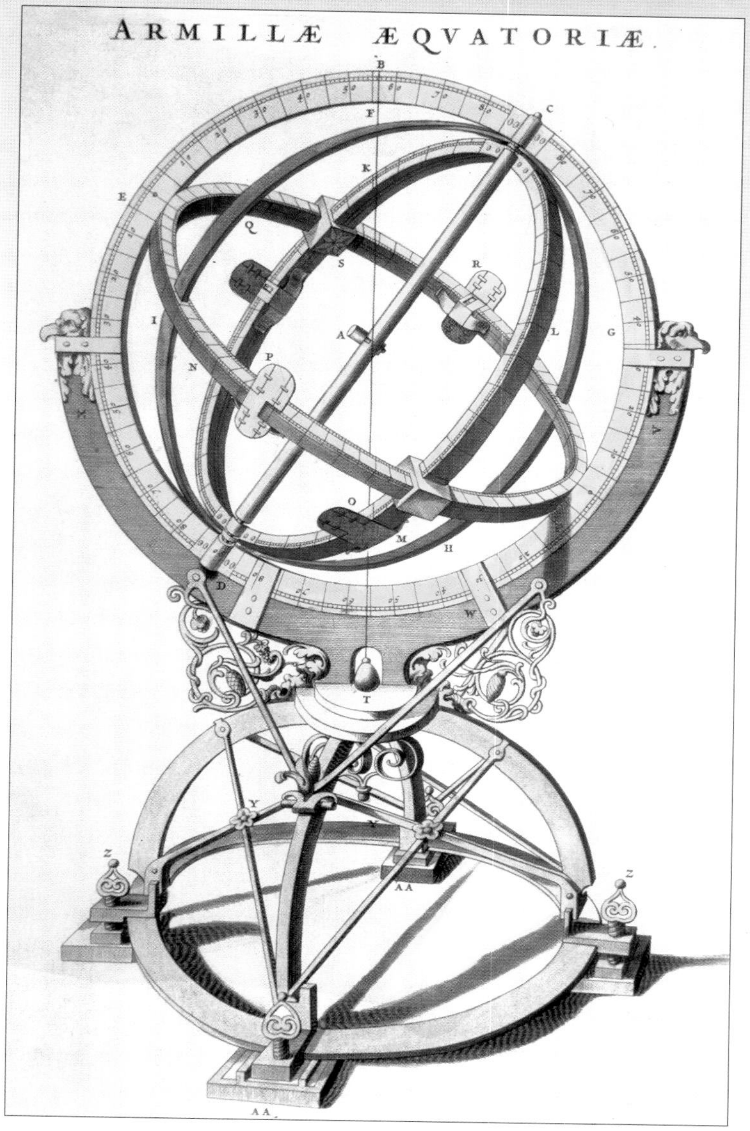

In addition to their mapmaking prowess, the Blaeus were also skilled manufacturers of scientific instruments. One such tool was the armillary sphere. First devised by the ancient Greeks, the most common type consisted of a small sphere representing the Earth within a series of graduated metal rings representing the spherical celestial shells of Aristotle's model of the Universe. With it, astronomers could teach astronavigation, determining the latitude and longitude of celestial bodies. This 1662 equatorial armillary sphere, made by Joan Blaeu, was a specialized type used to calculate the positions of stars in equatorial coordinates.

The Finest Line:

Frederick de Wit

Following in the long line of Dutch master mapmakers, and assuming the mantle of that country's foremost cartographer after the decline of the Blaeu and Jansson publishing houses in the 1660s and 1670s, was the engraver Frederick de Wit. Settling in Amsterdam—by then the undisputed world center of publishing and cartography—his popular and diverse body of work may be said to represent the very last flourish of the Dutch "Golden Age" of mapmaking.

De Wit's work was characterized first and foremost by the sheer variety of the material that he produced. He acquired at auction a considerable number of copperplates by several former masters (notably the Blaeus and the Janssons), and set about covering every conceivable aspect of cartography, among them atlases of the world, panoramic views, maritime charts, and large wall maps. Born in 1630 in Gouda, he may well have been active in the Dutch capital from the beginning of the 1650s, though the earliest extant dates on maps of his are from 1659. Before long, his business was thriving, with his beautiful maps being much sought after.

The quality of his work is attested by the fact that, despite being a Catholic, he enjoyed success in a city that, at the time, was very unsympathetic toward Rome, to understate the position. Municipal archives from the end of the 17th century describe him as being a "worthy citizen."

Lest we be tempted to think that de Wit was simply profiting from the skill of those who had gone before him, it is worth noting the historical accuracy that he was careful to invest in his work; thus, for example, in a new edition from the early 18th century of his 1660 *Atlas Maior*, the map of Scandinavia and the Baltic lands clearly shows the new city of St Petersburg (only founded in 1703) at the head of the Gulf of Finland. Such topicality was something that even his illustrious predecessors were not always scrupulous about when busily reissuing their maps.

One type of mapmaking for which de Wit became especially renowned was the plan-view of cities. These were often published bound together in handsome compendiums, such as the *Theatrum Praecipuorum Totius Europae Urbium* ("Theater of all the Principal Cities of Europe") of 1695. Taking a bird's-eye view of the towns, such works depicted individual streets and buildings and highlighted notable landmarks—such as the Leaning Tower of Pisa, a tourist attraction even in those days. They showed off to best advantage de Wit's mastery of detailed engraving, which employed very thin lines for extra clarity, and his outstanding use of color.

Other notable works were his sea atlases (the *Orbis Maritimus* of 1675 for example), where the precise engraving is used to good effect to show the irregularity of coastlines of such regions as Norway or South America. Even on such large-scale maps, small decorative details enhance the overall aesthetic appeal; for example, a map of the Kingdom of Naples (the whole of Southern Italy) includes a tiny, meticulously engraved eruption of Mount Vesuvius. Likewise, on a 1680 map showing all of North and South America, a vignette of warring Indian tribes adorns the region of Brazil. (The title of this latter map, "The Southern Ocean with the Island of California," reveals the contemporary misconception on which it was based, yet we cannot blame de Wit for this major geographical inaccuracy, nor does it detract from the excellence of its execution.)

De Wit's work attained such heights of popularity that it continued to be published long after his death in 1706 and the sale of the shop following the death of his wife five years later.

RIGHT *This 1680 double hemisphere world map by de Wit updated earlier similar maps of his from 1660 and 1670, giving greater detail and accuracy. For example, even on his 1670* Nova Orbis Tabula, *the island of New Guinea was incorrectly sited in the northern hemisphere, whereas here it is in the southern hemisphere. This work also demonstrates the rich decoration with which de Wit adorned his cartography, with nymphs, fauns, and gods (including Neptune, bottom right) shown around the margins.*

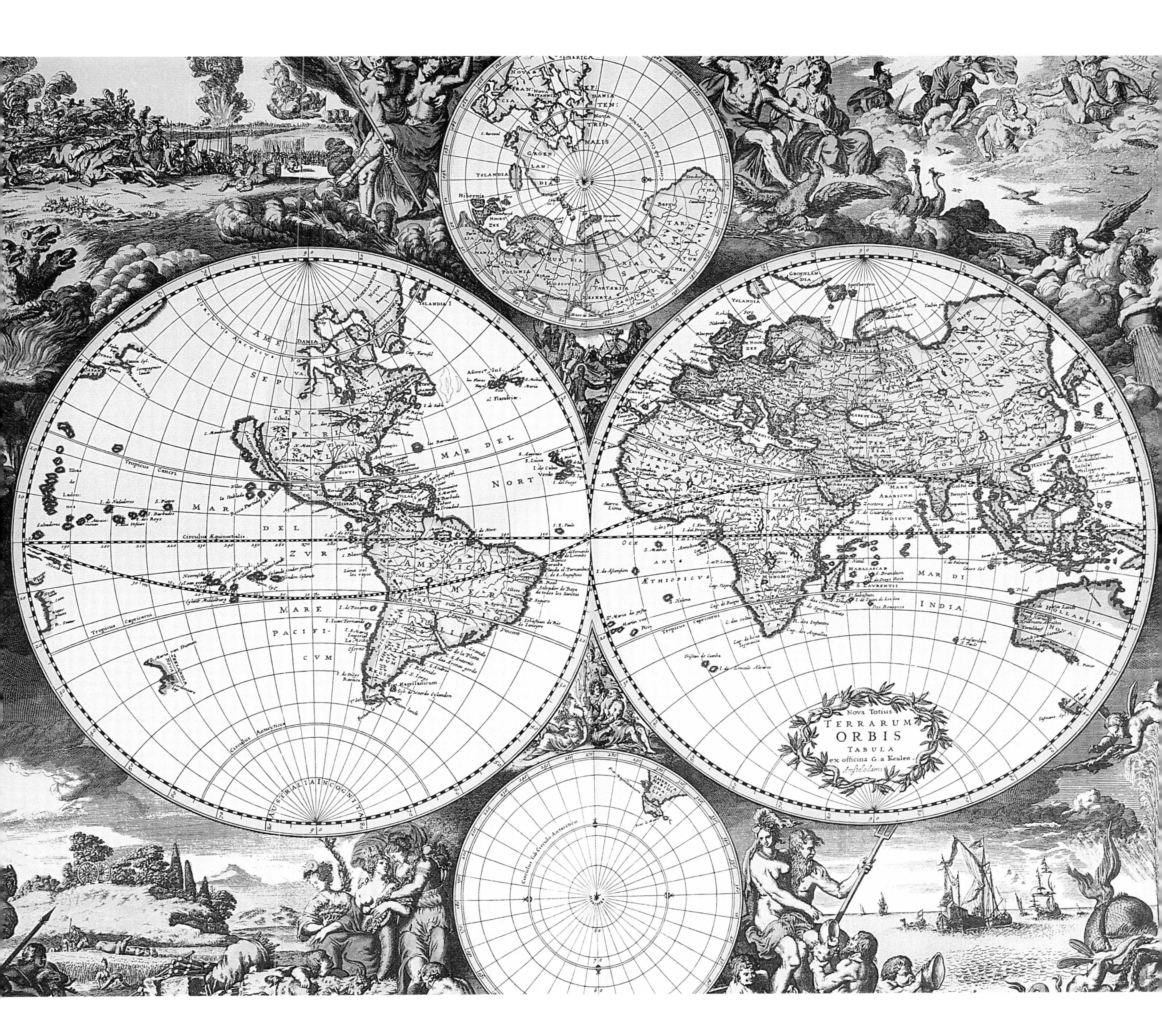
Nova Totius
TERRARUM
ORBIS
TABULA
ex officina G. a Keulen.
Amstelodami
MAR DEL NORT
MAR DEL ZVR
MARE PACIFICUM
MAR DI INDIA
AMERICA
Circulus Æquinoctialis
Tropicus Cancri
Tropicus Capricorni
Circulus Antarcticus
AUSTRALIA INCOGNITA
ÆTHIOPICUS
NOVA HOLLANDIA

"He Who Does Not Work, Will Not Eat":

John Smith and John White

Artist-explorers, propagandist-cartographers, expansionist-adventurers: the men behind the earliest European surveys of North America were motivated by a complex mix of political and personal considerations as they set out to map the new-found lands.

Some, such as Captain John Smith of Pocahontas fame, command enduring reputations that overshadow those of the maps they made. In life, Smith emerges as an ambitious and boastful man who reveled in his own brave adventures and wrangled with his fellow colonists for control of the American territory that he recorded in his maps. Today, immortalized by music, poetry and film, Smith survives in the modern imagination as a romantic hero, philanthropist and peacemaker. His famous admonishment to the Jamestown settlers, urging them to farm for their lives, (above), adds to the image of the tough leader of men.

It is difficult for us to know truly what sort of a man Smith was, and what motivated him in his American adventures as he struck out in the early years of the 17th century, leaving his compatriots and heading into the wilderness to map an expanse of land that stretched as far as the Chesapeake. Contemporary records point towards a dispute between the settlers, but was Smith's departure influenced by a self-sacrificing commitment to finding new lands for his fellow settlers to colonize, or merely by greed for land and power of his own? Whatever his motivation in surveying Virginia, the map produced by Smith and engraved by William Hole for publication in 1612 was considered definitive until well into the 1700s, and remains the most important source of information on Indian settlement to this day (see page 90).

Smith's map features a picture of the great Indian chief Powhatan, father to Pocahontas. He is seen wearing splendid headgear, and seated on a dais that elevates him above the people of his tribe. When John Smith was captured by one of Powhatan's scouting parties, the Chief was so impressed with the English man's spirit and bravery that he spared his life, and, following a trial ritual, made Smith a minor chief of his tribe. Pocahontas was 11 years old at the time of Smith's trial, and played a part in the ritual: Smith was afterwards convinced that she had played a role in saving his life, and later took her to England where she learned English and became a celebrity at the royal Court.

Smith's were not the first definitive European-made maps of Virginia, however. His map of Virginia, as well as the inset pictures of Indians that appeared on it, drew heavily on the work of another Englishman, John White, who first sailed to America some 27 years earlier. John White's map and John Smith's later adaptation both feature an intriguing mixture of English and Native American language place names and cartographic information. Were it not for maps such as these, the names and locations of many Native American settlements would be lost today.

When Sir Walter Raleigh's first expedition reached Roanoke Island in 1585, John White was the man responsible for pictorially recording the people, animals, plants, and territories they found in America. Together with the writer Thomas Hariot, John White traveled widely, making sketches, drawings, and miniature paintings of everything of note about the new world. Propaganda was a significant element of the work undertaken by White and Hariot, representing the colonies in a positive light to instill confidence into their financial backers, and to enthuse would-be colonists back in England. Many Englishmen and women had their first view of America through the eyes of John White —for most of these people, White's maps and sketches were as close as they would ever come to visiting the land across the ocean.

John White's maps were enormously influential in the late sixteenth and early seventeenth century, yet little is known about the man himself. His was a common name in 16th-century England, and, until the time of the first Roanoke voyage, very little is known about the soon-to-be-famous cartographer. He rises from obscurity in 1585 to travel across the ocean with Raleigh's adventurers. Following the failure

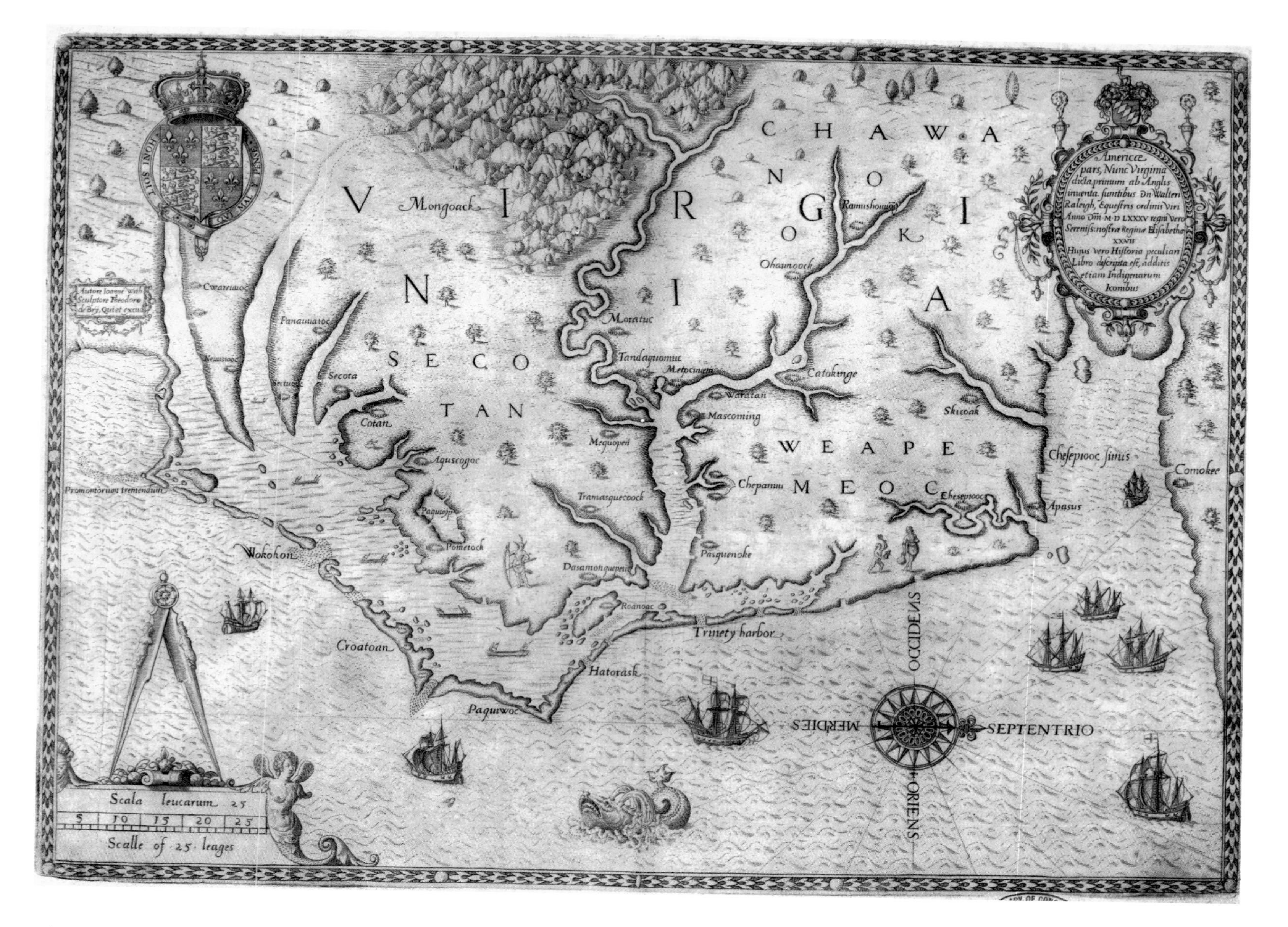

ABOVE Americae pars, nunc Virginia dicta *(part of America, now called Virginia); Theodor De Bry's engraving, based for the most part on two of John White's watercolor sketch maps, was published in 1590 in Thomas Hariot's* A briefe and true report of the new found land of Virginia, *reprinted by De Bry as part 1 of his* America. *North is to the right and the scale is about 1:1,700,000. A* scala leucarum *(scale of leagues) is provided: a league was not then a set distance and here it represents about two nautical miles. The map covers the coastal region from Cape Fear to Chesapeake Bay.*

of the second "Lost" Roanoke colony of which he was made governor, he disappears from record almost without a trace—the last evidence of his existence is a letter written from his Irish home in 1593. White was undoubtedly a skilled artist, and it is possible that he was the artist with Martin Frobisher's arctic expedition in the 1570s; perhaps his work on that expedition led to his being selected to join Raleigh's men a few years later. His daughter, Eleanor, married a builder named Ananias Dare, suggesting that White's origins were relatively modest. Yet in 1587 he received his own coat of arms—a sign that he was moving up the social ladder. His dates of birth and death remain unknown, yet we do know that Eleanor gave birth to his granddaughter, Virginia Dare, in 1587. Virginia, in her short life, won fame in her own right as the first child born to English parents on American soil, but nothing more is known about her following the disappearance of the ill-fated second colony some time between John White's departure from Roanoke in 1586, and his return to an abandoned island in 1587.

John White's sketch maps are not often seen today in their original formats—more frequently, we see the work of the engraver Theodore De Bry, who collated White's work for publication in the 1590s. Many

FAR RIGHT *Of course it was not only the English who were mapping the New World. Jacques Le Moyne de Morgues (b. ca 1533 d. 1588) was the first European to map Florida. It was Sir Walter Raleigh who persuaded the Frenchman to write about and illustrate his experiences, and it would be De Bry who published the book, in 1591. The waterfall he shows pouring into a lake in the "Montes Apalatci" may depict Native American references to Niagara Falls.*

of White's original sketches, together with much of Thomas Hariot's writing, were lost in the hasty departure of the first Roanoke settlers, but enough remains for us to see that de Bry's were not carbon copies of the White maps.

The best known of the Smith–de Bry maps is surely that of the east coast of North America, from the lower Chesapeake Bay in the north to just south of Cape Lookout. De Bry's engraving is based on a number of sources, including two of White's watercolor sketches—one covering the same area as the engraving; one showing part of North America and part of the West Indies—and a crudely-drawn pen-and-ink map, thought to be the work of Ralph Lane. In comparing the published engraving with White's original work, we can see that places are moved, omitted, re-named or added, as the European understanding of American geography evolved. Some of these changes are corrections; others are effectively errors, and may have been based on other source materials that have since been lost. The text in the top right hand corner of the map, beginning "Pars America," reads in translation:

> "Part of America, now called Virginia, first discovered by the English, taken up by Sir Walter Ralegh, Knight, in 1585 [sic] A.D., in the twenty-seventh [year] of the reign of our most serene Queen Elizabeth; the history whereof is truly described by this special book and also by the images of the natives included."

The images of natives alluded to here were also the work of White, in his role as visual chronicler of the Roanoke settlement. Small, inset images of natives are also used to decorate the Smith–de Bry map, and can be seen in the areas marked "Secotan" and "Weapemeoc"—both names of Algonquian tribes native to Virginia.

Virginia itself is one of the few English language place names to appear on this early engraving. The name honors Queen Elizabeth 1, known as England's "Virgin Queen," and was introduced in around 1584. Prior to this time, the area was known by either the Anglo-Indian "Wingadacon," a name that may have referred either to English attire or to trees, or the native name, "Ossomocumuck."

The Pocahontas Image

Predating Disney by about 140 years, this tobacco package label of 1860 is another example of how the John Smith legend became part of the American Grain. The chromolithograph depicts Pocahontas coming to the defense of Captain Smith and appealing to her father, Chief Powhatan, to spare his life. The depiction of the Native Americans here finds an echo in Theodor De Bry's engravings based upon John White's watercolor paintings of figures. In the reprint of Thomas Hariot's **A briefe and true report of the new found land of Virginia**, White's portraits of Native Americans are transformed in the the engravings. Originally in mannered but more or less realistic poses, the figures take on sculpture-like poses and more muscular physiques. Their hands and feet are smaller and, crucially, their features are Europeanized. Something similar is happening here to the face of the Indian "princess." The myth of the "Noble Savage" gets a makeover. Not all of the copper-plate engravings were by De Bry. Some were by his associate Gysbert van Veen, such as "A young gentle Woman Daughter of Secota"—tall, virginal, and European.

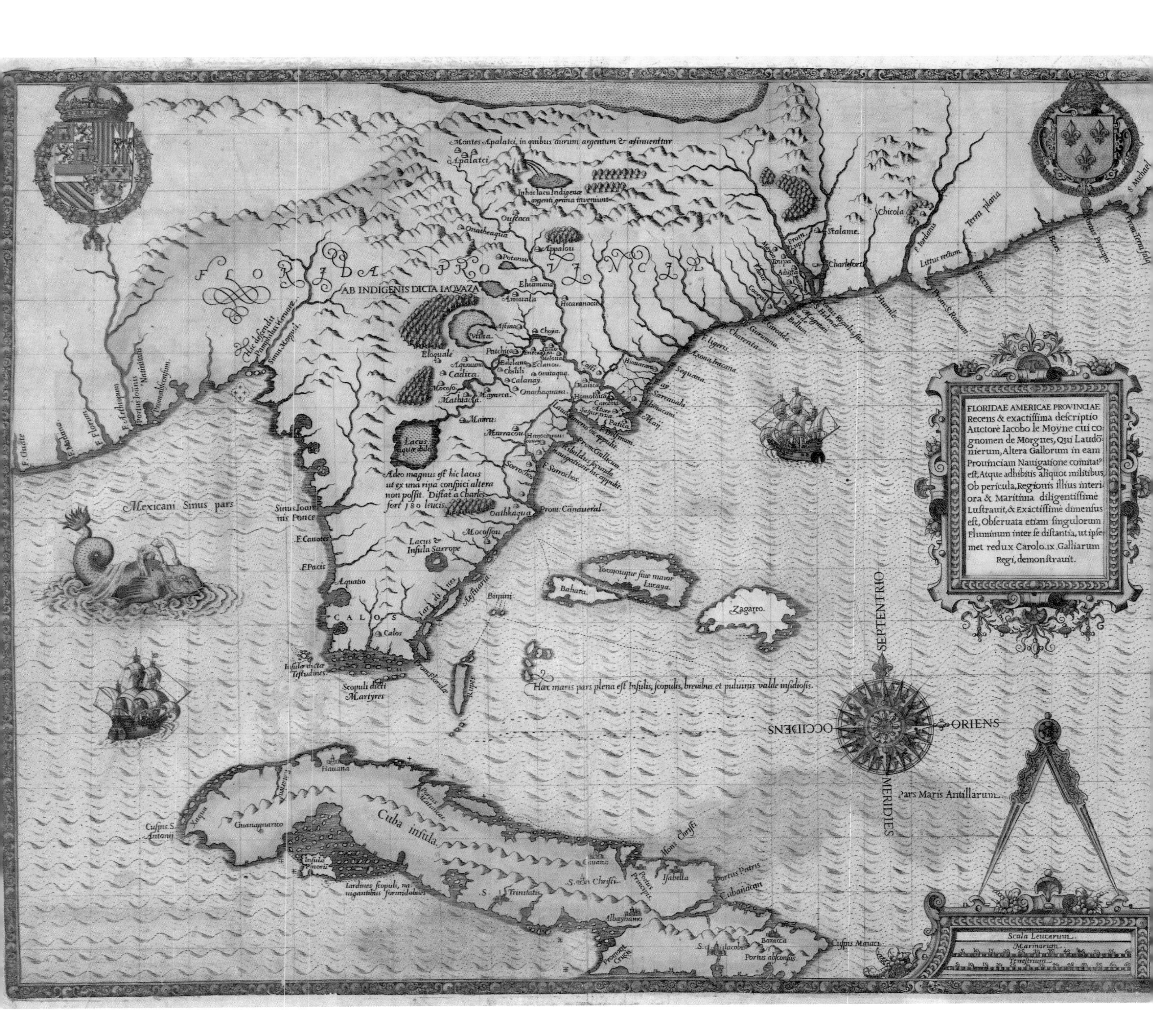
FLORIDAE AMERICAE PROVINCIAE Recens & exactissima descriptio Auctore Iacobo le Moyne cui cognomen de Morgues, Qui Laudōnierum, Altera Gallorum in eam Provinciam Navigatione comitat9 est, Atque adhibitis aliquot militibus, Ob pericula, Regionis illius interiora & Maritima diligentissime Lustrauit, & Exactissime dimensus est, Obseruata etiam singulorum Fluminum inter se distantia, ut ipsemet redux Carolo. IX. Galliarum Regi, demonstrauit.
Montes Apalatci, in quibus aurum argentum & æs inueniuntur
Apalatci
In hoc lacu Indigenæ argenti grana inueniunt
FLORIDA PROVINCIA
AB INDIGENIS DICTA IAQUAZA
Lacus aquæ dulcis
Adeo magnus est hic lacus ut ex una ripa conspici altera non possit. Distat a Charlesfort 180 leucis.
Mexicani Sinus pars
Sinus Ioannis Ponce
Lacus & Insula Sarrope
Prom: Cānaueral
Calos
Insulæ dictæ Testudines
Scopuli dicti Martyres
Hæc maris pars plena est Insulis, scopulis, breuibus et puluinis valde insidiosis.
Bahara
Zagareo
Charlesfort
Stalame
Chicola
SEPTENTRIO
OCCIDENS
ORIENS
MERIDIES
Pars Maris Antillarum
Hauana
Cuba insula
Guanaymarico
Cuspis S. Antonij
Iardines scopuli, nauigantibus formidabiles
Trinitatis
Albayhamo
Isabella
Portus Patris
Mons Christi
Cuspis Maiaci
Portus absconsus
Scala Leucarum
Marinarum
Terrestrium

Sanson d'Abbeville and de Champlain:

The First Accurate North American Maps

The most accurate 17th-century representations of Michigan and the Great Lakes were, without exception, produced by French mapmakers. Good quality cartography was dependent on detailed reports from pioneers and explorers, so French colonial interests in the region led directly to their production of superior maps. As late as 1701, British mapmakers were still consistently showing only one large lake at the end of the St Lawrence River.

The French had been aware of the existence of several distinct bodies of water since the publication of Samuel de Champlain's 1632 map, and Nicolas Sanson became the first cartographer to correctly note all five Great Lakes in 1650.

Louis XIII and XIV were deeply interested in the work of their cartographers, supporting them politically and financially. By the end of the century, Dutch maritime power had gone into decline and France was established as the center of geographical science, producing the clearest, most beautiful and most accurate maps of the day.

French cartography during this time was very often a family affair, and one of the greatest cartographic families of the century were the Sansons of Abbeville. Nicolas Sanson was born in 1600, and he, his sons, his grandsons, and even his son-in-law became leading lights of the world of mapmaking. Before Nicolas Sanson's rise to fame, however, the French navigator and soldier Samuel de Champlain had already played an important role in charting the new French territories.

Born in 1567, Samuel de Champlain was a trained mariner and a skilled navigator who made his first transatlantic journey in 1599, traveling as far as Mexico. By the time of his death in 1635, he had made a further seven voyages to and from North America. His adventures were many and varied: as well as charting vast areas of land (including his discovery of Lake Champlain, to which he gave his name), he found himself befriending the Montagnais Indians and fighting against the Iroquois and the British; taken prisoner and held in England for three years from 1629 to 1632, and, perhaps most significantly, founding and governing the French colony at Quebec.

De Champlain died in Quebec, having returned there in 1633 following the Treaty of St Germain-en-Lye, which saw Canada, Acadia, and Cape Breton restored to French control.

De Champlain produced a series of maps depicting "New France," updating details and altering particulars as he learned more about the geography of the prospective colony. Initially sent there in 1603 to report on the potential for colonial development in Canada, de Champlain gradually developed an unrivalled understanding of the region. His work served as the basis for the maps produced by the Parisian mapmakers of the period.

His first large-scale map of New France, entitled *Carte Géographique de la Nouvelle France,* was a highly ornate work. Published in 1612, it was the product of his early explorations in the St. Lawrence valley, combined with general information about New England. This early map is lavishly decorated with color illustrations of native flora, American Indians, ships, and sea-life, betraying de Champlain's hybrid role as cartographer, artist, and pioneer.

In 1632, he published another large-scale map of the same region. By this time, the French fur trade had expanded significantly and de Champlain had established his own trading company. This led to increased interaction with the native Hurons which, in turn, gave Sanson the opportunity to draw on their knowledge and improve his understanding of North American geography. De Champlain's 1632 map is essentially an updated and corrected version of his 1612 work.

The map shown right, *Le Canada,* is the product of Pierre du Val's modifications to a map that was begun by de Champlain as early as 1616. Although the original was on a smaller scale than de Champlain's 1632 map, it is very similar in content and was perhaps abandoned—it was certainly incomplete—at the time of his death. Only one copy of this unfinished map exists today, but the copper plate used for de Champlain's engraving was acquired by Pierre du Val, son-in-law to the acclaimed Nicolas Sanson and himself an influential mapmaker in the

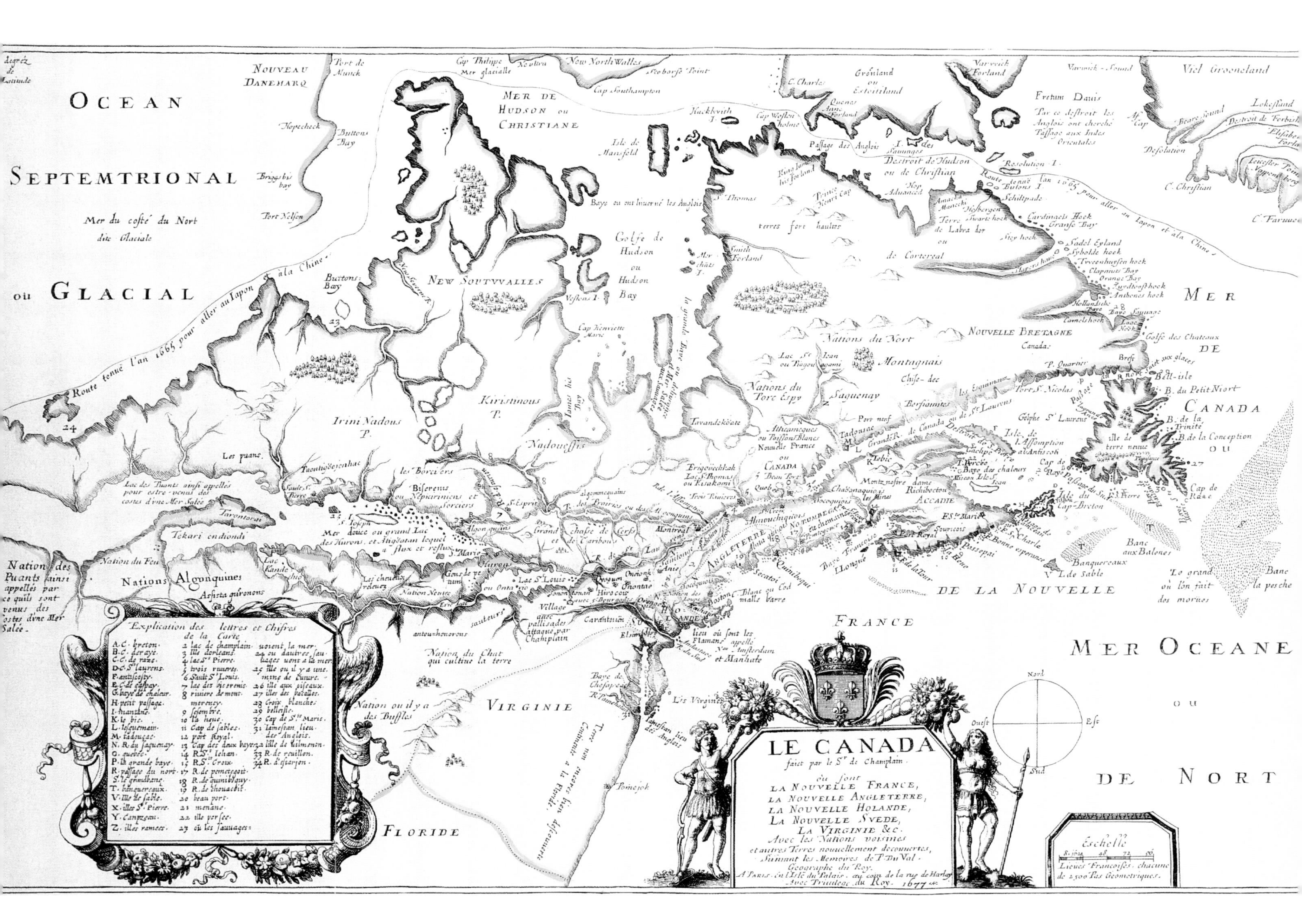

latter half of the 17th century. Finishing and coloring the map, he first published it in 1653 before producing this elaborate 1677 version, which retains its original attribution to "le Sieur de Champlain."

Nicolas Sanson, born in Abbeville, France in 1600, is often referred to as the father of French mapmaking. Certainly, he was father of a dynasty of distinguished cartographers, and his work was profoundly influential in the mid-17th century.

Sanson's interest in mapping grew out of his interest in ancient history, and he began to draw maps in the first instance as a means of illustrating his historical work. The quality and beauty of his work soon attracted the attention of Louis XIII, who is said to have enjoyed Sanson's company so much that he stayed with him when visiting

ABOVE Le Canada *(1677, Samuel de Champlain, modified by Pierre du Val). Published several years after de Champlain's death, this map is based on work that he begun as early as 1616. Samuel de Champlain first traveled to "Nouvelle France" in 1603, sent to survey the areas discovered by Jacques Cartier from 1534 to 1541. His was the first map to show all of the Great Lakes except Michigan.*

Abbeville. Like his sons Guillaume and Adrien after him, he was appointed *Géographe Ordinaire du Roi*—mapmaker to the king—and his duties included tutoring Louis XIII and XIV in geography. Louis XIII and later honored him further by naming him a Counselor of State.

Prolific in his work, Nicolas Sanson produced around 300 maps in his lifetime, and, when he died in 1667 his sons and grandson continued his business in partnership with A. H. Jaillot. His son in law Pierre du Val also helped to preserve the memory of his maps, re-engraving and republishing several of them after his death.

BELOW Mexique, ou Nouvelle Espagne, Nouvlle. Gallice, Iucatan &c.: et autres provinces jusques a l'Isthme de Panama, ou sont les Audiences de Mexico, le Guadalaiaira, et de Guatimala *(1656, Nicolas Sanson d'Abbeville). The Sanson dynasty was prolific in the production of high-quality printed maps, yet they rarely returned to earlier works with corrections or revisions. This map appeared in the collection* Cartes Générales de la Géographie Anciennes et Nouvelles, *a 1676 revision of the 1654* Cartes Générales de Toutes les Parties du Monde.

Sanson's maps charted the world, spanning territories from England to Africa, and from America to his French homeland. One of his most influential and significant works was "Amerique Septentrionale," (right). This was one of the first maps to draw on the "Jesuit Relations" for geographical information, acknowledging the value of the detailed reports produced by Jesuit missionaries in New France. The most accurate of its day, this map brought a number of "firsts" to American cartography: the first evewr portrayal of all five Great Lakes; the first suggestion that a Northwest Passage might be possible; and the first representation of the legendary "Quivira" to the East of New Mexico. Santa Fe appears for the first time on Sanson's map, and the Rio Grande ("Rio del Norte"), emptying into the Gulf of California.

Despite its many good qualities, Sanson's map was not perfect. Although the five Great Lakes appear, they are not accurately shaped. Lake Ontario and Lake Erie are particularly distorted, and Lakes Superior and Michigan are shown as open-ended stretches of water that fall off the edge of the picture. The entire north and west of the

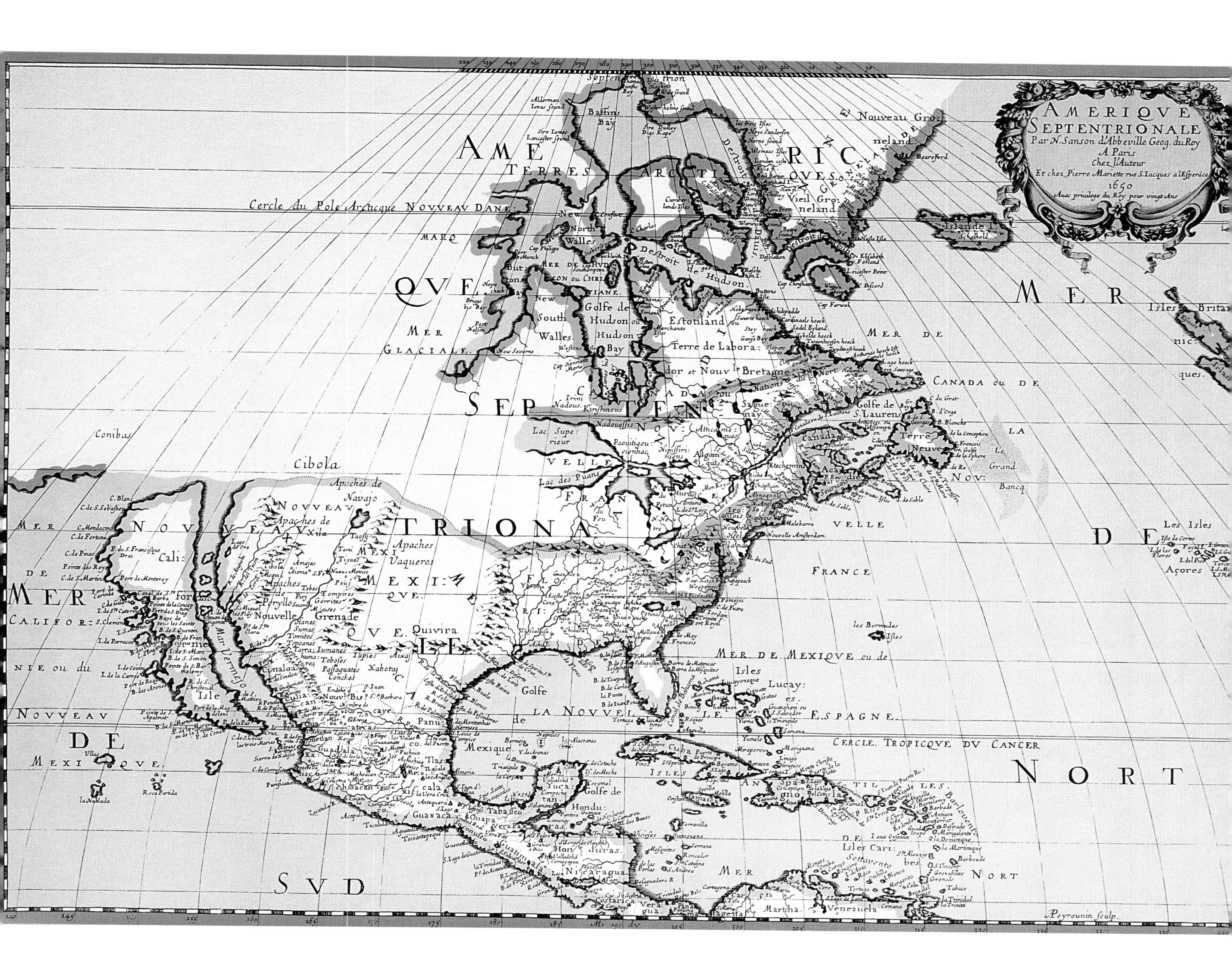

ABOVE Amerique Septentrionale *(1650, Nicolas Sanson d'Abbeville). This was the first map to depict North America using "sinusoidal" projection, where areas on the map are proportionate to the corresponding areas on a globe. The parallels and prime meridian appear as straight lines, and all other meridians curve increasingly outward from the prime. It was Sanson's concern with scientific exactness that enabled him to develop a reputation as a leading mapmaker.*

continent has been left blank as unknown territory, and the most serious fault in the map is the misrepresentation of California as an island, perpetuating a fallacy begun by Father Antoine Ascension in his map of 1602. This misconception proved long-lived, and was not formally dispelled until 1747, when King Ferdinand of Spain issued a royal decree to state that California was not an island after all.

"My Enemy's Enemy is My Friend":

The Chickasaw Map

The Chickasaw originally inhabited an area east of the Mississippi (roughly where southern Tennessee and northern Alabama are now) and were one of the southern tribal groupings whom the European incomers referred to as the "Five Civilized Tribes." This patronizing phrase expressed how relatively easily the peoples in question seemed to adapt to White culture. A remarkable map created by a Chickasaw draftsman in the early part of the 18th century reveals how alliances developed between indigenous peoples and settlers.

The Chickasaws occupied a strategically important position on the central reaches of the Mississippi river. From here, not only could they, as one contemporary settler observed "command all the water passages between New Orleans and Canada," but their homeland lay at the crossroads of important north–south and east–west traditional trading trails. This key location was undoubtedly a great advantage, but was also a source of considerable danger, particularly when settlers and their Indian confederates came into conflict over control of land and trade routes. The Chickasaws were a relatively small tribe (numbering only around 6,000 at the time of European arrival, compared to their far more populous neighbors and rivals to the south, the Choctaw, with some 21,000). Consequently, to preserve the integrity of their homeland and culture, they developed not only an indomitable fighting spirit but also a facility for skillful diplomacy and trade to forge alliances.

The British and French settlers, who started to arrive in southeastern North America from the end of the 17th century onwards in growing numbers, were quick to establish trading connections with native peoples. As rivalries, especially between France and Britain, increased and turned into open conflict, so these commercial links became strategic alliances. Before long, firearms were the most sought-after of all trade goods; with the French colony of Louisiana supplying guns to its Choctaw allies, the Chickasaws made common cause with the British colonies of the eastern seaboard in 1721, and were soon receiving munitions from Georgia and South Carolina. They were also careful to foster good relations with other, larger native groups (such as the Cherokee) who traded with the British, thereby increasing their network of potential allies in times of war.

In 1723 or 1724, a deerskin map was presented by the Chickasaws to Francis Nicholson, the British governor of Maryland, South Carolina, and Virginia. Unfortunately, the original—described at the time as being drawn by an "Indian cacique," a chief of a native group in territory once colonized by the Spanish—is now lost, and its content (along with a similar map) has only come down to us in a copy made by an English colonist. It is highly schematic, showing various peoples and their approximate relative sizes in a series of larger or smaller circles, connected by lines representing pathways, either trading paths or hostile warpaths. Stretching beyond the southeast, the map covers a total area of some 700,000 square miles (1,813 square kilometers) from Texas to New York. The main purpose was graphically to portray the vulnerability of the Chickasaws, surrounded by enemies, in a bid to get South Carolina to supply military aid. While not technically accomplished, the map reveals a sophisticated political worldview, and was deployed in the service of a pragmatic and subtle *Realpolitik* by the threatened native nation. A later deerskin map of 1737 was directed at the French, showing recent alliances between the Chickasaws and the pro-French Choctaw and Alabamas and appealing for peace. This overture clearly failed, as a French force was dispatched from Montreal and Fort Michilimackinac (Michigan) to attack the Chickasaws in 1739. With their diplomacy thwarted, the Chickasaws resorted to their other traditional strength and annihilated the French force.

Tragically, despite their diplomatic adroitness and military prowess (and in betrayal of the advantages their alliance had brought the

RIGHT *From 1720-1725 the Chickasaws fought against France and her allies the Choctaws. The French hated the strong relationship between the Carolina English and the Chickasaws because when France and England were at war the Chickasaws atacked French shipping on the Mississippi River, disrupting the connection between Louisiana and its sister colonies in Canada.*

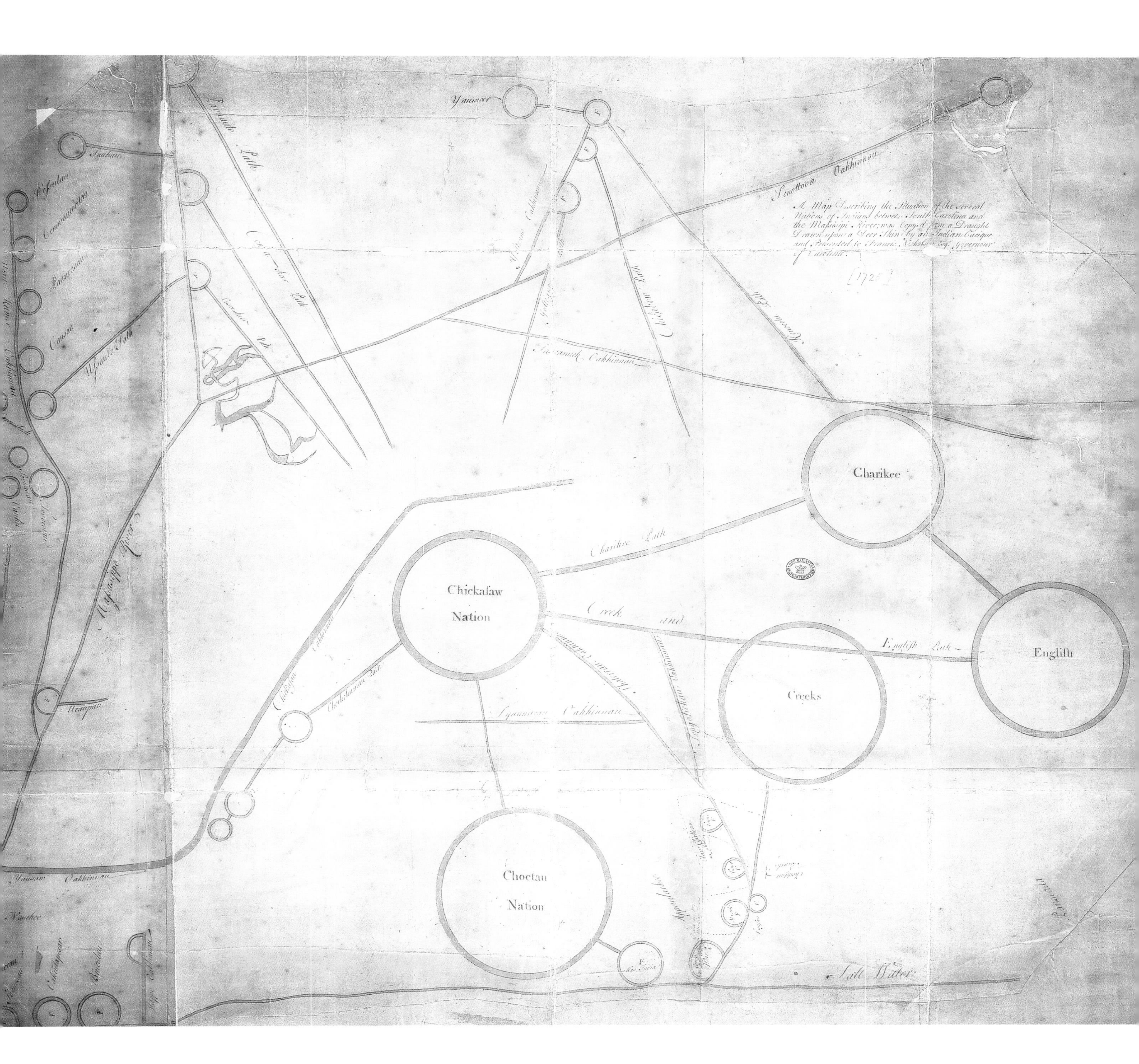

Senottova Oakhinnau
A Map Describing the Situation of the several Nations of Indians between South Carolina and the Massisipi River; was Copyed from a Draught Drawn upon a Deer Skin by an Indian Cacique and Presented to Francis Nicholson Esqr. Governour of Carolina.
[1723]
Yaumeer
Charikee
Charikee Path
Chickasaw Nation
Creek and English Path
English
Creeks
Choctau Nation
Salt Water

Artifacts of the Creek Confederacy

These items belonged to the Yuchi, (see map, right), part of the Creek Confederacy who were removed along with the Creeks in 1836 from their ancestral lands in Georgia to Indian Territory, in present-day Oklahoma. Even as late as the mid-20th century the Yuchi retained their own language. The items shown were all collected by the ethnologist Frank G. Speck in 1904 and 1905. They include, from left to right: a green cotton hunting jacket; a flageolet of cedar wood used by young men during courting; a cane tube used to impart curing power by medicine men; a bow of bois d'arc; a pounder used by medicine men to prepare emetics taken by men at the Green Corn Dance; a ballstick for a team game; and a girl's calico dress. Mute witnesses to a lost culture.

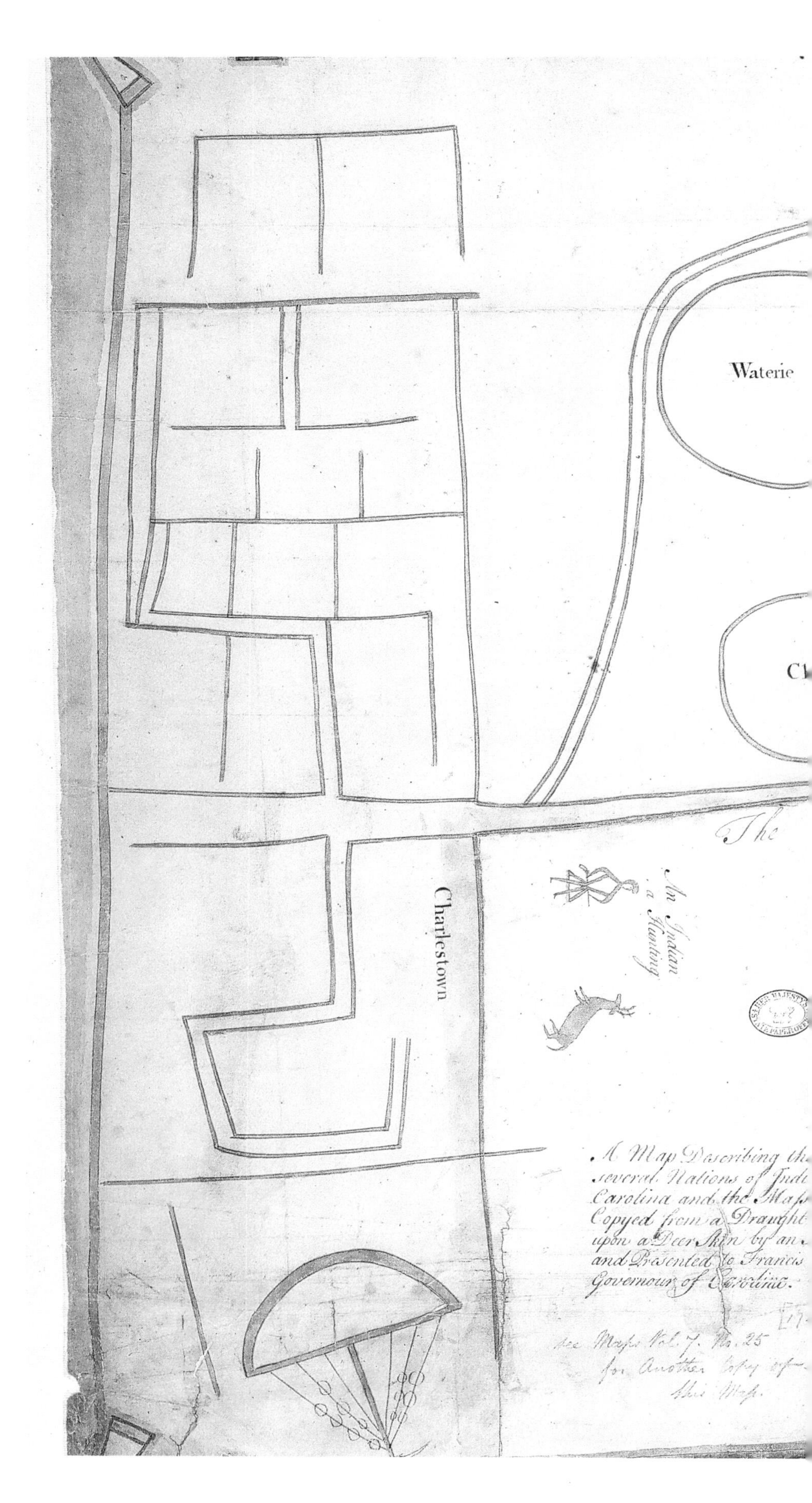

RIGHT *Both this map and the one on the previous page are described as being supplied to Governor Francis Nicholson. This version includes a rough plan of Charleston with a ship in the harbor, and covers North and South Carolina and the southern Appalachian mountain range. The tribes and their locations differ.*

colonists), the "civilized" Chickasaw fell victim to President Andrew Jackson's genocidal Indian Removal Act of 1830 and were forcibly relocated west of the Mississippi. However, to this day, the Chickasaws of Oklahoma believe that "the spirits of all Chickasaws will go back to the Mississippi, and join the spirits of those that had died there."

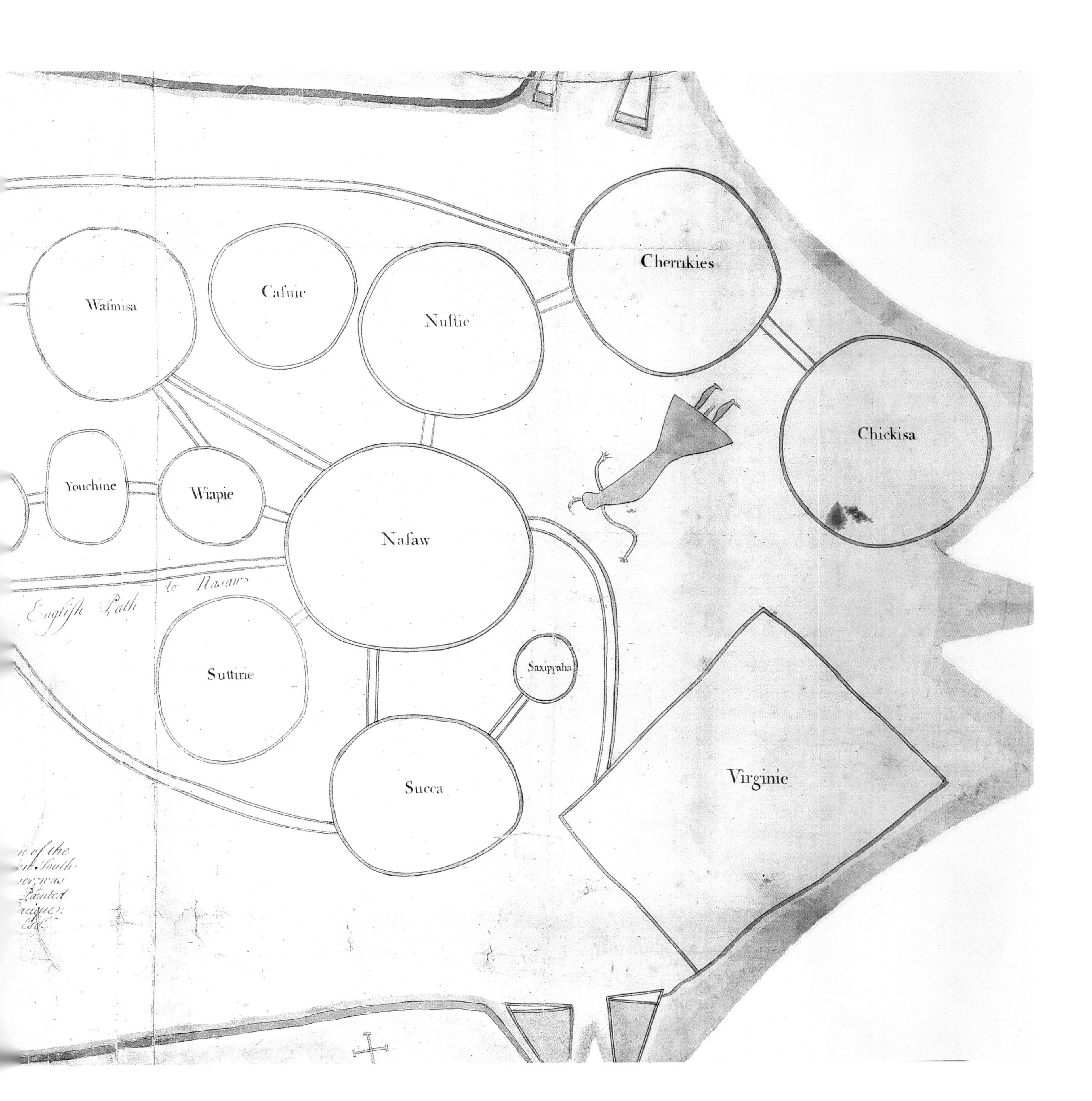
Cherrikies
Wafmisa
Cafuie
Nuftie
Chickisa
Youchine
Wiapie
Nafaw
English Path to Nasaw
Suttirie
Saxippaha
Succa
Virginie

East is East:

Japanese Isolationism

Japan's first real exposure to Western influence came in the middle of the 16th century, when a ship from China (a longstanding trading partner) foundered off the southern island of Kyushu in 1542 with two Portuguese travelers aboard. Initially, the Japanese cautiously welcomed the commercial growth stimulated by European merchants, and especially the technological innovations they brought–including the introduction of Western mapping. Before a century had passed, the rulers of Japan closed its doors almost totally to European trade.

The Japanese resented and feared the activites and influence of Christian missionaries and would enter a period of isolation lasting two centuries. Before reaching this point however, once the southern port of Nagasaki was opened to foreign commerce in 1570, the Portuguese began to trade freely with Japan. It was not long before the other great seafaring nations of the time, the Netherlands and England, appeared on the scene. A shipwrecked English adventurer, Will Adams, settled in the country and helped Japan develop her own merchant fleet from 1600 onward. The main commodities that the foreign merchants imported were raw silk and finished textiles from China, and—most lucratively—new European models of firearms that transformed warfare in Japan. These they bartered in exchange for precious metals such as gold, silver, and copper from the country's recently developed mines.

Increased trade brought not only goods, but also ideas. Jesuits from Spain and Portugal were the prime movers in teaching important practical subjects, for example mathematics, astronomy, engineering, and mining. These missionaries had arrived shortly after the first wave of European traders; in 1548-51, Francis Xavier traveled to the then-capital Kyoto and to Kagoshima to win converts, particularly among the influential feudal class (*daimyo*). Yet the Jesuits' proselytizing zeal and disruption of the traditional social system caused deep resentment, and, after several bouts of persecution had failed to quell Christianity, matters came to a head.

The final catalyst was the major Christian-inspired Shimabara rebellion that broke out near Nagasaki in 1637, and prompted the shogun Tokugawa Iemitsu to order the expulsion of almost all foreigners. Japanese citizens were also banned from journeying overseas. Henceforth, the only Western nation sanctioned to trade with Japan would be the Netherlands. Trade with these foreigners was now to take place under strictly controlled conditions. The agents of the Dutch East India Company were moved from their original base at Hirado and confined to the small artificial island of Deshima in Nagasaki harbor, whose construction the shogun had ordered as a base for the Portuguese in 1634.

Before the expulsion occurred, one of the new branches of learning that the Jesuits introduced was Western cartography. The traditional Japanese concept of geography regarded the world as consisting only of the three Buddhist cultures of Japan, China, and India. Exposure to Western mapmaking revealed that a much wider world lay beyond, and local artists soon began to copy maps produced in Europe onto folding screens to satisfy a demand from rich patrons.

Scholars have determined that these early Japanese cartographers were especially influenced by Petrus Kaerius' large wall-map of the world of 1607 (see page 116); the vignettes of ten world rulers that Kaerius placed around the margin of this map make a reappearance in Japanese paintings of this period.

Yet even after 1640, the know-how that the Japanese had gleaned meant that western cartographic knowledge continued to spread. Improvements in printing saw the production and dissemination of the first truly "modern" Japanese map in 1645. Known as the *Kon-yo Bankoku Zenzu* ("general world map"), it was based on the so-called "Great Map of Ten Thousand Countries" published in Beijing in 1602 by Matteo Ricci, (1552-1610) an Italian Jesuit missionary who had been instrumental in introducing the principles of Western cartography to China. Under his tutelage, the Chinese had learned for the first time about latitude and longitude, as well as the distribution of the Earth's oceans and land masses, and in doing so began to adopt a less Sinocentric view of the world.

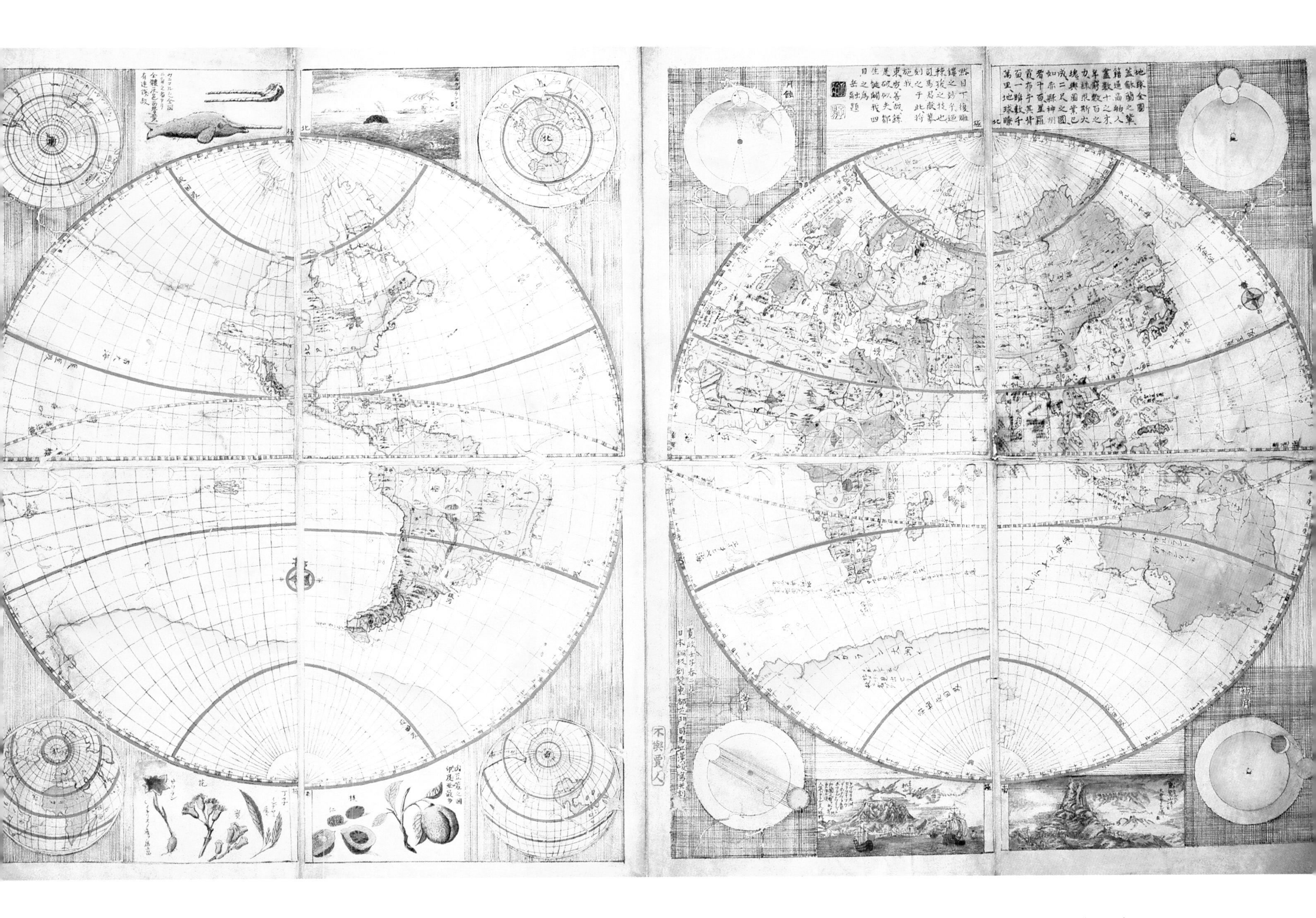

ABOVE *The world map known as the* Chikyu Zenzu *produced in the late Edo period (1792) by the Nagasaki scholar Shiba Kokan marked a new departure in Japanese cartography, incorporating as it did the recent geographical knowledge of Dutch and other Western explorers. Thus, for the first time in Japan, this copperplate printed map showed New Guinea and Australia.*

Ricci's own map drew heavily on works by Ortelius and Mercator, and first arrived in Japan shortly after the start of the Tokugawa Shogunate (1603). The *Bankoku Zenzu* was extensively copied and reprinted, and for a long time remained the standard map of the world in Japan. Even in the late 18th century, a direct spin-off, Nagakubo Sekisui's *Kaisei Chikyu Bankoku Zenzu* "Revised World Map") was widespread in Japan.

In 1720, a small but significant relaxation in the import of foreign goods came when Tokugawa Yoshimune eased the embargo on Western books (except those on the subject of Christianity) to promote scientific and technical learning. Arriving through Deshima, Dutch atlases ensured that some current geographical knowledge was available to Japanese artists. It is through this line of influence that one of the most celebrated maps to emerge from Japan during its long

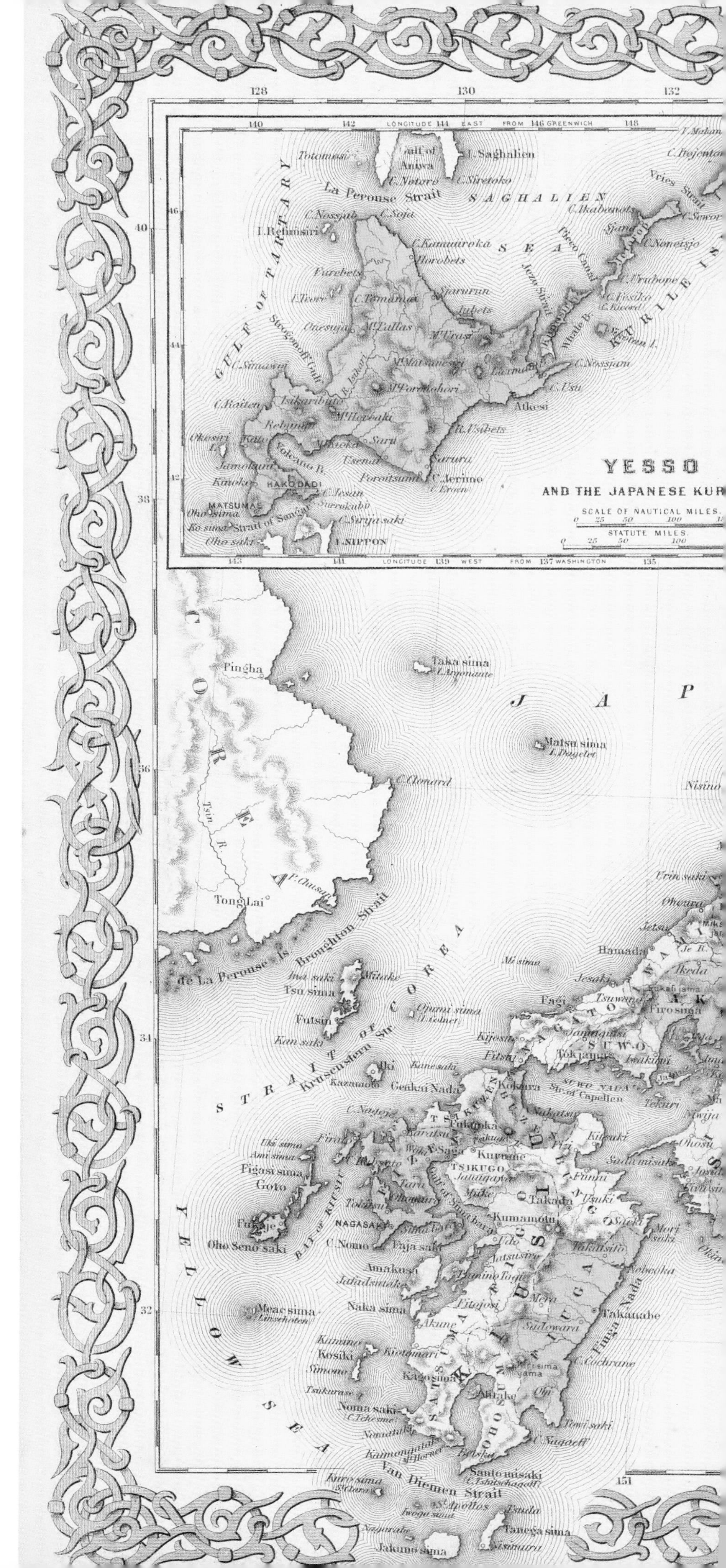

RIGHT *In 1855, just one year after US Commodore Matthew Perry had compelled Japan to resume trade and diplomatic contacts with the West after more than two centuries, the New York firm of G.W. Colton & Co. responded to the new American popular awareness of Japan by producing this map. The legend acknowledges the mapmaker's debt to the "surveys and reconnaissances of the US Japan Expedition." Coincidentally, the pastel, hand-colored lithographic technique recalls the fine delicacy of Shiba Kokan's map from the previous century. Perhaps it was not a coincidence: Colton maps hace a great variety of styles, so this may have been a conscious echo of Eastern watercolor tradition.*

period of isolation, Shiba Kokan's *Chikyu Zenzu* ("Complete Map of the World") of 1792, came into being. Shiba Kokan (1747-1818) was an artist-scholar who had steeped himself in Dutch culture. He was an adherent of a small but important school of thought known as *Rangaku* ("Dutch learning") that was centered on Nagasaki.

Prevented by the travel ban from visiting the Netherlands, Shiba Kokan nevertheless produced a remarkable series of landscape paintings showing a strong Dutch influence, which were based on engravings by the Amsterdam printer Jan Luiken. To help him draw perspective accurately, he even constructed his own camera obscura. Kokan was also responsible for making the first Japanese copperplate etchings, whose subjects he derived, almost inevitably, from illustrations in a Dutch encyclopedia.

The *Chikyu Zenzu* was a faithful copy of a world map issued by the Amsterdam publishing house of Covens and Mortier earlier in the century. It was accompanied by a text by Kokan, *Chikyu Zenzu Ryakusetsu* ("Explanation of the Complete Map of the World"). Although it was already long obsolete in Europe, it represented a major advance over the Japanese map of 1645. New Guinea and Australia appear for the first time, and although the huge, spurious southern continent of *Terra Australis Magellanica* (which had been imagined since ancient times, and appeared on the maps of Ortelius, Plancius, and others) is still present, it is far smaller than before. Another oddity that reveals its provenance from an earlier period is the curious shape of North America, with California still shown as an island.

Kokan was an inquisitive polymath, and it is largely thanks to his desire to enquire beyond imposed cultural restrictions that mapping in Japan continued to improve in the 19th century, before the country finally emerged from isolation with the Meiji restoration of 1867. Inspired by his map, scholars began to ask for access to more recent and accurate publications. By 1810, a newly revised Japanese world map—the *Shintei Bankoku Zenzu* by Takahashi Kageyasu—had appeared, fully up-to-date and reflecting the latest geographical findings.

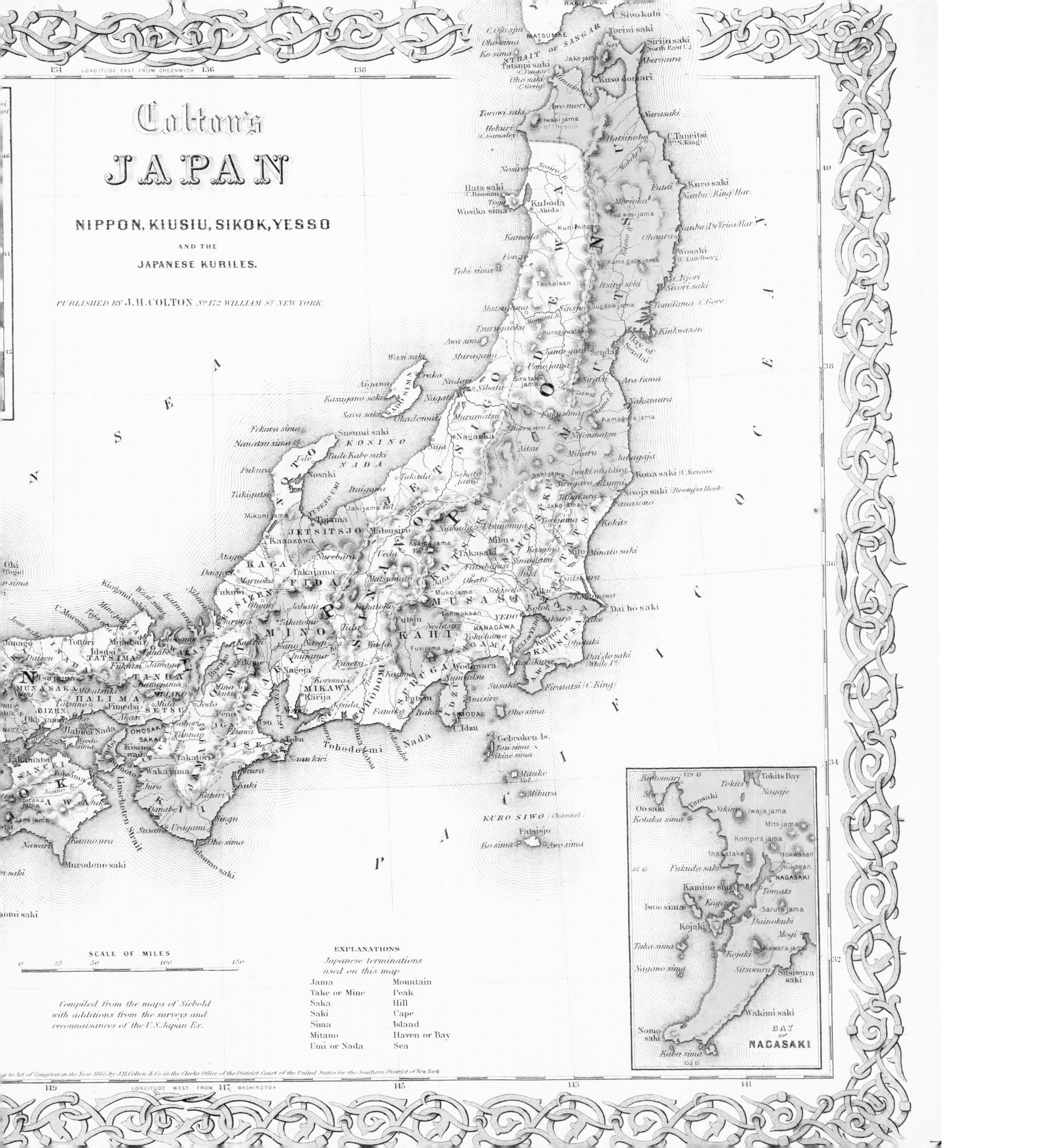

Colton's
JAPAN
NIPPON, KIUSIU, SIKOK, YESSO
AND THE
JAPANESE KURILES.
PUBLISHED BY J.H.COLTON No. 172 WILLIAM St. NEW YORK.
LONGITUDE EAST FROM GREENWICH
LONGITUDE WEST FROM WASHINGTON
SCALE OF MILES
0 25 50 100 150
EXPLANATIONS
Japanese terminations used on this map
Jama — Mountain
Take or Mine — Peak
Saka — Hill
Saki — Cape
Sima — Island
Mitano — Haven or Bay
Umi or Nada — Sea
Compiled from the maps of Siebold with additions from the surveys and reconnaisances of the U.S. Japan Ex.
BAY OF NAGASAKI
NAGASAKI
PACIFIC OCEAN
SEA OF JAPAN
STRAIT OF SANGAR
MATSUMAE
YEDO
KANAGAWA
SIMODA
OHOSAKA
KURO SIWO (Channel)
Gebroken Is.
Fatsisjo
Sendai
Niigata
Kanazawa
Nagoja

Terra Incognita:

Australia Becomes an Island Continent

In 2002 Australians celebrated the bicentennial of the circumnavigation of their country by Matthew Flinders, an event that literally put Australia on the world map. In Britain the occasion went unnoticed, met with the same apathy that greeted Flinders two centuries ago when he finally returned to England after nearly seven years as a prisoner of the French.

The first stirrings of initial interest in Australia among the British public came in 1771 at the end of the first Pacific voyage of Captain James Cook (1728-79), which began in 1768. After visiting Tahiti, he discovered and sailed around the islands of New Zealand, making contact with the warlike local Maori people. He then proceeded to Australia, where scientists from HMS *Endeavour* went ashore and botanist Joseph Banks (1743-1820) collected specimens and made drawings of the local flora and fauna. (Banks later helped with the settlement of New South Wales when it became a British penal colony.) Cook named their landing place Botany Bay.

Cook then sailed up the east coast of Australia, where his ex-Whitby collier ran aground on the Great Barrier Reef. He put ashore to extract a lump of coral from the *Endeavour's* hull and repair the ship, at the same time claiming the whole of the land for England on behalf of his monarch King George III. Cook had at last given identity (in the eyes of Europeans) to the *Terra Australis Ingognita*—the Unknown Southern Land. He continued northward to Cape York, Australia's northernmost point, and then westward across the Indian Ocean, round the Cape of Good Hope and northward through the Atlantic Ocean to home.

On Cook's second expedition of 1772-75 he renewed his search for the non-existent great southern continent, this time with two ships, HMS *Resolution* and HMS *Adventure*. But the voyage only added New Caledonia and South Georgia to his list of discoveries; he did not even see Australia on this trip.

His third fatal expedition of 1776-79 took HMS *Resolution* and HMS *Discovery* to complete the charting of North America's Pacific coastline as far north as the Bering Strait and to search (unsuccessfully) for a northwest passage around northern Canada. He visited the Sandwich Islands (now Hawaii) where he intervened in a dispute with the local people about a stolen boat, and was killed on the beach at Kealakekua by the Polynesians.

There are notable commentators who have in all seriousness presented Cook with the laurels as the greatest explorer of all time. An impossible judgement of course: but one that relies in part on the outstanding accuracy of the charts he produced. He did not work alone; the Polynesian navigator Tupaia in his employ was an important contributor. Perhaps Cook's greatest cartographic achievement was the coastal survey of "New South Wales," the eastern coast of Australia, until then unmapped by Europeans.

Despite Cook's monumental achievements, nobody had yet established that Australia is in fact an island. That task fell to Matthew Flinders. He was born in Donington, Lincolnshire, in 1774 and as a teenager joined the Royal Navy in 1789 as a trainee navigator. In 1791-93 he accompanied the infamous Captain William Bligh on HMS *Providence* on a voyage to Australia, where he helped chart some of the waters off Van Diemen's Land (now Tasmania). He returned to Australia in 1795 on HMS *Reliance*, whose ship's surgeon was George Bass (1771-1803). Bass was also born in Lincolnshire, in the town of Aswarby, and the two men became firm friends.

In 1796 Flinders and Bass with a servant boy named Martin set out in their 8-foot (2.5-meter) open boat *Tom Thumb* to explore the coast south of Sydney, where they rowed up the Georges River and established that the land was suitable for settlement. The following year

RIGHT *On the map based on Matthew Flinders' circumnavigation of 1801-03, New Holland and New South Wales were finally combined under the name of Australia. Captain James Cook had charted much of the east coast on his first expedition, but Flinders continued the survey right around the new continent.*

Bass, accompanied by six sailors, went exploring on his own in a slightly more substantial whaleboat. On his way from Port Jackson to Western Port on the southern coast he encountered seven escaped convicts and promised to pick them up on his return, which he did two weeks later. During his trip along 300 miles (480 kilometers) of uncharted coastline, he surmised that there must be a strait of water between Van Diemen's Land and mainland Australia.

Flinders had come to the same conclusion and in 1798 the two sailors persuaded Governor Hunter to provide a larger boat, the *Norfolk*, and more men to mount a mini-expedition to carry out a hydrographic survey. The boat, the first to be built in the colony, was constructed by convicts on Norfolk Island. On Tasmania they discovered the Tamar River, sailed along the western coast to the Derwent River (site of present-day Hobart, the capital), and then right

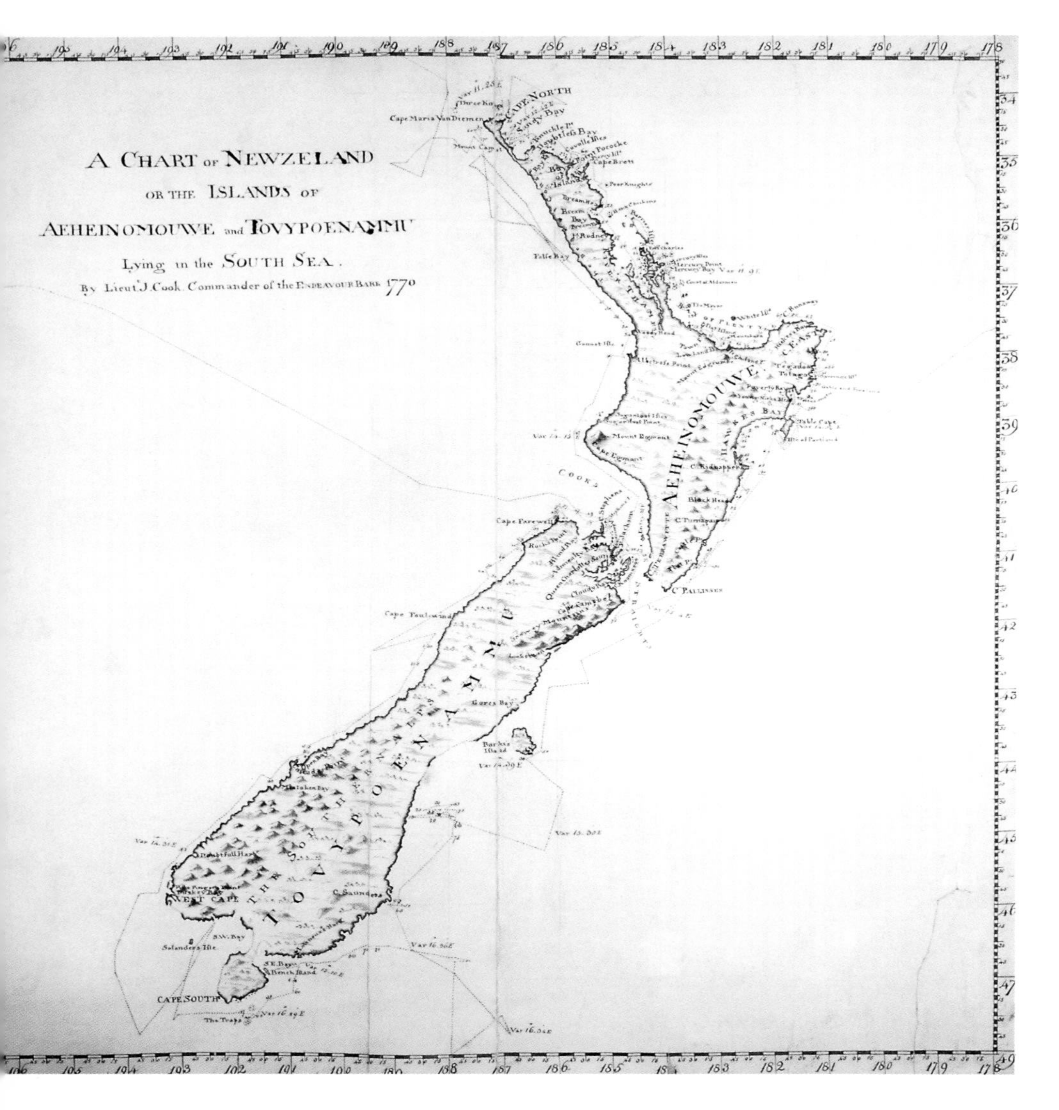

ABOVE *So good was Captain Cook's map of New Zealand, which he called Newzeland, that it served virtually unchanged for more than 100 years. He described the local Maori people as warlike, but was careful not to get involved with them. He was not so lucky on his third expedition to the Pacific islands in 1776-79, and was killed by Polynesians in the Sandwich Islands (now Hawaii) after intervening in a dispute over a stolen boat.*

RIGHT *Early explorers of new lands such as Australia had a field day devising names for the places and features they discovered. In 1803 Frenchman Nicolas Baudin named Brué Reef for the signaling mate of one of his ships, whom he described as "a young man of good disposition and with a zeal for geography." Adrien Brué (1786-1832) turned out to be one of the finest map publishers of the early 19th century, as seen in his map of Australia dated 1826.*

round the island back to Sidney. Flinders gave his friend's name to the Bass Strait, a waterway that allowed ships to sail directly along the coast of the mainland instead of having to detour south around Tasmania as previously. Their map of Tasmania was published in 1800. Bass left Sydney in 1803 to sell a cargo in South America, and is thought to have died in the mines there, for nothing more was heard of him.

Flinders had returned to England in 1800 where the British Government, represented by Joseph Banks, asked him to repeat a voyage of circumnavigation but this time right around Australia—if this was possible. He returned to Australia in 1801 as captain of the appropriately named HMS *Investigator,* which had a crew of 80 men and two cats, one of which was named Trim. First Lieutenant on the ship was Flinders' brother Samuel, and the famous Scottish botanist Robert Brown (1773-1858), nominated by Banks, was also on board. They first sighted land at Cape Leeuwin, the most southwesterly point of Australia. Flinders mapped the south coast from there to Spencer Gulf, discovered Kangaroo Island, and in Encounter Bay met French explorer Nicolas Baudin before sailing on to Sydney. In 1802 he continued mapping from Port Jackson (Sydney) following Cook's route along the east coat to the Gulf of Carpentaria. He then continued west and south, taking his leaking craft right around Australia and back to Port Jackson a year later, in 1803. The thoroughly rotten *Investigator* was beyond repair and abandoned. During the voyage, Brown collected specimens of nearly 4,000 new species of plants.

Flinders set sail for England later that year as a passenger in HMS *Porpoise,* which promptly struck a reef. He returned to Sydney in the ship's cutter, to set sail again in command of HMS *Cumberland,* which appears to have been no more seaworthy than the *Investigator.* In the western Indian Ocean he had to put in to Île de France (now Mauritius) for repairs, but the French authorities accused him of being a spy, confiscated his charts and papers, and interned him, his crew, and his cat Trim for nearly seven years. He was finally released when the British claimed the island, and he arrived back in England, ill and totally forgotten, in 1810. He related his experiences in a book with the lengthy title of *Voyage to Terra Australis Undertaken for the Purpose of Completing the Discovery of that Vast Country,* which was published a day before he died in 1814. The two-volume work included a folio atlas, in which Flinders for the first time inscribed the world Australia; the name was officially adopted in 1824.

Australians themselves commemorated the great man many times with the town of Flinders, south of Melbourne; Flinders Bay at Cape Leeuwin; the Flinders Group of islands off the Cape York Peninsula; Flinders Island in the southern Bass Strait; the Flinders Ranges in South Australia; and Flinders Reefs in the Coral Sea. Yet it was Cook who captured the English imagination.

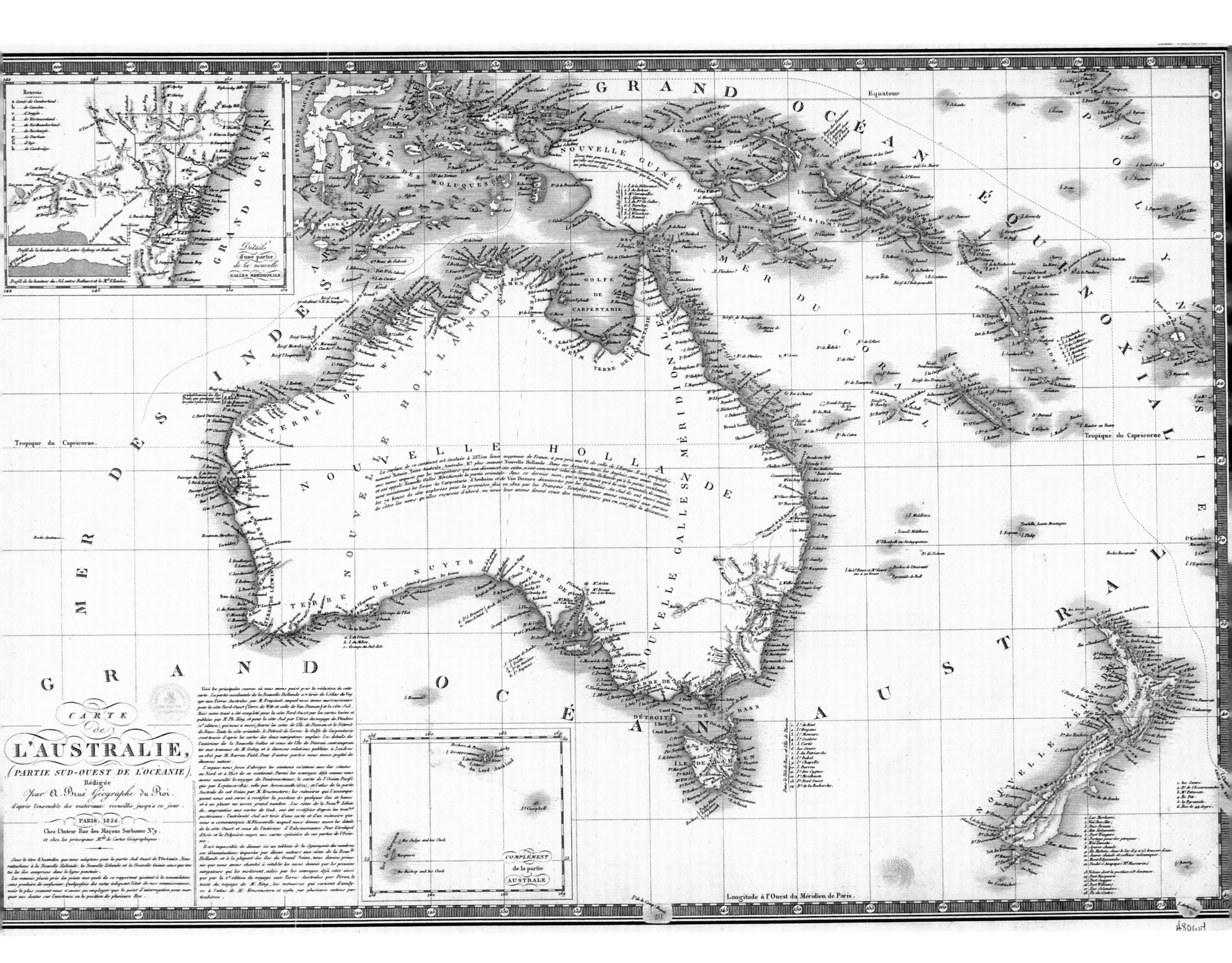

CARTE de L'AUSTRALIE,
(PARTIE SUD-OUEST DE L'OCÉANIE)
Rédigée par A. Brué Géographe du Roi.
d'après l'ensemble des matériaux recueillis jusqu'à ce jour.
PARIS, 1826.
Chez l'Auteur Rue des Maçons Sorbonne N.o 9.
et chez les principaux M.ds de Cartes Géographiques.
GRAND OCÉAN
GRAND OCÉAN ÉQUINOXIAL
POLYNÉSIE
MER DES INDES
GRAND OCÉAN AUSTRAL
NOUVELLE HOLLANDE
NOUVELLE GUINÉE
MER DES MOLUQUES
GOLFE DE CARPENTARIE
TERRE D'ARNHEIM
TERRE DE WITT
TERRE DE NUYTS
TERRE DE LEUWIN
NOUVELLE GALLES MÉRIDIONALE
MER DU CORAIL
DÉTROIT DE BASS
NOUVELLE ZÉLANDE
Equateur
Tropique du Capricorne
Longitude à l'Ouest du Méridien de Paris.
Détails d'une partie de la Nouvelle GALLES MÉRIDIONALE
COMPLÉMENT de la partie AUSTRALE

From Sea to Shining Sea:

Lewis and Clark

Once the American colonies had won their long sought-after freedom from the British crown and established the United States of America, their focus began to turn westward—to the vast expanse of land between the populated eastern seaboard and the Pacific Ocean. It was a long, hard way.

In 1803 President Thomas Jefferson arranged for the purchase of the Louisiana territory from the French. This area had originally been ceded to Spain in 1762, but under a secret treaty of 1800 Napoleon Bonaparte had won the area back for the French, and he had grand plans for establishing a new empire based around this huge area. However, pressing military matters intervened, and by 1803 the need for money to fund the European war meant than Bonaparte was willing to negotiate. The eventual agreed price was 60 million francs, around $15 million.

The Louisiana Territory at this time was far larger than the State of Louisiana today; it stretched for 828 thousand square miles (2,140,500 square kilometers) from the Mississippi River to the Rocky Mountains, and at a single stroke Jefferson had almost doubled the size of the fledgling United States. It was one of the crowning achievements of his presidency.

This new territory was practically unexplored and Jefferson immediately set about rectifying this. He supervised the creation of the Corps of Discovery to both explore and map the new area and, hopefully, to discover the fabled "Northwest Passage"—a navigable route from one side of the continent to the other. The expedition was to be led by two men, Meriwether Lewis and William Clark.

Meriwether Lewis was born in Virginia in 1774 and had joined the army in 1794. He had seen much service in the Northwest Territory fighting against Native Americans, and had grown accustomed to their culture and languages. In 1801 Thomas Jefferson appointed Lewis to be his private secretary, and when the decisions came to be made who should lead the expedition, Lewis was the obvious choice.

As his co-leader Lewis chose William Clark, an old army comrade. Clark was also born in Virginia in 1770, and had fought the Native Americans in the Ohio Valley before enlisting in the regular army. Clark prepared for the expedition by studying astronomy and cartography, as he was to be the expedition's principal mapmaker.

The two leaders gathered a group of around 40 soldiers to make up the rest of the Corps of Discovery, and on 14 May 1804 the party set out from St. Louis up the Missouri River. They made the journey in a specially commissioned 55-foot (16.8-meter) boat, with enough room for 12 tons of cargo.

The first stage of the expedition took the party up the Missouri and across the Great Plains, where they encountered a number of different tribes of Native Americans, including the Missouris, the Omahas, the Yankton Sioux, the Teton Sioux, and the Arikaras. Out of all of these meetings the only hostility they encountered was from the Teton Sioux.

They reached what is now North Dakota by October, and, having built themselves a fort, decided to winter in the area with a tribe of friendly Native Americans—the Mandan. Over the course of this winter they employed Toussaint Charbonneau, a French fur trader, and his wife Sacagawea, a member of the Shoshone tribe, to act as guides and interpreters for the expedition.

When spring came they set out once more on the Missouri, and followed the river upstream to its headwaters, traveling this time by canoe as their boat had been sent back to St Louis laden with samples of flora and fauna.

After passing through present-day Montana they finally came to the headwaters of the Missouri and, by a stroke of luck, managed to find a Shoshone village whose chief was Sacagawea's brother. Here they bought horses and set out to cross the imposing mountain range ahead of them.

It was immediately obvious to the expedition that this was no minor range of mountains with a river the other side as Thomas Jefferson had

ABOVE *We begin with a quick visual trick. Looking like an antique map, this reconstruction of the route was published for the Bicentennial Commemoration in 2003-2006. In addition to the preparation, exploration, and homecoming, it includes details of Indian reservations and the Louisiana Purchase.*

hoped. Instead the Rockies formed an imposing barrier running from north to south. Lewis and Clark spent 11 days crossing the mountain range and were very short of food when they reached the other side. Luckily for them, they were welcomed by the Nez Percé tribe and were able to continue their expedition in good shape.

The route to the Pacific Ocean from here was relatively straightforward, and the party followed the Clearwater, Snake, and Columbia rivers until they reached the coast in November 1805. Although they had hoped to be picked up by a passing ship this did not happen and they had to spend the winter on the Pacific Coast before setting off eastwards in March 1806.

The return journey was much more straightforward, as, once they had crossed the barrier of the Rockies, they were following the flow of the Missouri River, and they reached their final destination of St. Louis in November 1806, to the utter surprise of the inhabitants, who had given them up for dead long ago.

One of the primary focuses of the Lewis and Clark expedition was mapping of the new territories. Thomas Jefferson had been very explicit in his demands that:

> "Beginning at the mouth of the Missouri, you will take observations of latitude and longitude, at all remarkable points on the river. Your observations are to be taken with great pains and accuracy; to be entered distinctly and intelligibly for others as well as yourself."

Although Lewis was the one Jefferson was talking to, Clark was principally responsible for the creation of maps of the expedition, along with the other records. The maps they were equipped with, including Nicholas King's map created especially for the expedition, were woefully inaccurate, and throughout the expedition Clark took continual surveys of the area around them, while Lewis was responsible for cataloging the different flora and fauna they encountered, as well as recording their meetings with various Indian tribes.

Two Million for a Swamp?

The PRAIRIE DOG sickened at the sting of the HORNET
or a Diplomatic Puppet exhibiting his Deceptions!

Events at around the time of the Lewis and Clark expedition included the purchase of land from France and Spain to consolidate the territory of the fledgling United States. In this cartoon by satirical artist James Akin (ca 1773-1846) President Thomas Jefferson is depicted as a prairie dog being stung by a hornet (Napoleon) and made to vomit up "two millions" in gold coins. The character on the right is a happy dancing French diplomat with orders from French minister Charles de Talleyrand in his pocket and maps of east and west Florida in his hands. In 1804 Jefferson carried out negotiations with Spain in secret and peruaded Congress to agree to an appropriation of $2 million for the purchase of Florida. The deal was not formalized until the Adams-Onis Treaty (called the Transcontinental Treaty) of 1819, which also defined the border between the US and the Spanish territory of Mexico. Spain retained Texas, California, and New Mexico. John Quincy Adams was American Secretary of State and Onis was the Spanish foreign minister. Jefferson had done a much better deal with the Louisiana Purchase in 1803, when he almost doubled the size of US territory by paying the French only $19 per square mile (less than 3 cents per acre), hence the French joy at Jefferson being stung by the Spanish over the Florida Purchase. No wonder the "balloon" from the Frenchman's mouth says "A gull for the people."

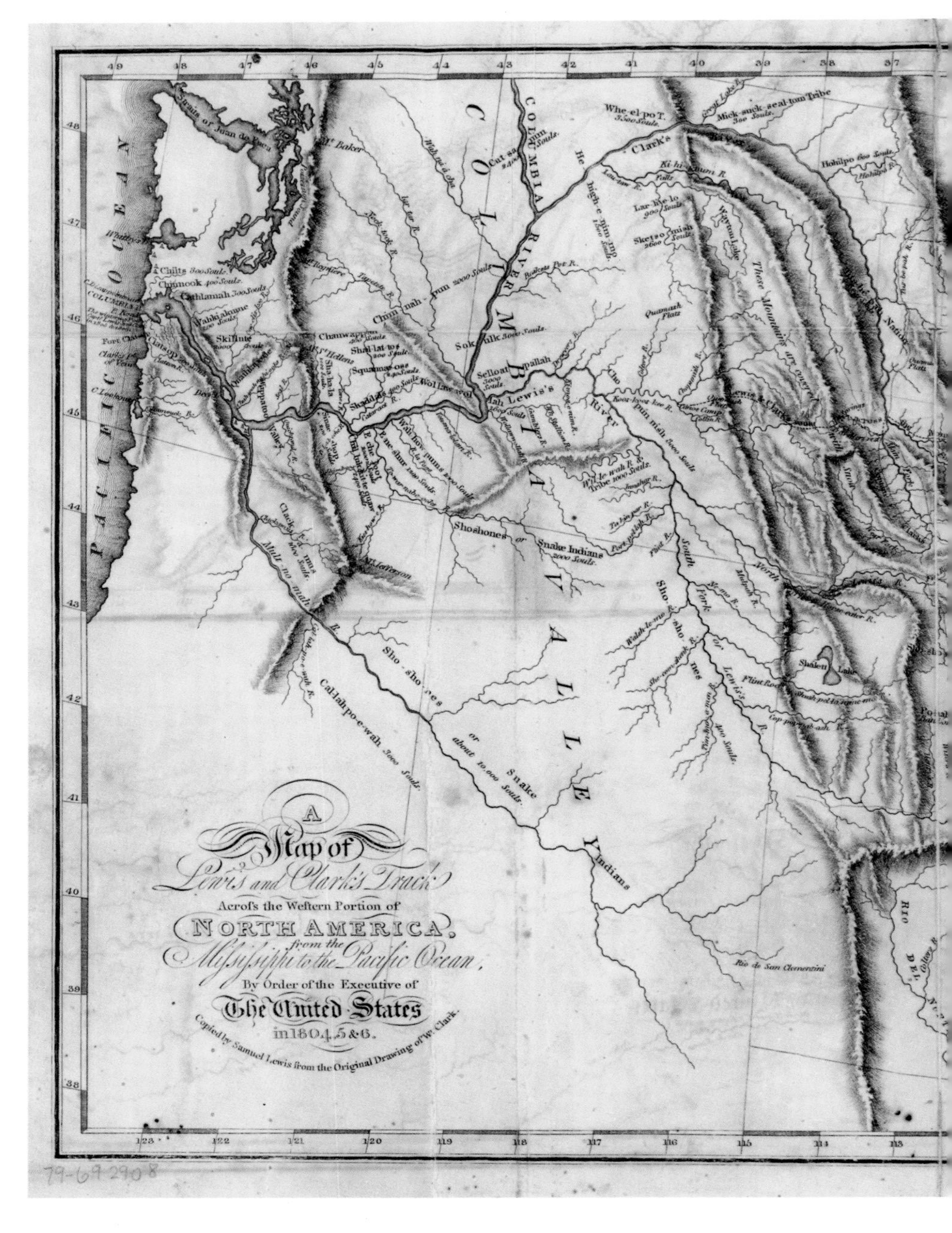

ABOVE *Copied by Samuel Lewis from William Clark's original drawing, this detailed chart is titled:* "A map of Lewis and Clark's track, across the western portion of North America from the Mississippi to the Pacific Ocean: by order of the executive of the United States in 1804." *Samuel Lewis was a cartographer, and should not be confused with the Meriwether Lewis who accompanied Clark on the expedition. The map shows the tortuous route up the Missouri River as far as its source, the journey across the Great Divide, and the final leg down the Columbia River to the Pacific Ocean. The total journey, by boat, canoe, and horseback, had taken 17 months from May 1804 to November 1805. The return trip—by the same route but downstream on the long Missouri River—was much quicker, taking only nine months (March 1806 to November 1806).*

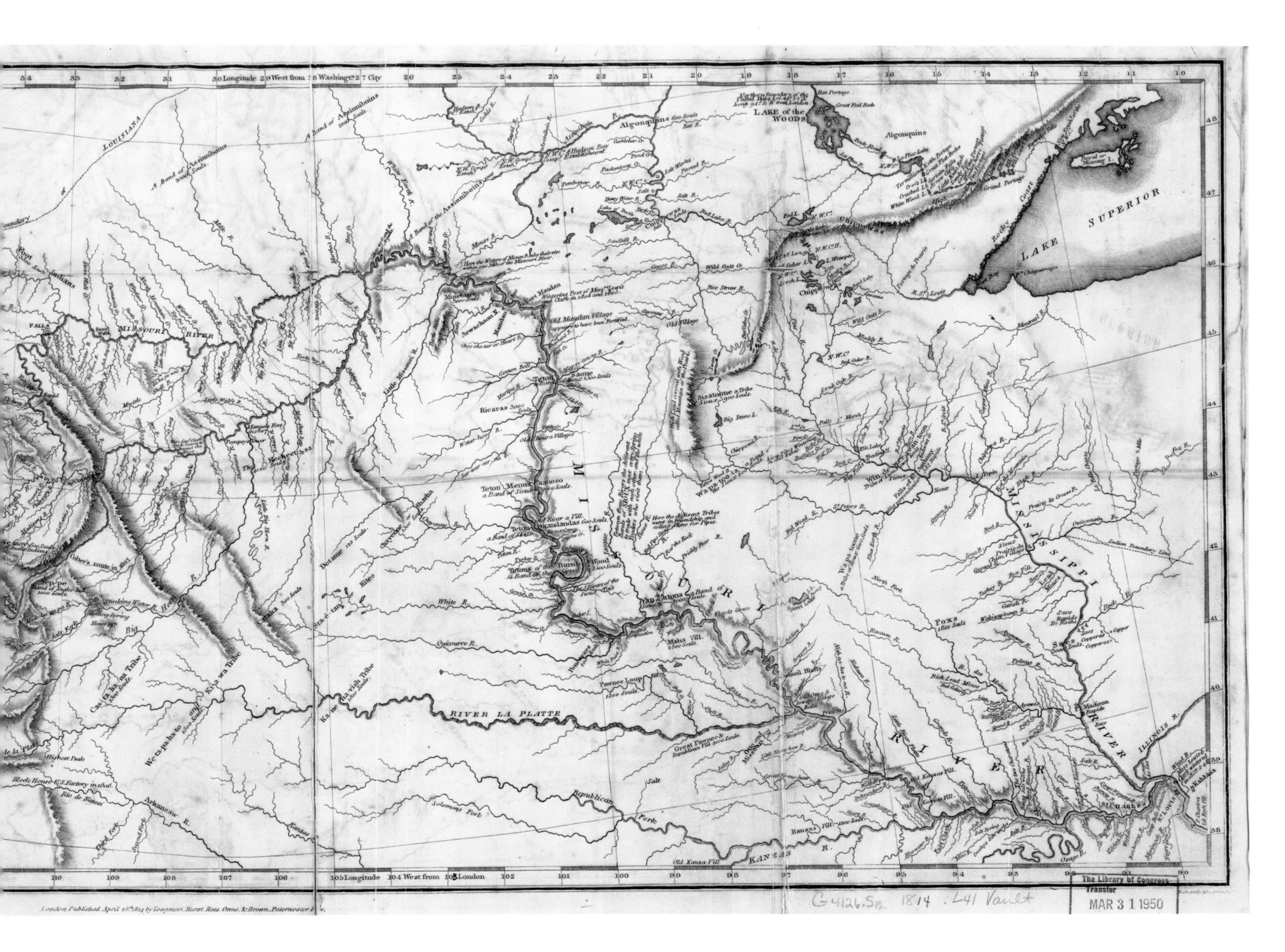

In fact, many of these tribes helped them in their cartographic enterprise, as Lewis records in his journal:

> "I now prevailed on the Chief to instruct me with respect to the geography of his country. This he undertook very cheerfully, by delineating the rivers on the ground."

Once their mission had been completed Clark collated all this information and constructed a new map of the western territories of the United States, also incorporating information from a number of later expeditions of discovery, including William Dunbar's exploration of the Ouachita River, Thomas Freeman's survey of the Red River and James Wilkinson's journey down the Arkansas River. By 1810 Clark had finished his map and Bradford and Inskeep in Philadelphia eventually published it in 1814. Although Lewis and Clark were not the first white men to cross the continent, beaten by Alexander Mackenzie (1764-1820) a decade earlier, and although the Clark Map still contained a large number of inaccuracies, it was by far the best available map of the new territories. It may have put an end to the hopes of finding a Northwest Passage connecting the east and west coasts of the North American continent, but it also inspired a new generation of explorers and colonists who would go west in search of land and prosperity.

Rule Britannia:

The Royal Navy "Conquers" Antarctica

The British Imperial Conference met at the end of 1926 and resolved to assert British title to areas of Antarctica. As they had been discovered by officers of the British Crown, they were part of the Dominions and should be taken under control. This would be declared ceremonially, by formal taking of possession on the spot. The resulting Committee on British Policy in the Antarctic needed maps of the area.

The Admiralty Hydrographic Office's South Polar Chart was by far the best overview available. However, as originally published, it could be regarded as off-message. The first edition had been drawn in 1839, based on the information then available, and it had been updated ever since as news of discoveries came in. This meant that as well as the tracks of Cook and James Clark Ross it recorded discoveries by foreign naval expeditions, for example the French led by d'Urville, the Russian led by Bellinghausen, and the US led by Wilkes. It may be a coincidence that in 1927 the Hydrographic Office produced the version of the chart reproduced here, overlaid with stencilled captions and boldly drawn tracks of British Naval Explorers, celebrating 150 years of British naval achievement. Small corrections are given at the bottom of the chart, as more information came into the Hydrographic office from various sources, such as merchant shipping and whalers.

James Cook's circumpolar voyage in 1772-5 was undertaken in search of a great Southern Continent, which some scholars thought scientifically plausible and some mariners claimed to have already seen. He circumnavigated the Antarctic, sailing farther south than anyone before, and demonstrated that if a land-mass existed it must be much smaller than expected and unsuitable for colonization. He discovered South Georgia (named for the king) and the South Shetland Islands.

James Clark Ross, who had discovered the North magnetic Pole, led a three-year expedition to explore the Antarctic. He gave his name to the Ross Sea and the Ross Ice Shelf, the Queen's name to Victoria Land and his ships' names to Mounts Erebus and Terror. Two of the other 19th-century naval officers whose names appear on the list of tracks are less well known. Henry Foster commanded a voyage in 1828-31 to chart Deception Island in the South Shetlands and measure magnetism and gravity. Thomas Moore had sailed with Ross and was dispatched in 1844 to make magnetic observations. He found the north-east part of Enderby land, an island originally discovered and named by British whalers. George Nares led the first British world oceanographic expedition. In the course of the four-year voyage, HMS *Challenger* became the first steamship to cross the Antarctic Circle and came within 1,400 miles of the South Pole. The expedition found open sea where American explorers had reported seeing a large landmass.

Other nations had sent expeditions all around Antarctica after the 1895 International Geographical Congress, but the British always had the South Pole as the main aim. This was the personal ambition of the President of the Royal Geographical Society, Sir Clements Markham. Robert Scott made three expeditions to the Antarctic, naming Edward VII Peninsula in his first season in 1901. Ernest Shackleton's *Nimrod* expedition reached the South Magnetic Pole in January 1909 but the glory of reaching the geographical Pole (90 degrees south) was still to be won. Scott's third voyage famously ended in heroic tragedy in January 1912, when he and his companions died in the ice. They had succeeded in carrying the Union flag to the South Pole, only to find that Amundsen's Norwegian party had been there before them.

RIGHT *South Polar Chart, British Admiralty Hydrographic Office, 1927, overprinted with tracks of British Naval explorers in South Polar Regions 1772 to 1922. The limits of Antarctica are very tentatively drawn, with discontinuous coast, faint lines labeled "appearance of land" and broken lines indicating the edges of the ice. Soundings are sparse, sporadic and very deep. The Polar projection gives the chart the appearance of a target, with the South Pole at the bull's-eye."Scott 1912" occupies the center, but the stencilled tracks and captions deflect attention from the galling truth printed below: "Amundsen 14th to 17th December 1911, Scott 13th Jan 1912."*

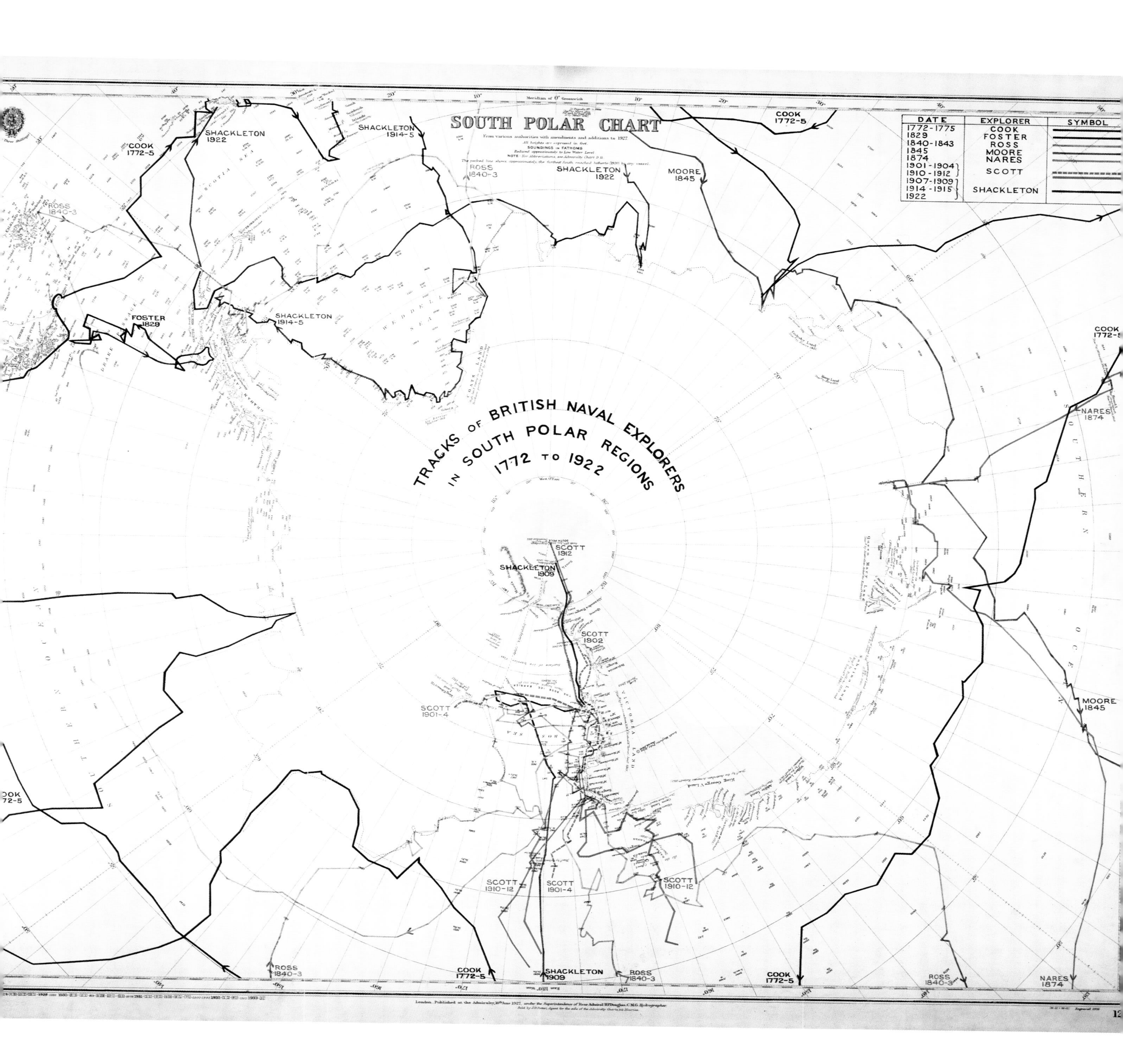
SOUTH POLAR CHART
TRACKS OF BRITISH NAVAL EXPLORERS IN SOUTH POLAR REGIONS 1772 TO 1922
DATE
1772-1775
1829
1840-1843
1845
1874
1901-1904
1910-1912
1907-1909
1914-1915
1922
EXPLORER
COOK
FOSTER
ROSS
MOORE
NARES
SCOTT
SHACKLETON
SYMBOL
SHACKLETON 1922
COOK 1772-5
SHACKLETON 1914-5
ROSS 1840-3
SHACKLETON 1922
MOORE 1845
COOK 1772-5
ROSS 1840-3
FOSTER 1829
SHACKLETON 1914-5
NARES 1874
SCOTT 1912
SHACKLETON 1909
SCOTT 1902
SCOTT 1901-4
MOORE 1845
SCOTT 1910-12
SCOTT 1901-4
SCOTT 1910-12
ROSS 1840-3
COOK 1772-5
SHACKLETON 1909
ROSS 1840-3
COOK 1772-5
ROSS 1840-3
NARES 1874
WEDDELL SEA
ROSS SEA
VICTORIA LAND

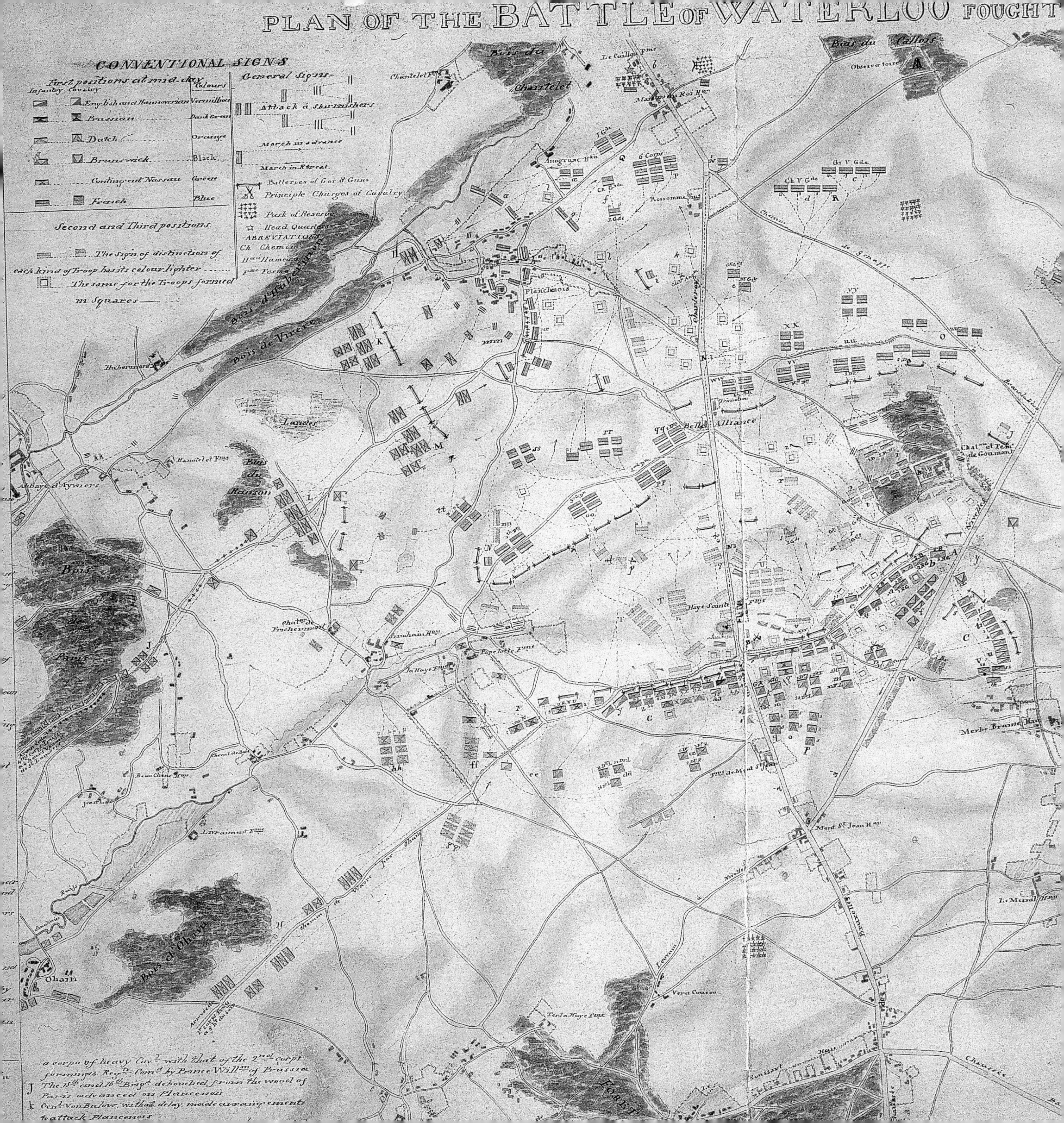
PLAN OF THE BATTLE OF WATERLOO FOUGHT
CONVENTIONAL SIGNS
First positions at mid-day
Infantry Cavalry
Colours
English and Hannoverian Vermilion
Prussian Dark Green
Dutch Orange
Brunswick Black
Contingent Nassau Green
French Blue
Second and Third positions
The sign of distinction of each kind of Troop has its colour lighter
The same for the Troops formed in Squares
General Signs
Attack á Skirmishers
March in advance
March in Retreat
Batteries of 6 or 8 Guns
Principle Charges of Cavalry
Park of Reserve
Head Quarters
ABREVIATIONS
Ch Chemin
Hau Hameau
Fme Ferme
Bois du Chantelet
Chantelet Fme
Le Caillou Fme
Bois du Callois
Observatoire
Maison du Roi Hau
Anogrune Hau
6 Corps
Rossomme Fme
Planchenois
Bois d'Hubermont
Bois de Virere
Hubermont
Landes
Belle Alliance
Chau et Fme de Goumont
Hameau et Fme
Bois du Rossau
Abbaye d'Aywiers
Bois de Paris
Chau de Frichermont
Smohain Hau
Papelotte Fme
la Haye Fme
Haye Sainte
Merbe Braine Hau
Fme de Mont St Jean
Mont St Jean Hau
Cheval de Bois
Beau Chene Hau
Livraimont Fme
Chemin de Wavre par Ohain
Bois d'Ohain
Ohain
Vera Coucou
Ter la Haye Fme
Le Mesnil Hau
Chaussée
Forest
Arrivée des 1r corps Brigs a 8 h. du soir
a corps of heavy Cavy with that of the 2nd corps forming 14 Regts Comd by Prince Willm of Prussia
J The 15th and 16th Brigs debouched from the wood of Paris advanced on Plancenois
k Genl Von Bulow, without delay, made arrangements to attack Plancenois

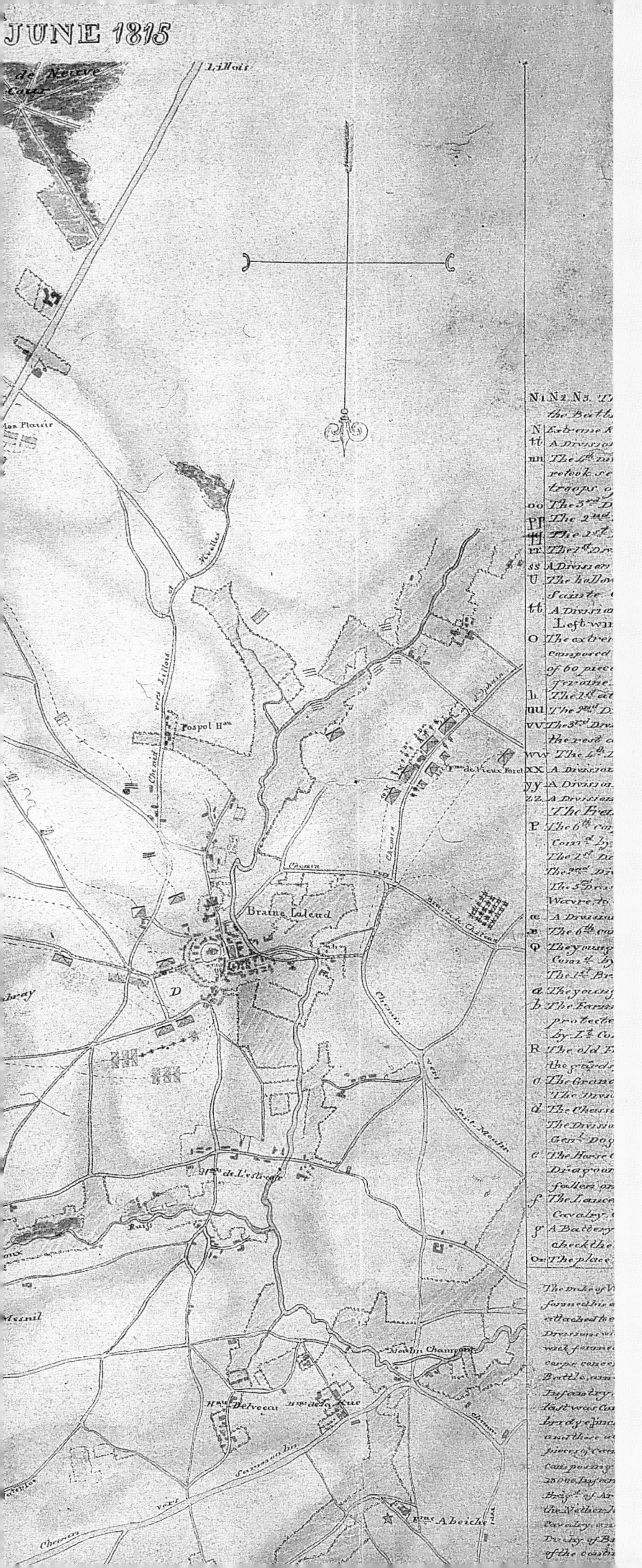

MILITARY MAPS

An excellent sketch map of one of the most historically significant battles in history, Waterloo, June 18, 1815. Captain John Thornton's sketch map includes a reference table to stages of the battle, concentrating on three French deployments. The legend shows symbols for the English and Hanoverian, Prussian, Dutch, Brunswick, "contingent Nassau" (Netherlands, i.e. Dutch-Belgian), and French troops. Scale is 6 inches to one mile. Captain Johnson didn't have complete information at his fingertips; the unit symbols for the French, for example (off the map to the right) include: "PP: the 2nd division commanded by Lt. Genl. —"

The Imola Plan:

Da Vinci's New Viewpoint

Leonardo da Vinci (1452–1519) was one of the greatest artists, scientists, and inventors of the Renaissance. His most famous works such as The Last Supper and the La Gioconda (Mona Lisa) still attract hundreds of thousands of visitors to this day. Da Vinci was also an extraordinarily talented cartographer, and is widely believed to be the author of the plan of the town of Imola shown here. This is not, as it may seem, just a town plan: it is a military map.

Leonardo was born in the small town of Vinci, just outside Florence, in 1452, and at the age of 15 was apprenticed to the painter Andrea del Verrochio in Florence. He stayed with Verrochio for ten years, before setting up on his own in 1477. In 1482 he was drawn to Milan and entered the service of Duke Ludovico Sforza. Da Vinci's work for the duke not only included painting and sculpture, but also engineering projects, including designing weapons, buildings, and machinery. It was during this period that he also produced his first anatomical studies. Following the French conquest of Milan and the flight of the Sforza family in 1499, da Vinci was left without a patron and, following a few years back in Florence, he became a military engineer for the infamous Cesare Borgia in 1502.

Cesare Borgia (1476-1507) was the youngest son of Pope Alexander VI, and his father appointed him as Archbishop of Valencia and a cardinal in 1493. By 1502, Cesare had resigned from the priesthood and become a secular lord, having been made Duke of Valence by the French King Louis XII in return for his help in obtaining the French King's divorce. His father entrusted him with the task of subduing the central Italian territory of Romagna, which owed nominal allegiance to the papacy. One of the first territories attacked by Cesare was the town of Imola, which surrendered in December 1499.

Da Vinci is recorded to have been in Imola from September to December 1502, at the same time as the famed politician Niccolò Macchiavelli, author of *The Prince*. At the time the city was an armed camp for Borgia's forces, and Leonardo appears to have been commissioned to carry out a survey of the existing fortifications of the city and also to suggest improvements and additional defenses.

There is some controversy over whether this map is an original da Vinci work or whether it is actually largely an earlier work by an engineer called Danesio Manieri, who was employed in Imola by the Duke of Milan. Most historians tend to lean towards da Vinci as the true author of the map, and it is usually acknowledged as the first surviving example of a European plan of a town. Most city maps of the time were drawn at an angle from above, giving a "bird's-eye view" of the location. This da Vinci map developed a whole new style of cartography. The map itself is set in a circle, divided into eight sectors, with individual streets, district, fortifications, and the Senterno River all marked out carefully. Cesare Borgia had attacked the weak points of the garrison walls from the east. Da Vinci's notebooks include several assessments of distances around Imola, such as: "Castel San Piero is seen from Imola at four points from the West towards the North West, at a distance of seven miles. Faenza stands with regard to Imola between East and South East at a distance of ten miles." In October 1502 Borgia had been trapped in the city by a revolt, and some have linked the map, da Vinci's notes, and this specific event.

Following his creation of the Imola map, da Vinci also worked on plans of Arezzo to assist the military campaigns of Cesare Borgia. However, in August 1503 Pope Alexander VI died while Cesare Borgia was incapacitated in Rome. These two concurrent events undermined his political power and he never returned to his conquests in the Romagna. Da Vinci returned to Florence and began work on what is probably his most famous painting—*La Gioconda*, the Mona Lisa.

RIGHT *The parts which are now black close to the river were yellow ocher in the original. Da Vinci's approach owes something to the technique of architect Leon Battista Alberti (1404-1472), who made measurements of Rome using polar coordinates. The ichnographic view of the city shows it as a* system, *not the sum of its important structures. With a plan view, one can* plan*—including city defense.*

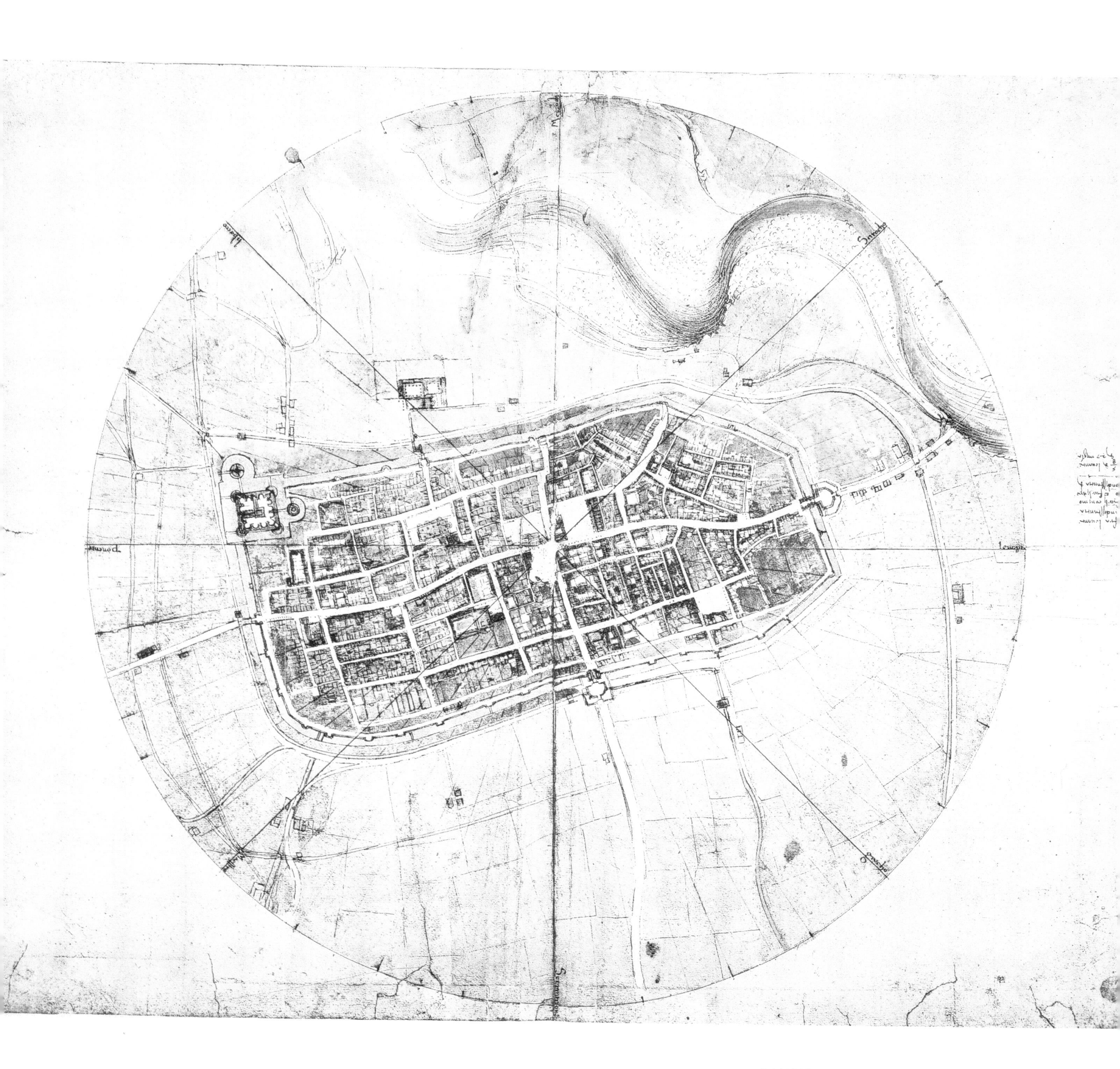

Jedediah Hotchkiss:

"Make Me A Map of the Valley"

On March 26th 1862 the spring campaign of America's Civil War was just beginning in the Shenandoah Valley when Major General Thomas J. "Stonewall" Jackson discovered that there were no maps of the region. He sent for 34-year-old Jedediah Hotchkiss, a schoolteacher and self-taught cartographer, and said, "I want you to make me a map of the Valley, from Harper's Ferry to Lexington, showing all the points of offence and defense in those places. Mr. [Alexander Swift] Pendleton will give you orders for whatever outfit you want."

Born and educated in Windsor, New York, Jed Hotchkiss celebrated his graduation from the local academy by hiking through the hills and valleys of Appalachia. One trek took him to Mossy Creek in the Shenandoah Valley, where he met Daniel Forrer and accepted the position of tutoring the family's children. With time on his hands, Hotchkiss taught himself the principles of mapmaking and engineering, subjects only available in military schools. In 1853 he went to Pennsylvania, married an educated woman, returned to the Shenandoah Valley, and eventually settled in Churchville. In 1859 he opened the Loch Willow Academy and with his wife and small staff began the 1860-61 term with 54 students.

Being absorbed in the new school, Hotchkiss paid little attention to the national crisis until April 17th 1861, when Virginia seceded from the Union. When an assistant teacher organized a local infantry company, several pupils joined it. Weeks later, more students enrolled in a cavalry company. In June Hotchkiss acknowledged the inevitable, dismissed the few remaining students, and closed the school. He offered to serve the Confederacy as a civilian, but the army did not know what to do with him until the government began looking for maps. Those they found were old and inaccurate, so they sent Hotchkiss into the wilds of western Virginia to survey Camp Garnett at Rich Mountain. A week later Federal forces struck the camp and drove the defenders into the hills. That night, only the men who followed Hotchkiss over the mountain escaped. When General Robert E. Lee arrived on the field and discovered the talents of Hotchkiss, he put him to work drafting the first Confederate maps. As the years passed, Hotchkiss sketched himself into the chronicles of America's Civil War and became the foremost topographical engineer in the Army of Northern Virginia.

On March 26th 1862, after General Jackson ordered a map made of the Valley, Hotchkiss wrote in his journal: "I secured for myself, a wagon and two horses, and the driver that I detailed … remained in my service during all the war up to the defeat of Gen. [Jubal] Early at Waynesboro, on the 2nd of March, 1865." On that day Hotchkiss packed up all his maps, which numbered in the hundreds, and joined General Lee a month before the surrender at Appomattox Court House. Instead of turning the maps over to the government, he retained the collection and for the next 33 years refined them. Today, the best Confederate maps made during the war are certainly those of Jed Hotchkiss.

Wherever Stonewall Jackson went, he took Hotchkiss, his adjutant and topographical engineer. When on March 23rd 1862 Jackson fought and lost the opening battle at Kernstown, Hotchkiss had just joined the brigade and not made his first map. Three days later Jackson began asking for maps, and Hotchkiss reconstructed the battle at Kernstown (opposite). Having witnessed the dispositions of Union and Confederate forces, he accurately detailed the topography and later the flow of the battle. From Kernstown, Jackson received the first of many pieces that fit into the tangle of mountains, gaps, and riverbeds in the Shenandoah Valley. For the next ten weeks Jackson advanced and retreated up and down the Valley, sometimes dodging superior Union forces and sometimes striking their flanks through passes scattered among the mountains. Dressed in civilian clothes, Hotchkiss regularly

RIGHT *After Stonewall Jackson lost the opening battle of the Shenandoah Valley Campaign in the spring of 1862 he turned to Jedediah Hotchkiss and asked for maps. Hotchkiss recreated the Battle of Kernstown and presented Jackson with the first map of the Valley.*

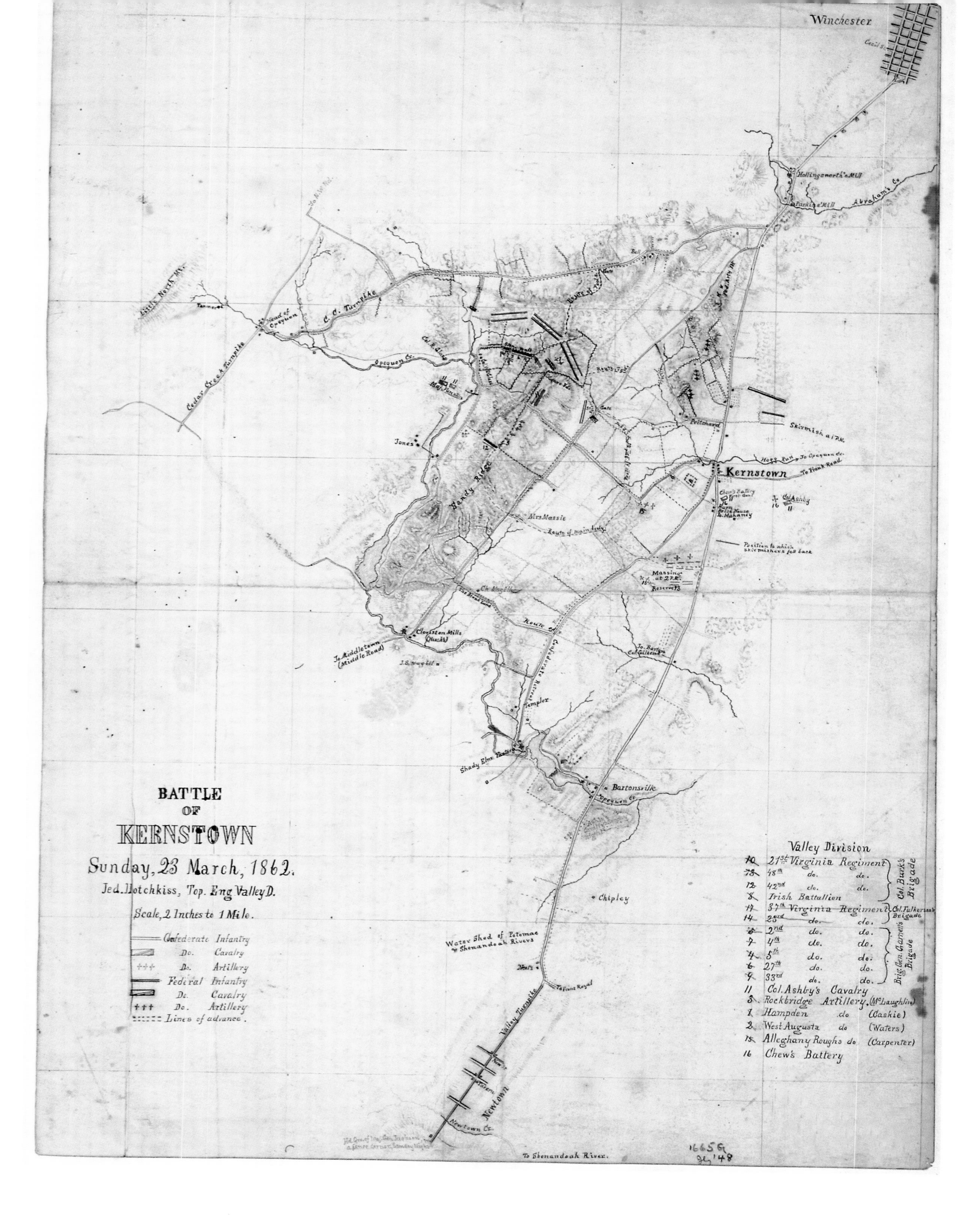
Winchester
Hollingsworth's Mill
Abraham's Cr.
C. C. Turnpike
Head of Opequon
Cedar Creek Turnpike
Opequon Cr.
Sandy Ridge
Jones
Pritchard
Skirmish at 7 A.M.
Hoge Run
To Opequon Cr.
Kernstown
To Front Road
Mrs Massie
Position to which skirmishers fell back
Massing at 2 P.M.
Reserves
Clevlston Mills
To Middletown (Middle Road)
Templer
Shady Elm
Bartonsville
Opequon Cr.
Chipley
Water Shed of Potomac & Shenandoah Rivers
To Front Royal
Valley Turnpike
Newtown
Newtown Cr.
To Shenandoah River.
BATTLE
OF
KERNSTOWN
Sunday, 23 March, 1862.
Jed. Hotchkiss, Top. Eng Valley D.
Scale, 2 Inches to 1 Mile.
Confederate Infantry
Do. Cavalry
Do. Artillery
Federal Infantry
Do. Cavalry
Do. Artillery
Lines of advance.
Valley Division
10 21st Virginia Regiment
13 48th do. do.
12 42nd do. do.
8 Irish Battallion
Col. Burk's Brigade
17 37th Virginia Regiment
14 23rd do. do.
Col. Fulkerson's Brigade
5 2nd do. do.
7 4th do. do.
4 5th do. do.
6 27th do. do.
9 33rd do. do.
Brig. Gen. Garnett's Brigade
11 Col. Ashby's Cavalry
3 Rockbridge Artillery. (McLaughlin)
1 Hampden do (Caskie)
2 West Augusta do (Waters)
15 Alleghany Roughs do (Carpenter)
16 Chew's Battery

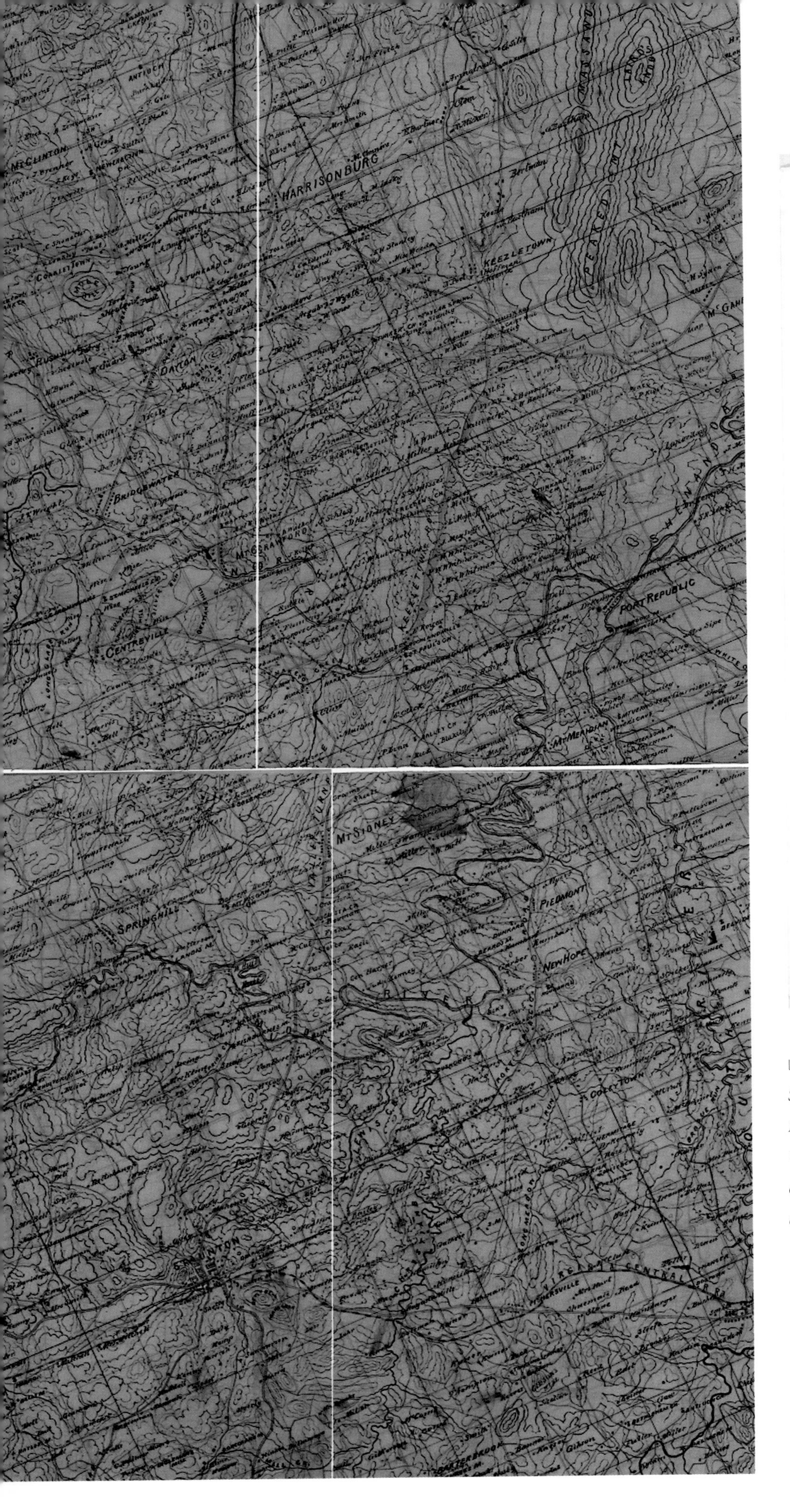

LEFT AND ABOVE *Hotchkiss made many of his maps in rough draft, such as the Shenandoah River Valley from Winchester to Staunton (detail, left) with the Blue Ridge mountains to the east, and then recopied and refined them when in camp. While circulating through the Valley, Hotchkiss crossed paths with women engaged in the dangerous work of spying. Union captors never quite knew what to do with a woman suspect. Such encounters were romanticized after the War.*

roamed the Valley, sometimes far from Jackson's camp. He sketched from hillsides, beside creeks, in hidden passages, and often in sight of the enemy. He inked uncharted farm roads and followed creeks flowing off the mountains until they branched into deeper waterways and flowed into the Shenandoah. He talked to farmers and learned their byways, and he spoke to Jackson's cavalrymen, sometimes sending them into remote areas to verify his observations. Nothing escaped his eye. Because of Hotchkiss, Jackson never lost sight of the enemy, but the enemy regularly lost sight of him. On occasion Hotchkiss returned to camp to warn the general, "My reconnaissance westward showed that [your position] could be easily turned by several roads and that it would be difficult ... to speedily fall back from that point." A few days later he would return to report that he had discovered the enemy in a vulnerable position that could be attacked by advancing on previously unmarked roads.

The Shenandoah Valley Campaign, which began on March 23rd, climaxed on 8-9th June in a brilliant victory for Jackson's Confederates at Cross Keys and Port Republic. Hotchkiss had prepared detailed maps of the area, and Jackson observed how the rivers converged and where the bridges were located. On June 7th, having just escaped a Union pincer movement, Jackson found his forces squeezed between Major General John C. Frémont's division to the west and Major General James Shields' division to the east. On June 8th, while one of Jackson's brigades engaged Frémont's force at Cross Keys, another brigade engaged Shields' division east of Port Republic. Jackson burned the bridge at Port Republic, stranded Frémont's division on the opposite side of the river, turned his entire force on Shields' division, and drove the enemy out of the Valley.

Hotchkiss shed his civilian clothes, became a major in Jackson's II Corps, and plotted many of General Lee's most famous campaigns. On May 2nd 1863, after Jackson sustained a mortal wound at Chancellorsville, Hotchkiss remained with the Army of Northern Virginia. During the summer of 1864, he guided and sketched General Early's route during the second Shenandoah Valley campaign, which took Confederate forces into Pennsylvania and to the very outskirts of Washington.

When the war ended, Hotchkiss gathered his maps, went back to the Valley, and opened another academy. Several months later a federal detective stopped at Hotchkiss' home in Staunton and demanded his maps. Hotchkiss refused, but agreed to talk with General Ulysses S. Grant. Hotchkiss had been thinking about the historical value of the maps while Grant only saw the practical value. During negotiations Grant picked the maps he wanted. Hotchkiss agreed to copy them but retained his originals.

The Hotchkiss imprint on American history remains to this day. He provided more than half of the Confederate maps that appear in *The Official Military Atlas of the Civil War*. Oddly enough, when Hotchkiss wrote the 692-page *Civil War History of Virginia*, he never used any of his maps, but all the key figures that had fought for the Confederacy and wrote reminiscences filled their volumes with his sketches.

Air War

HARPER'S WEEKLY.
A JOURNAL OF CIVILIZATION.

Vol. VI.—No. 283.] NEW YORK, SATURDAY, MAY 31, 1862. [SINGLE COPIES SIX CENTS. $2 50 PER YEAR IN ADVANCE.

Entered according to Act of Congress, in the Year 1862, by Harper & Brothers, in the Clerk's Office of the District Court for the Southern District of New York.

ALLOON VIEW OF THE ATTACK ON FORT DARLING, IN THE JAMES RIVER, BY COMMANDER ROGERS'S GUN-BOAT FLOTILLA, "GALENA," "MONITOR," ETC.—[See Page 3

Harper's Weekly, v. 6, May 31st, 1862, acquired by the Library of Congress in 1948 with the purchase of the papers and maps of Major Jedediah Hotchkiss. It illustrates a "Balloon view of the attack on Fort Darling in the James River, by Commander Rogers's [Rodger's] gun-boat flotilla, 'Galena,' 'Monitor,' etc., May 16." After the fall of Yorktown, the Confederate ironclad **Virginia** at Norfolk was scuttled to prevent her capture. This opened the James River to Federal gunboats. The five Union gunboats encountered submerged obstacles and deadly accurate fire from the fort at Drewry's Bluff and were turned back. There is something satisfying about Hotchkiss having kept the bird's-eye view: this, after all, was how he "saw" the entire conflict.

Pinned Down by the Rains:

The Battle of the Modder River

Before the outbreak of the Second Boer War, as it is known to the British, or more acceptably, the South African War of 1899-1902, in October 1888 Captain W. A. J. O'Meara, Royal Engineers, was making a clandestine survey, by eye only, of the approaches to diamond-rich Kimberley from Cape Town. River crossings were of obvious importance, and he paid particular attention to the railway bridge over the Modder River.

O'Meara then moved north to Kimberley where he was trapped when the town was besieged on October 14th 1899. Also in the town was Cecil Rhodes, the mining magnate and politically active colonialist, who made a great fuss to be rescued by the British. Lieutenant General Lord Methuen, commanding 1st Infantry Division, was sent to relieve Kimberley. He fought the Boers at Belmont on November 23rd and at Graspan on November 25th, taking their positions on both occasions but failing to prevent their withdrawal to reform and fight again. This they did, choosing the Modder River as their next battleground.

The station master at Modder River was helpful. He informed the British that the Boers had blown the railway bridge and were in occupation of the north bank. Reconnaissance by the 9th Lancers and by Rimington's Scouts put their number somewhere between "thousands" to the Scouts' figure of 400 men. This led Methuen to abandon his idea of making a great encircling foray to the east; the Modder had to be secured here first.

The scouting left a good deal to be desired. So did the map. Methuen had a sketch headed "Modder River Railway Bridge" and dated October 19th 1899 made by Captain O'Meara and, as the date was after the siege had begun, drawn in Kimberley and smuggled out. It was made from memory alone and failed to show the course of the Riet southwards on the British right, or the resort of Rosmead with its dam on the British left. The title of the sketch suggests it was never intended to do more than show the bridge, and, since the mapmaker had been on the site, two months' of wet season rains had swollen the rivers. This possibility appears to have escaped the British general; but what no map could tell Methuen was the plan developed by his enemy, Vecht-general Koos De la Rey.

The Boers had been dismayed by the losses they had sustained at Belmont and Graspan. True, the British losses had been greater, but the Boers fought to live rather than to die; dying was for the professionals. Taking position on hill-tops had not proved satisfactory. Firing down from the kopjes failed to make the most of the Mauser rifle's flat trajectory; indeed, the British Naval Brigade had suffered most when enfiladed by fire from men who had been forced down from their kopje by shell-fire. Moreover, the smokeless gun-powder now in use would allow them to take up positions which might be vulnerable if known, but were secure when undetected. De la Rey persuaded his superior, Assistant Commandant-General Piet Cronje, and General Marthinus Prinsloo that they should entrench on the southern bank of the river with fall-back positions ready dug on the northern bank. The artillery was less cunningly arranged. There were four 75 mm guns to the north-west of the bridge but the rest were on "Twee Rivier" or south at Bosman's Drift, the first ford on the Riet where it crosses the Free State border.

Against a frontal attack the Boers were excellently arrayed, and a frontal attack is what Methuen gave them. At 4.30 a.m. on Tuesday November 28th the British moved forward. On the left was the 9th Brigade under Major-General R. Pole-Carew, with the 1st Loyal North Lancashires, the 2nd Kings Own Yorkshire Light Infantry, and the 1st Northumberland Fusiliers, with the newly arrived 1st Argyll and Sutherland Highlanders in reserve. On the right were the Guards Brigade, the 2nd Coldstream, 3rd Grenadiers, 1st Scots, and the 1st

RIGHT *O'Meara's map with his handwritten notes and subsequent annotation by Lord Methuen. O'Meara's sketch of the railway crossing itself was accurate. Methuen marked a ford to the right of the bridge that was missing on the original, which would have allowed him to turn the Boer flank.*

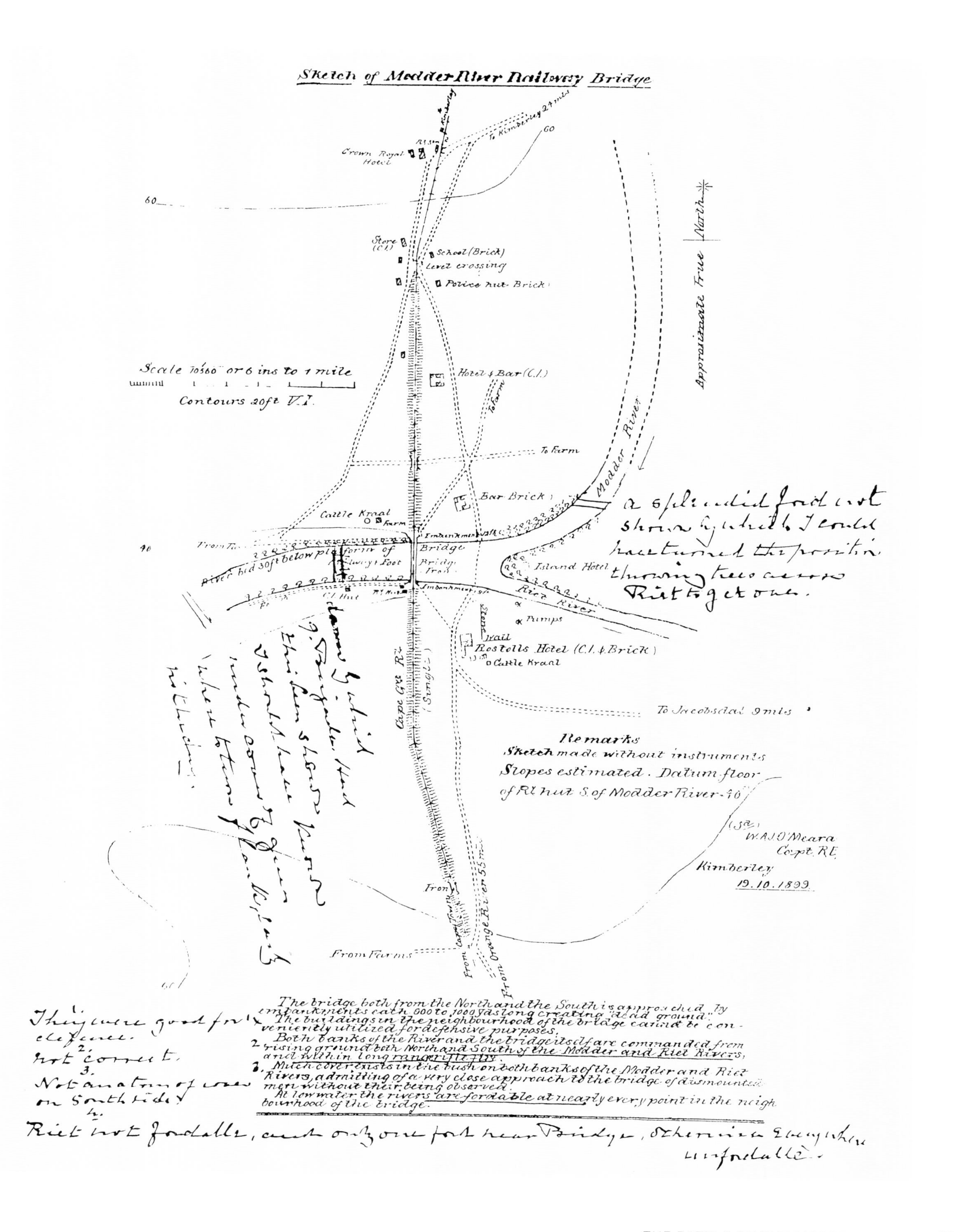

Sketch of Modder River Railway Bridge
To Kimberley 24 mls
60
Crown Royal Hotel
Store (C.I.)
School (Brick)
Level crossing
Police hut Brick
Scale 1/10560 or 6 ins to 1 mile
Contours 20ft V.I.
Hotel & Bar (C.I.)
To Farm
Bar Brick
Modder River
Approximate True North
Cattle Kraal
Farm
Embankment
Bridge
Bridge Iron
Island Hotel
Riet River
Embankment
Pumps
Well
Rosetta Hotel (C.I. & Brick)
Cattle Kraal
Cape Gt. Rd.
(Single)
To Jacobsdal 9 mls
Remarks
Sketch made without instruments
Slopes estimated. Datum floor of Rl. hut S. of Modder River-40
(Sd.) W.A.J. O'Meara Capt. R.E.
Kimberley 19.10.1899
From Farms
From Orange River 55 mls
a splendid ford not shown by which I could have turned the position throwing trees across Riet to get one.
The bridge both from the North and the South is approached by embankments each 600 to 1000 yds long creating "dead ground".
1. The buildings in the neighbourhood of the bridge cannot be conveniently utilized for defensive purposes.
2. Both banks of the River and the bridge itself are commanded from rising ground both North and South of the Modder and Riet Rivers, and within long range rifle fire.
3. Much cover exists in the bush on both banks of the Modder and Riet Rivers, admitting of a very close approach to the bridge of dismounted men without their being observed.
4. At low water the rivers are fordable at nearly every point in the neighbourhood of the bridge.
1. They were good for defence.
2. not correct.
3. Not an atom of cover on South side
4. Riet not fordable, and only one ford near Bridge, otherwise everywhere unfordable.

Coldstream in reserve. The artillery in support of the 9th was the naval battery of 12-pounders, on a low crest some 4,800 yards (4.4 kilometers) from the Modder. The 18th Field Battery was on the railway and 75th Field Battery was with the Guards. An hour after starting the infantry was approaching the Modder crossing, not a sign of the enemy to be seen.

Methuen was up with the troops as they came down the gentle slope towards the river and remarked to Major-General Sir Henry Colvile, commander of the Guards Brigade, that the Boers were not there. Colvile responded that if they were they were sitting tight when, at some 1,000 yards (910m) from the Modder, the Boers opened fire. Immediately the British hit the dirt. Colvile tried to have the 1st Coldstream work round the right flank but they stumbled up against the unexpected Riet River, and though they worked along the left bank for some way they were soon pinned down like everyone else. And there they had to stay, baked by the sun, steadily becoming more and more thirsty, bitten by ants and shot if they raised themselves above a totally prone position. It is also true to say that the entrenched Boers, with the river at their backs, were pinned down in equal measure and unable to indulge their inclination to make tactical retreats when things got hot. The artillery attempted to engage the Boer guns with some success, but the Free State Artillery commander, Major F W R Albrecht, had prepared alternative positions for his pieces and moved them about to confuse his adversaries.

On the left the 9th was similarly stuck, except that they could see some of their enemies in a clutch of farm buildings on the near bank forward of the hamlet of Rosmead. Here, after midday, the Argylls managed to insinuate themselves along a gulley and cut the Boer front line. The Free Staters were worried by this and their fire slackened. The KOYLI then charged the farm and ejected the Free Staters who fell

BELOW *Later in the war the British built a blockhouse to protect the river crossing. When the Boers opened fire, there was hardly any cover. Men threw themselves flat in the scrub, behind an anthill if they were lucky. A string of shells from one of the "pom-poms" killed the entire crew of the Scots Guards' Maxim machine gun. It remained the one object visible above the plain all day, untouched.*

back across the river by way of a drift and the dam to Rosmead. Pole-Carew followed with some of the Loyal North Lancashires and Argylls. He sent for reinforcements, but got very few. The 62nd Battery, which had come all the way from the Orange River that day and had to cut dead horses from their traces on the way, was able to add its efforts to the naval guns, but by now Methuen was wounded and Colvile was in command. Communication failed and Pole-Carew stood where he was, Rosmead in his hands and with De la Rey's Transvaalers blocking further progress.

As darkness came on, the Free Staters drifted quietly away. De la Rey was willing to hold on, but Cronje, his senior, was for withdrawal. By dawn on Wednesday the Boers were gone. The Modder was crossed without another shot fired. It had cost Methuen 70 dead and over 400 wounded, while the Boers suffered about 50 killed and an unknown number wounded. And now they were positioned, or so it seemed, on the hills ten miles south of Kimberley, ready to fight again.

Methuen later annotated the map he had had from O'Meara, but without an intention to publish it or use it to defend himself against criticism; it was found subsequently among his papers. On the Modder River to the right of the bridge he marked a ford which, if he had known of it, he said could have permitted him to turn the Boer flank. On the left he marked another "draw," which, under the cover of the guns, could also have been used. In the dry season there would have been many others, which is, perhaps, why they did not imprint themselves on O'Meara's memory. Indeed, the surveyor notes in his precise handwriting that the rivers are fordable almost everywhere and two months previously they evidently were. There are also comments on the amount of cover available, but that again was subject to seasonal variation. The misinterpretation arose not from faulty data, but the lack of local knowledge that would have told Methuen how things would change with the onset of the rains.

BELOW *The modern railway bridge over the Modder River, photographed in September when the water is low. The new bridge, beneath which a pier of the old one can be seen, is higher than the structure of a century ago. Unobtrusive reconnaissance in the flat, open country would have been hard to carry out.*

The Wrong Drift:

Colenso

On Thursday, November 2nd 1899, the last train left the little town of Ladysmith in Natal, among its passengers Lieutenant General Sir John French and Colonel Douglas Haig. The Boers had the place besieged. With Kimberley in Cape Colony besieged as well, the British commander, General Sir Redvers Buller, found himself obliged to divide his army and send Lieutenant General Lord Methuen to relieve Kimberley, while he tackled Ladysmith's problems. Methuen's advance was stopped at Magersfontein on December 11th.

By this time Buller had devised his approach to attacking the Tugela line, where the river wound its way beneath the towering kopjes to present a moat and a wall for the Boers to oppose the relief of Ladysmith, which lies beyond the hills. Buller's initial plan was to march to the left to outflank the Boer force on the hills north of Colenso, for the approach to the river from the south was across flat, open land. This he could see for himself, and when he arrived he lacked any more detailed map than Russel's five-inch to the mile school map. He intended to hold the Boers at Colenso with a simulated advance by the 6th Brigade, while he made a night march to Potgieter's Drift, a ford some 18 miles (29 kilometers) upstream, overlooked by Spioenkop from the north-west. But when news came of Methuen's defeat he canceled the operation as being too risky. The frontal attack on Colenso would be undertaken, in spite of Buller's previous misgivings.

The problem of mapping was tackled as soon as the army reached this front and Major G. S. Elliot, Captain Elton, and the range-takers of two Royal Field Artillery batteries made the best survey they could manage, given the presence of the enemy. A map called "Sketch Map of the Country Around Ladysmith" had been published by the War Office in that year to replace an earlier version that omitted the "drifts" or fords across the River Tugela, but whether the new edition was available to Elliot is not known. What is interesting is that the major errors of cartography appear both on the War Office maps and on the new map made by Elliot. Was he influenced to see only what he expected to see?

The relief force was based about eight miles (13kilometers) to the south at Frere and the surveyors spent two days moving from one observation point to another, eleven in all, which they identified with letters of the alphabet along the route, shown by dots and crosses on their map. The points of which they took bearings were shown by dots within circles. The map also had remarks such as, "Said to be $2^{1}/_{2}$ ft. deep" at Robinson's Wagon Drift, west of Colenso. This was, as the map was annotated, the result of guides giving assistance to provide the information and the names. Between points E and F on the route taken, a stream is shown with a continuous line, but as it goes north its course becomes uncertain and a series of dashes, also specifically mentioned on the key as indicating "doubtful," suggests that it runs into the Tugela east of Bridle Drift and west of the bend known as "The Loop." On the War Office map this rivulet is called Doorn Sp.—a contraction of "Doornkop Spruit."

Commandant General Louis Botha was in command of the Boer defense, and he had banked on Buller attacking here. He had, in spite of the obvious danger to them, persuaded the Wakkerstroom commando under Joshua Joubert to occupy Nhlangwini. The Krugersdorp, Vryheid, and Heildelberg commandos held the ground north of Colenso, including the lofty Fort Wylie, while the Swaziland Police and Ermelo and Middleberg commandos held the land west to Vertrek and beyond. They had ten 75 mm guns, a 5-inch howitzer and a pom-pom in well-concealed positions in the hills. Along the river trenches had been dug.

On December 14th at 10pm the orders were issued, too late to reconnoiter the ground. Major General A. F. Hart's 5th (Irish) Brigade would advance on the left, the west, to cross the Tugela at Bridle Drift,

RIGHT *Buller (left) and his staff observing the battle (above). Intelligence Division War Office map No. 1449, "Sketch Map of Country Around Ladysmith," published not long before the battle, scale one inch to two miles. The Tugela drifts, or fords, are marked but not named. Confusing two of them proved fatal.*

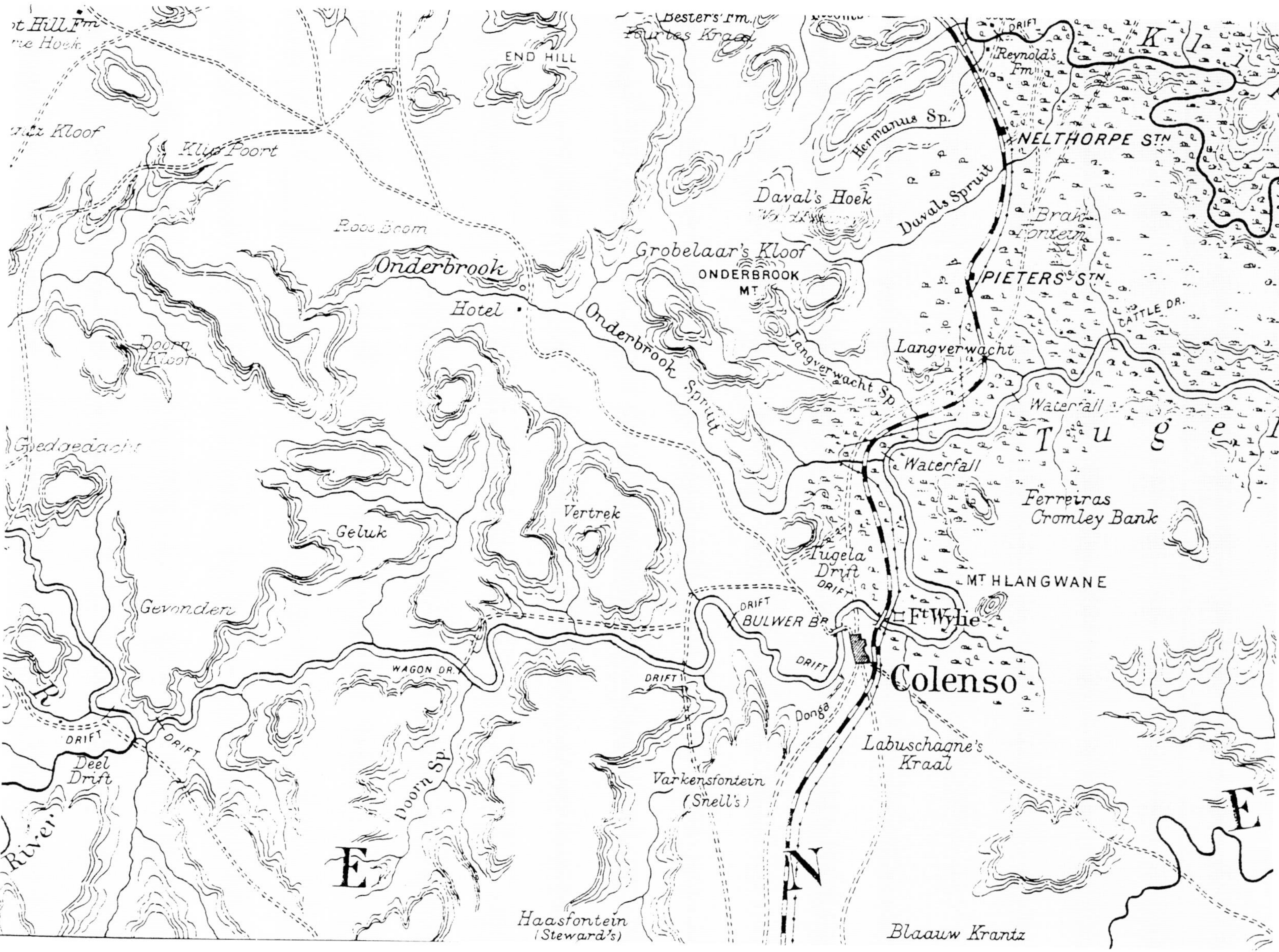
END HILL
Klip Poort
Roos Boom
Onderbrook
Hotel
Bester's Fm.
Hermanus Sp.
NELTHORPE STN
Reynold's Fm
Daval's Hoek
Duvals Spruit
Grobelaar's Kloof
ONDERBROOK MT
Brak Fontein
PIETERS STN
CATTLE DR.
Onderbrook Spruit
Langverwacht Sp.
Langverwacht
Waterfall
Doorn Kloof
Geluk
Vertrek
Gevonden
Ferreiras
Cromley Bank
Tugela Drift
MT HLANGWANE
DRIFT
BULWER BR.
Ft Wylie
Colenso
WAGON DR.
Donga
Labuschagne's Kraal
Deel Drift
Doorn Sp.
Varkensfontein (Snell's)
River
Haasfontein (Steward's)
Blaauw Krantz

ABOVE AND RIGHT *The Loop, (above) marked by the modern planting of river-side trees curving from the right, from a Boer gun position on Vertrek. Hart led his men to the Punt Drift within the Loop and into withering fire. Boer gun position (right) near the Ladysmith road's junction with the side-road to Grobbelaarskloof, with the dark green curve of the Loop in the distance.*

to the west of the great meander of "The Loop," and to turn back eastwards to roll up the Boer line. Major General H. J. T. Hildyard's 2nd (English) Brigade was tasked with taking Colenso village itself and then, reaching the far bank of the river by way of two drifts and the Ladysmith road that left over Bulwer Bridge, attacking the Colenso Kopjes. Meanwhile, on the right, Colonel Lord Dundonald's Mounted Brigade was to attack Hlangwane Hill (Nhlangwini). Colonel C. J. Long's 14th and 66th Field Batteries and Lieutenant Ogilvy's naval guns would be on the right and Colonel Parson's 63rd and 64th Field Batteries on the left, where Major General N. Lyttleton's 4th Brigade was in support. Major General G. Barton was in command of the supporting 6th Brigade on the right.

The wording of General Hart's orders was crucial to the events that followed. He had been given a guide, an African man, and a Natal-resident interpreter, Mr Norgate, to help him locate "the Bridle Drift immediately to the west of the junction of the Doornkop Spruit and the Tugela." This appears to be clear enough, looking at Elliot's map. The Loop has the stream entering the river at its western end and immediately west of that is the Bridle Drift. The problem was that Elliot, probably because he simply could not see the ground clearly enough while keeping out of Boer rifle range, had not created an accurate map. The Doornkop Spruit, on his map, flows past a "Little Cliff on Bank of Stream" and is joined by another stream from the west to make a single watercourse running down to the Tugela. In fact, these streams do not join up; the little stream carries on to the western side of the Loop while another, today named Doringspruit, enters the river on the eastern end of the Loop. Not only are there two streams and confusion about their names, but there are two drifts, or fords. The intended crossing, the Bridle Drift, is west of the nameless stream's entry point, but the Punt Drift, within the Loop, is the one "immediately west ... of the Doornkop Spruit." The scene was set for confusion and disaster.

Hart's Brigade was given some bracing drill before moving off at about 4am. Hildyard's were not so quick off the mark, and Long's artillery, which started at 3.30 as ordered, was soon racing far ahead of the infantry. At 5.30 the 4.7-inch guns and four 12-pounders under

Commander A. H. Limpus began shelling the heights beyond the river from a position west of the railway. For three-quarters of an hour the bombardment continued without reply from the invisible Boers, who were probably unharmed. On Hart's left 1st Royal Dragoons saw the Boers across the river and sent warning to the General, who decided to ignore them unless they attacked. Hart approached the Tugela, keeping the Doornkop Spruit on his right, the east, but it became evident that he was not as far west as the map suggested he should be. He asked his African guide where the drift was and, assuming the general line of the advance was what was intended, the man naturally pointed him towards the Punt Drift within the Loop, while others warned him he was entering a salient. Leaving the 1st Scottish Borderers to hold the bank on the west, he led 1st Connaught Rangers, 2nd Royal Dublin Fusiliers, and 1st Royal Inniskillin Fusiliers on towards the ford. At that moment the Boer shelling started, three shells causing 33 casualties immediately. Hart liked to keep his troops under his close control and they marched on in mass of quarter columns, the formation that had proved so costly at Magersfontein and was equally costly here. The guide, wisely, fled.

Hart was now enduring the crushing, scything fire of Mauser rifles and Krupp field guns, or, to be precise, the Irish Brigade were, while, apparently bearing a charmed life, the General himself strode about unharmed, pushing his men deeper into trouble and even preventing the Iniskillings from creeping out on the left and attempting the river closer to the correct place. Trooper T. Sullivan, a Queenslander, was attached to the Irish that day:

> "We found the Boers entrenched on the banks of the River Tugela. As we approached they opened a withering fire on our brigade, so the order was given to swim the river and in we jumped. I left my horse behind and joined the rush. To our surprise we found the Boer entrenchments surrounded by barbed wire, so we set to work to cut a gap in the fence. We were all but through when the Boer cannon opened on us and the order was given to retire. While I was cutting my last bit of wire a shell hit the iron post on my right, and took the head clean off a Connaught Ranger. Then I thought it might be my turn next so I got back to the bank."

From the height of Vertrek, from a gun emplacement on the Ladysmith road and from trenches on the north bank, the Boers poured an unrelenting fire on the Irish.

Buller ordered Lyttleton to extract Hart from the trap in which he had entwined himself and Hildyard, who had captured Colenso village without serious problems, to cover the Commander-in-Chief's mission to extract Long's guns which had become isolated on the right. At 8am the whole action was called off and the general order to retire was given. The British struggled to extract themselves, in daylight and without significant covering fire, from their positions. Some managed it in three hours, others took all day and yet others were still out there when evening came and the Boers crept over the river to take them prisoner. Buller had lost seven officers and 138 men killed, 43 officers and 719 men wounded, and another 220 missing or taken prisoner; trivial numbers by the standard of later wars or the American Civil War, but enough to convince those at home in Britain that this was the third massive disaster of "Black Week."

Hart remarked that during the action he did not see a single Boer and that the attack was a complete fiasco, but apparently felt no great responsibility for what he had allowed to befall his men. Buller, however, held him responsible, but left him in command of the Brigade all the same. It was Buller who lost his post as Commander-in-Chief, replaced by Lord Roberts ... for the want of an accurate map.

Serving with the Boers was Colonel J. Y. F. Blake, Commander of the Irish Brigade. After the war he wrote:

> "The Boers generally acknowledge General Buller as by far the ablest commander the English had in the field. True it is, he made mistakes on the Tugela, but ... had [he] been in supreme command, I firmly believe the war would have been brought to an end within six months after the relief of Ladysmith."

2,000 Dead:

Thiepval Sacrifice 1916

Between Beaumont Hamel at the northern end of the attacking front and La Boiselle in the center, the British front line was established where the French had arrived in September 1914. Clothing the western slope down to the river, Thiepval Wood offered some shelter and the front line ran along its eastern edge. On July 1st, 1916, 10 battalions of the 36th Ulster Division advanced at speed from the wood and into a terrible trap. For four miles on either side of them there was no help, nothing to distract the German machine guns and artillery.

To the north the river curves away eastwards around St Pierre-Divion and to the south, below the fortification known as the Leipzig Redoubt, a shallow valley cuts into the higher ground north of Orvillers towards Mouquet Farm beyond the Pozières-Thiepval road. Mouquet Farm itself had been converted into a fortress, and between Thiepval and St Pierre-Divion a huge fortification had been built, the Schwaben Redoubt (grid square 19), together with the complex of trenches that characterized the whole of the German front line.

Immediately in front of Thiepval, where the château (25) then stood and where now the massive memorial to the missing dominates the skyline, the 1st Salford Pals and the Newcastle Commercials launched the initial assault. Not a man of the six waves of troops even reached the German wire. The final waves were wisely held back from futile sacrifice and, manning their trenches once more, laid down fire on the enemy. To the south, from Campbell Avenue, men of the 32nd Division had worked their way forward before the bombardment lifted and then rushed the Leipzig Redoubt, taking part of the position. Attempts to reinforce them were prevented by machine-gun fire and, in a rare instance of flexibilty in the use of the artillery, two howitzers were detailed to cover the consolidation of the gains.

In the wood the men of the Ulster Division were in fine fighting spirits. July 1st is, by the old calendar, the anniversary of the Battle of the Boyne and these Orangemen took that for an excellent omen. Before zero hour they crept forward towards the German lines and when the bugle sounded all earlier orders to advance in line were forgotten. As the barrage was raised they rushed forward and before the Germans could emerge from their shelters the Ulstermen had swept into the first line of trenches. From there they pressed forward into the Schwaben Redoubt itself. Here the resistance was tough, for the defenders had not been caught in their dug-outs. Meanwhile the Germans at Thiepval, no longer faced with frontal attack, could turn their guns on the flank of the Ulsters, and as the next wave, four Belfast battalions, came through they found themselves under heavy enfilading fire. Major George Gaffikin of the West Belfasts waved his orange sash and cried "Come on, boys! No surrender!" and the men hurled themselves forward.

The fight for the Schwaben Redoubt was long and vicious. Whole packs of grenades were hurled down dug-outs to explode among the defenders. The stove-pipes that projected from the shelters were used to post grenades into the living quarters. Captain Eric Bell of the Tyrone Volunteers, a trench-mortar officer, actually resorted to throwing his mortar bombs at the enemy. He died leading an ad-hoc formation of infantry in attack later in the day, and was posthumously awarded the Victoria Cross. By mid-morning the position was taken and some 500 prisoners were in British hands. The troops were now leaderless and out of touch with their divisional command. A brief foray towards Thiepval found an unmanned trench which might have enabled them to attack that position from the rear, but they lacked orders and officers to take the initiative.

What the Belfasts had been ordered to do, and what they expected to be doing had they not become embroiled in the fight for the Schwaben Redoubt, was to take the second line that ran south from

RIGHT *The British front line is marked by the blue dashes, the German trenches are in red; this is Thiepval as at May 16th, 1916—so far as the British knew it. It is possible to comprehend how seductive Thiepval Wood looked "on paper" as a point of attack. A painstaking (and challenging) comparison of this map with the German one on page 169 reveals that British knowledge of the enemy trench system was partial.*

St Pierre
Divion
The Mound
Hamel
75.3
Mill
Mill Road
24
23
19
20
Road
150
Crucifix
Hamel
Bridge
75.1
Thiepval Road
Cemetery
145
Thiepval
Wood
29
30
25
26
snil
Thiepval
Passerelle
de Magenta
133.0
135.8
Row of
Apple Trees
BRITISH FRONT LINE
122
70
35
36
31
32
Yatman
Bridge
74.2
Mill
89.3
140
130
120
110
Authuille
Bridge
69.3
Campbell
Avenue
Authuille
89.1
Cemetery
Black Horse Bridge
5
6
1
2
Black Horse Road
72
Ancre
Kelly
Authuille Wood
100

Grandcourt with Feste Staufen, Stuff Redoubt, lying on the line of advance through the Schwaben Redoubt. They pressed on, but fatally ahead of schedule. Felix Kircher of the German 26th Field Artillery was there to see them.

> "At 9 o'clock, I was down in a dug-out in the Feste Staufen when someone shouted down to me in an amazed voice 'the Tommies are here.' I rushed up and there, just outside the barbed wire, were ten or twenty English soldiers with flat steel helmets. We had no rifle, no revolver, no grenades ... we were purely artillery observers. We would have had to surrender but, then, the English artillery began to fire at our trench; but a great deal of the shells were too short and hit the English infantrymen and they began to fall back."

There were no fighting troops to resist the Belfasts, but their own artillery fire stopped their attack. The few who remained could see German troops gathering in Grandcourt ready to counter-attack, and made their way back to their comrades. There, in the newly captured strongpoint, the Ulsters found themselves isolated. The attacks on each flank had failed and these men, who had been the only troops to penetrate to the second German line, had no support against the gathering strength of the enemy.

As the counter-attacks started the reserves were sent to relieve the Ulsters. Thiepval, however, remained in German hands and any attempt to cross no-man's-land from the wood was doomed. Private J. Wilson of the 1/6th West Yorkshires describes the experience.

> "We went forward in single file, through a gap in what had once been a hedge; only one man could get through at a time. The Germans had a machine-gun trained on the gap and when my turn came I paused. The machine-gun stopped and, thinking his belt had run out, or he had jammed, I moved through, but what I saw when I got to the other side shook me to pieces. There was a trench running parallel with the hedge which was full to the top with the men who had gone before me. They were all either dead or dying."

BELOW *The Castle and the surroundings of the Church tower at Thiepval, photographed in 1915. Thiepval is now the site of the largest British memorial.*

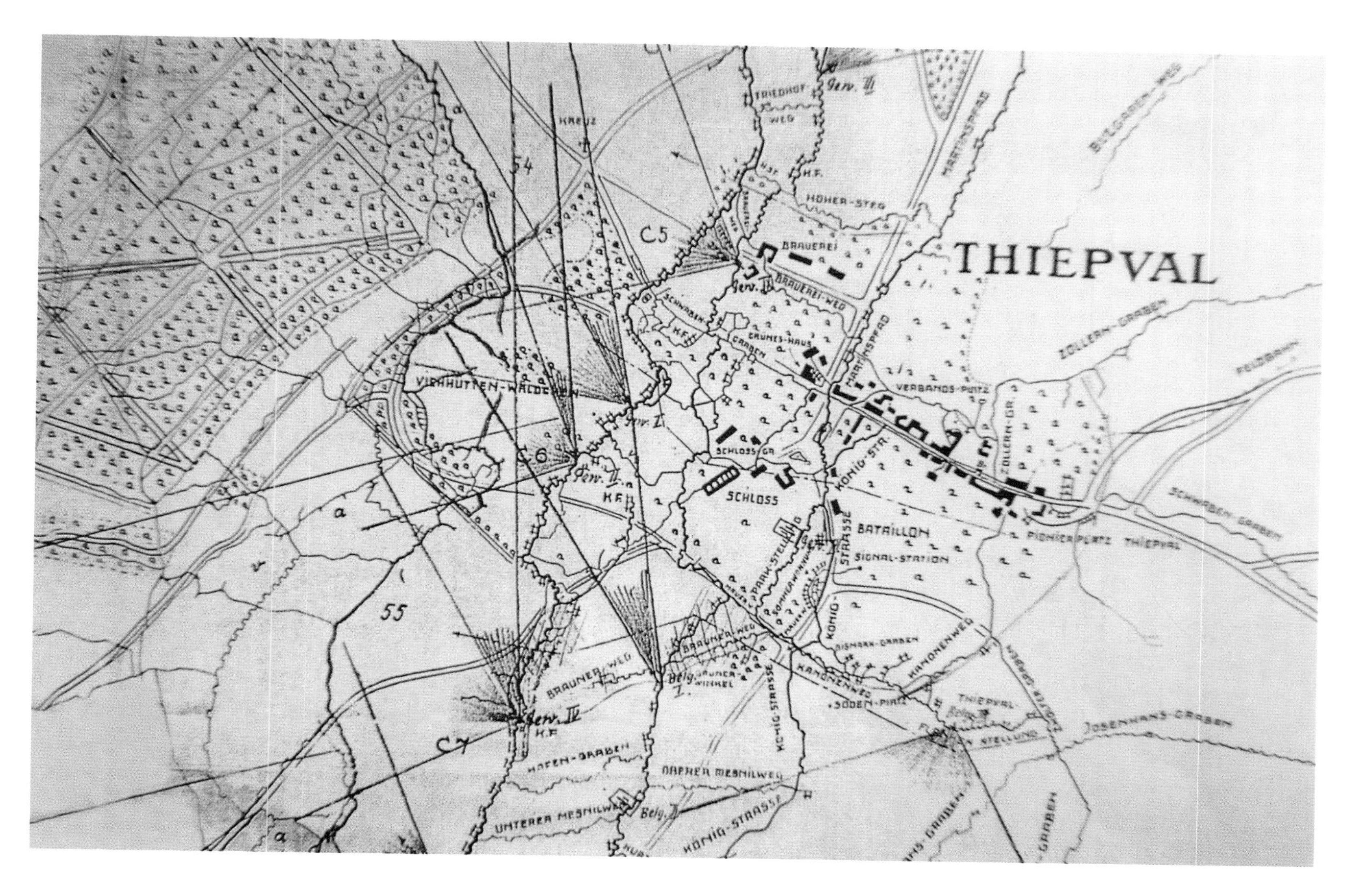

ABOVE *Difficult to interpet and even read, this rare map shows the German record of their machine gun arcs of fire. This would have been a life-saving document for the British. With it, perhaps there would have been no VC, fewer military crosses for the men of Bushmills, and fewer than 2,000 of their comrades-in-arms might have fallen. Immediately in front of Thiepval, where the chateau then stood, the 15th Lancashire and 16th Northumberland Fusiliers launched the first attack. Not a single man of the six waves of troops reached the German wire.*

The German attacks on the Schwaben Redoubt were unremitting. As each was driven off, the casualties mounted and the ammunition grew less. By the evening the Ulsters were forced to retire to the old German front line, where they were at last relieved by the West Yorkshires. The adversaries were to occupy these same positions for the next three months. The Ulsters had lost some 2,000 dead and 2,700 wounded. 165 were taken prisoner.

Like so many of the regiments of the New Army, those making up the Ulster Division consisted of numerous groups of men who had enlisted in the same unit; neighbors and brothers fought side by side. From the little town of Bushmills, close by the Giants Causeway, many of the men joined the Royal Irish Rifles, 12th Battalion and another Bushmills man, 2nd Lieutenant Sir Edward MacNaughten (Bart) was seconded to the Battalion from the Black Watch. On July 1st they were on the left flank of the Ulster line, on the river. Twenty-three Bushmills men died that day, of whom only six have known graves. Alex Craig was 27. His younger brother, Samuel, was 20. John McGowan was 19, his brother James 18; both died, one helping the other. Bushmills, with a population not much in excess of 2,500, lost nearly a quarter of all its men killed in the First World War on this one day. Private Robert Quigg won the Victoria Cross on July 1st, and the other decorations won by Bushmills men by 1918 would number two Military Crosses, seven Military Medals, two Distinguished Conduct Medals and one Croix de Guerre. Throughout Britain towns and villages suffered similarly. Whole village football teams were gone. All the children of a school class were dead or wounded. The impact was enormous.

Taking the Ridge:

Passchendaele

In the summer of 1917 the British, under Field Marshal Sir Douglas Haig, undertook a massive campaign in Flanders. In France the Germans had retreated to the daunting fortifications of what the British called the Hindenburg Line.. In April the Allies mounted a great pincer attack with the British and Canadians assaulting Vimy Ridge and the enemy lines in front of Arras while the French threw themselves against the heights of the Chemin des Dames. For the time being the burden of fighting the Germans would rest upon the British alone.

The Ypres front was dominated by a ridge of hills, not particularly high at about 170 feet (50 meters), but in this flat countryside lofty enough to give the army that held it an overview of their opponents and all their movements. The underlying geology, as it had been mapped by the Belgians some twenty years earlier, was a layer of clay, impervious to water, about 115 feet (35 meters) deep, beneath which were water-bearing sands down to which engineers drilled to make wells. Over the clay, and forming a line of hills that encircled the ancient town of Ypres from south to north, were a series of layers of sand and clay deposits crowned with a final icing of well-drained soil. The cultivated fields depended on the soil being drained into the network of streams that came down from the ridge, and so it is today. But once the drainage system was disrupted the terrain became water-logged, and if the underlying clay was pierced water welled up from below even when the weather was dry.

The Third Battle of Ypres began on July 31st and the rain began at the same time. By late August the British were still well short of the ridge and Passchendaele village that crowned it. The terrain had prevented the excavation of trench lines, and the Germans depended on a complex of pill-boxes, small concrete forts, between which their troops manned shell-holes. The traditional front line was thus replaced with a more flexible defense, into which the British were supposed to be drawn before being destroyed by shell and machine-gun fire. The British responded by reorganizing their infantry into specialist sections within platoons—rifle-grenade men, hand-grenade men, Lewis-gunners—and introducing new tactics to pin down and outflank German resistance. The British also raised the science of the artillery bombardment to new heights, the attackers following a creeping barrage so closely that they were among their enemies before the counter-barrage could be brought to bear on them. The shellfire, predictably, made the countryside a complex of bog and mud-traps.

The Canadians returned to the Ypres Salient on October 18th, under a new commander, a Canadian, Lieutenant-General Sir Arthur Currie. They could hardly recognize the ground. The villages they had known were all gone, the woods had disappeared, the streams were now broad bogs. Only the faint trace of the roads to Zonnebeke and Gravenstafel served as reference points in the blighted landscape. They took over the line from the Zonnebeke road in the south to a point astride the Gravenstafel-Mosselmarkt road, the sector of the last great ANZAC attack. It was to be their task to secure the longed-for objective, Passchendaele and its ridge.

A huge effort was at once undertaken to build roads and tracks to get guns and supplies forward. Their work was hampered by continual shelling and by the introduction of a new gas by the Germans, diphenyl chlorarsine, "Blue Cross," which penetrated the current design of gas-mask and caused uncontrollable sneezing and vomiting. The Germans had also been rethinking their defensive tactics. The British approach of leap-frogging their units, one consolidating as the next passed through, behind a creeping barrage that made any German counter-attack so costly, had worked well at Menin Road Ridge and Polygon Wood against the German scheme of lightly defended front

RIGHT *8th September British map; red crosses are "wire entanglements or other obstacles." Red lines are German trenches: "Important ones are shown by thick line, old or disused by dotted line." There is no "conventionally" defined front line. As Haig acknowledged in his Christmas Day despatch: "Where the Menin Road crosses the crest of the Wytschaete-Passchendaele Ridge formed … was the key to the enemy's position, and here the most determined opposition was encountered."*

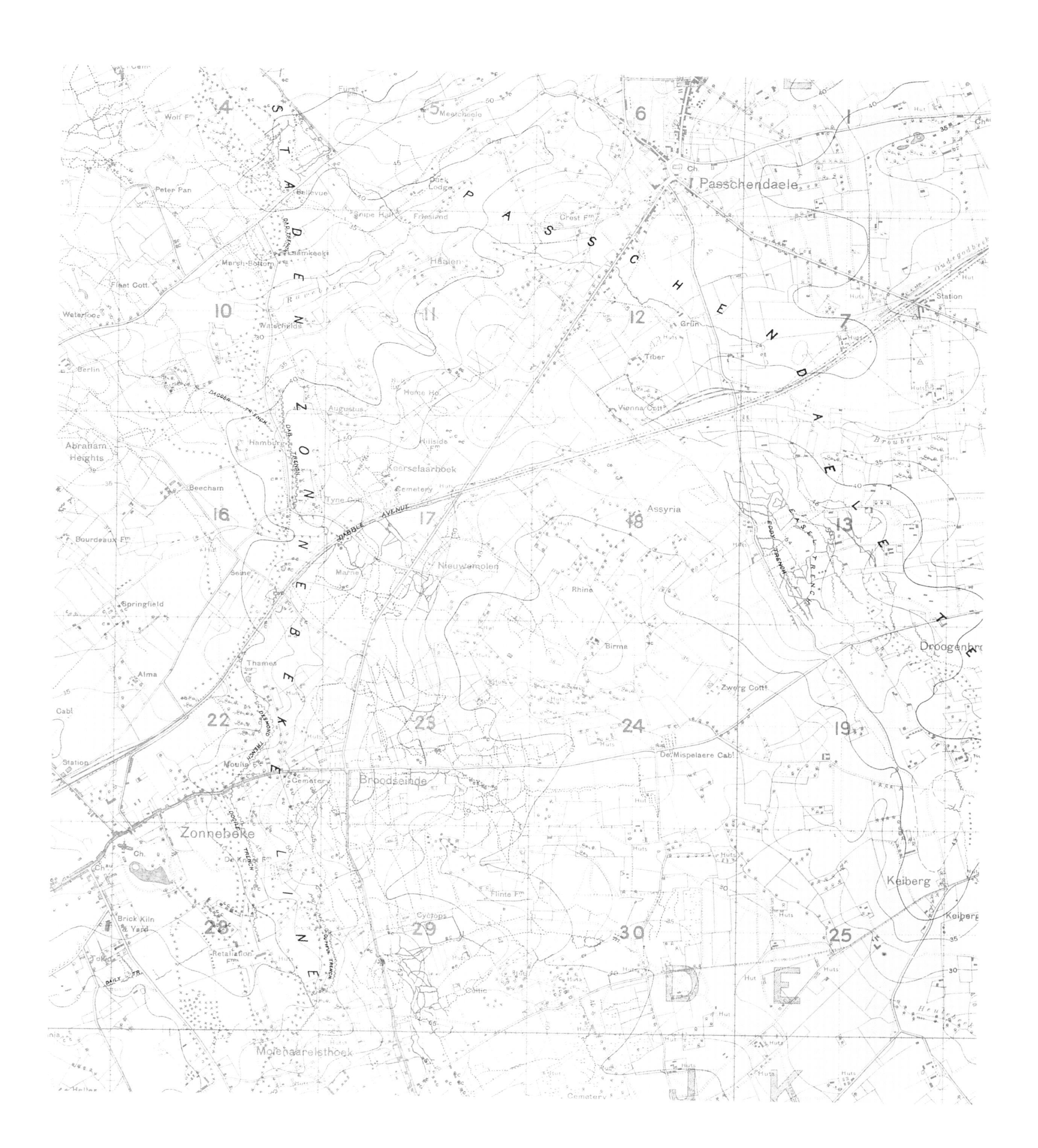

Passchendaele
Station
Wolf Fm
Peter Pan
Bellevue
Snipe Hall
Friesland
Crest Fm
Meetcheele
Marsh Bottom
Flat Cott.
Waterloo
Watchfields
Berlin
Gravenstafel
Augustus
Home Ho.
Tiber
Vienna Cott
Abraham Heights
Hamburg
Hillside Fm
Keerselaarhoek
Cemetery
Beecham
Tyne Cott
Assyria
Bourdeaux Fm
DABBLE AVENUE
Seine
Marne
Nieuwemolen
Rhine
Springfield
Birma
Droogenbroodhoek
Thames
Alma
Zwerg Cott.
Station
Moulin
Broodseinde
De Mispelaere Cab.
Zonnebeke
Keiberg
Flinte Fm
Brick Kiln & Yard
Cyclops
Retaliation Fm
Molenaarelsthoek
Cemetery
STADEN ZONNEBEKE LINE
PASSCHENDAELE

lines. A heavily defended front line had failed at Broodseinde. The concept of the "forefield" was now introduced: the lightly defended front line would be separated from the main defensive positions by as much as 500 to 1,000 yards, the intervening ground to be saturated with artillery fire as soon as the outposts had been withdrawn. The three regiments of the 11th Bavarian Division on the Passchendaele Ridge each had one battalion in the main defensive position with the remaining two in successive lines to the rear, awaiting the Canadians.

The German front was plastered with shellfire for four days along its whole length. Particular attention paid to those pillboxes and blockhouses known of on the intended line of attack was disguised by equal fury falling elsewhere, but the detail on the British mapping failed to cover all the installations created by the energetic German defenders. General Currie decided to solve the problem of the Ravebeek bog by leaving it alone, a decision which certainly prevented useless casualties.

At 5.40am on October 26th the Second Battle of Passchendaele began. On either side of the Menin Road the British 7th and 5th Divisions were frustrated by the marshes that now protected Gheluvelt. Along the ridge and across the Broodseinde-Passchendaele road the Canadian 46th Battalion, with the Australian 18th Battalion on their right, went forward in a mist that, as the day drew on, turned to steady rain. They took their objectives, but the Canadians paid heavily with 70 percent casualties. It was a mistake to allocate this sector to two different formations, as confusion was inevitable. Decline Copse, straddling the railway, was reported as taken by both Canadians and Australians. Each therefore withdrew to leave it to the other and the Germans moved back in. They were not ejected for another 24 hours, after a night attack by the Canadian 44th Battalion.

Against the Bellevue Ridge the 43rd Battalion from 9 Canadian Brigade made good progress, clearing the pillboxes with grenades, but on their right the 58th were checked by the Laamkeek blockhouse. From the hill above, the Germans were able to direct a telling shellfire on the attackers, and they were forced to fall back, but some of them held on. Lieutenant Robert Shankland of the 43rd, with the added strength of men of 9 Machine Gun Company, clung on to a position on the Bellevue Spur, occupying the former German positions. By noon the 52nd had come up in support to reunite the outpost with the brigade and went on to secure the rest of the spur. They took the pillboxes one by one, rifle grenades providing cover while small parties of soldiers crept their way round to hurl hand grenades through the loopholes. It was tough, dirty work, but by evening defenses that had repelled the British and the New Zealanders were secured. Robert Shankland was awarded the VC. To their left the 63rd (Royal Naval)

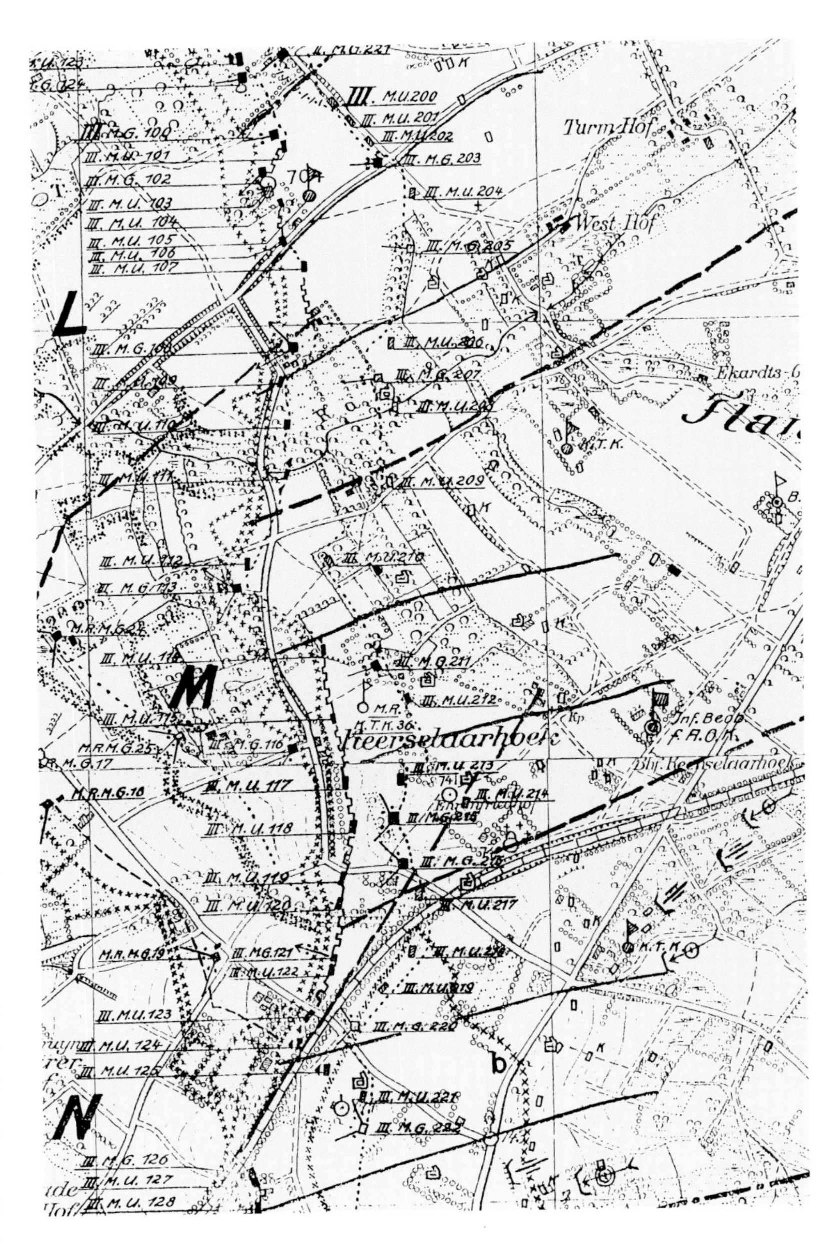

ABOVE *German map of machine-gun posts on the western side of the Passchendaele Ridge, now held at the* Bayerisches Haupstaatsarchiv, *Munich.*

Division had gained some ground but, just as down on the Menin Road, anything other than a hill was now a lake.

In the dark the work of finding and tending the wounded went on. Private F. Hodgson, 11th Canadian Field Ambulance, was at Tyne Cot:

> "We had two pillboxes there ... the doctor and his helpers were in one and we stretcher-bearers were in another about a hundred feet away ... There were three squads of us. Three squads of eight—because it took six of us at a time to get one stretcher out through the mud. That day we drew lots to see

RIGHT *Private Reginald le Brun in a shellhole with his Vickers machine-gun, ready to give covering fire to the Canadian attack.*

who should go first. My squad drew the last carry ... Away we went with our wounded man, struggling down the track. After a few hundred yards we were caught in a barrage. We put the stretcher-case in a depression on the ground. He was very frightened, the wounded boy ... He died before we could get him to the dressing station. On the way back we passed the remains of our No.1 Squad. There was nothing but limbs all over the place. We lost ten of our stretcher-bearers that day."

The Canadians had lost 2,481 men over the three days of their first strike at Passchendaele, of whom 585 were killed. Currie declined to rush into further attacks and three more days passed as roads were repaired and fresh preparations made. On October 30th the assault began again. The German reply to the artillery barrage did not come for eight precious minutes, by which time the Canadians were well on the move. On the right the 85th Battalion took Vienna Cottage and reached its final objective but lost half its men in doing so, and to their left the 78th were overlooking the eastern slope of the ridge. The 72nd were heading for Crest Farm, moving along the side of the ridge and fanning out to the left to take it. Their forward patrols found the Germans preparing to retreat from Passchendaele village. Below them the Ravebeek was flooded, limiting their ability to establish a secure line, and beyond it Princess Patricia's Canadian Light Infantry were having a tough time, so the final line established curled back from Crest Farm. Private J. Pickard of the 78th was a signaller, but the shellfire destroyed telephone cable as fast as it could be laid, so he became a runner:

"It started to rain in the afternoon, but it went well that day. I was back and forward to the line as acting-runner, and every time we'd got a bit nearer to Passchendaele. They stopped eventually at the foot of a lane leading into the village. You could tell it had been a lane by the ruined cottages on either side ... It was a place they called Crest Farm. They had to fight hard to get it and the place was thick with bodies. But we took it, and we held the line."

The PPCLI had to work their way through the strongpoints of Snipe Hall, which they took at night before the attack in a surprise move, and Duck Lodge, before they could seize the Meetcheele crossroads under heavy fire, and there they dug in, well up the Bellevue Spur and an obvious threat to the German possession of the ridge. They nearly did not make it. A pillbox beside the road had them under a punishing fire and two men tackled it. Lieutenant Hugh Mackenzie drew their fire, and died doing so, while Sergeant George Mullin worked his way round to attack the post alone, killing the two machine-gunners with his revolver. Both men were awarded the VC. Below the spur to the left the 5th Canadian Mounted Rifles struggled through the swamp of Woodland Plantation and by mid-afternoon they were well up the ridge. The third Canadian VC of the day was won here when Major George Pearkes held Vapour Farm and Source Farm against German counter-attacks. The action proved as costly to the Canadians in a single day as the previous three-day attack: 884 killed and 1,429 wounded.

German counter-attacks over the next few days were determined and frequent, but largely unsuccessful. Currie took a seven-day break before the next operation. It came on November 6th, with the rest of Plumer's force raising a ruckus on the flanks, but leaving the hard fighting to the Canadians. By 8.45am the village was in Canadian hands. The action was swift, decisive, and costly. Casualties amounted to 2,238, of whom 734 were killed or died of wounds.

In a final effort in torrential rain four days later, the line was pushed to the north, the Canadians fighting alongside the British. The final position was a vulnerable salient above Passchendaele, which the Germans pounded unmercifully for four days, although the Canadians succeeded in pushing forward down the eastern side of the ridge to make the position more secure. On the 14th their relief commenced. On November 15th the Third Battle of Ypres was declared over.

Blitzkrieg in the West:

Plan D in Belgium

In 1939 the British Expeditionary Force under General the Viscount Gort became part of the Allied army expecting to defend France and Belgium against a German attack. In eastern France the forts of the Maginot Line presented an insuperable barrier to invasion, in southern Belgium the forests of the Ardennes precluded, it was thought, a swift incursion by the enemy and so that left the Belgian route, the same as that used by the Schlieffen Plan in the First World War, as the obvious line of attack for Hitler's forces.

The BEF was tiny; only five divisions were available in September and another five were to follow, so the resistance would depend on the French, with more than eighty divisions at their disposal.

The major problem was Belgium, which had declared her neutrality. It was impossible to discuss or plan anything with the government of the probable battleground. The impasse was eased somewhat in January 1940 when a German officer was captured when his aircraft came down near Maastricht. It turned out he carried plans for an invasion of Belgium and the Netherlands which were passed on to the Allies and led to Plan D, the scheme for setting up a defensive line east of Brussels. The French Seventh Army was to hurry north to the Dutch border, the French First Army would occupy Belgium as such and the BEF, as part of it, would cover the Belgian capital in the center. Detailed arrangements were made and the movement was marked on a large map at Lord Gort's headquarters. Then they waited.

At 6.45am on Friday May 10th, two hours after the German attacks had begun on the Netherlands and the Belgian borders, the Belgians at last invited the French and British to enter their country. The French cavalry units rushed on as quickly as possible to throw a screen in front of the approaching allies as they took up position on the Dyle and elements of 2nd and 3rd D.L.M. (light mechanized divisions) reached their positions near Liège just in time the next day. On Saturday, as the air forces of three countries were attacking the bridges further east, the Belgian 7th Infantry Division collapsed, leaving the road open for 4 Panzer.

The BEF was on the move to take up its positions east of Brussels. In the vanguard were the 12th Royal Lancers, who were to form the forward screen alongside, and to the north of, the French cavalry. They were slightly delayed by a Belgian official who was unhappy with their lack of documentation for entering his country, but overcame this by driving through his barrier. The reception given to the troops by the Belgians was enthusiastic. Major Lord Sysonby of 1st/5th The Queen's Royal Regiment wrote home:

> "When we stopped for a cup of coffee in the big town half way [possibly Ath] we were surrounded by a jabbering mass of people who were overpoweringly enthusiastic at our appearance... The standard greeting seemed to be "Goodbye Tipperary" said with a heavy accent to which we usually replied "Good night Leicester Square" which nobody seemed to understand but took in very good part."

Lieutenant Christopher Seton-Watson of 2 Regiment, Royal Horse Artillery reported:

> "Fruit and packets of biscuits and chocolate were thrown into our trucks ... Wheeler and I were presented with huge sprays of white lilac by a charming peasant girl, who threw her arms round his neck and kissed him on both cheeks. He completely failed to register the embarrassment to be expected from a respectable married man."

There was a great fear of spies and fifth-columnists. Sysonby wrote:

> "This part of the world is literally honeycombed with spies and Basil Hunt has had a busy day chasing them. The first was a man who was seen by Corporal Simmonds letting off pigeons out of a loft at dawn this morning. After that we have had a

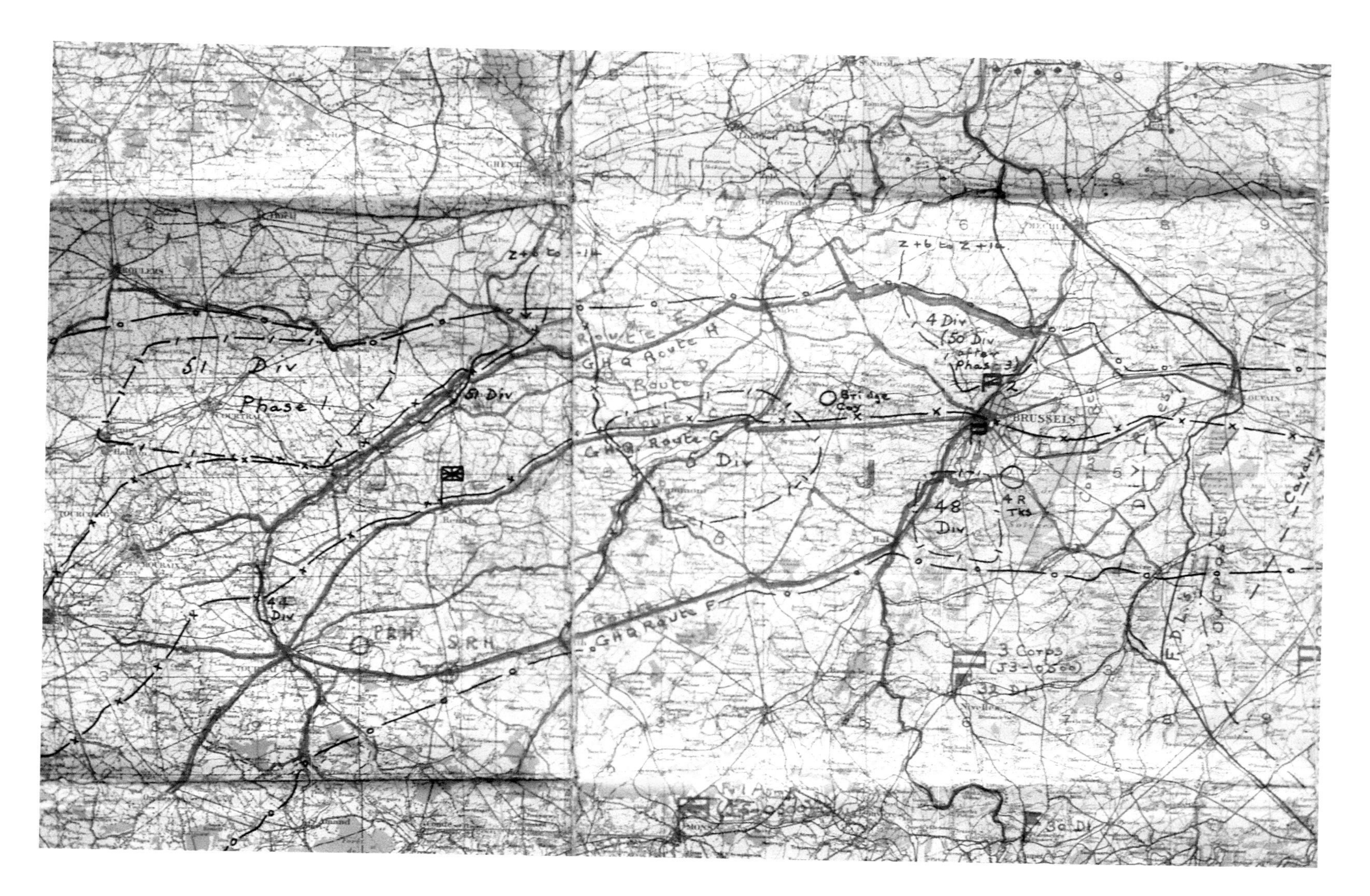

ABOVE *A rare photograh of Lord Gort's HQ map for Plan D, now held in a private collection. The Belgian Army was to fall back on the Dyle and the lower reaches of the Albert Canal to protect Antwerp (Antwerpen), and the French were to hold the Gembloux gap between the Dyle and the Meuse at Namur (Namen) and the Meuse itself, and the British were to defend the upper Dyle. The German armored thrust would leave the plan in tatters.*

series of people cutting the grass or ploughing the fields in extraordinary patterns which are obviously signals to enemy aeroplanes. The air, as usual, is alive with rumours, the best one being that Divisional Headquarters is surrounded by enraged German parachutists. So far we do not know why they are enraged! … Today [April 15th] we had our usual parachute scare and were warned that the Germans might adopt any type of disguise including that of priests so when we saw 80 priests advancing down the road we felt confident we had caught a large contingent of them. We surrounded them and covered them from every possible angle and with every possible weapon; we then followed them into a wood. I myself, revolver in hand, advanced into the wood debating how many of them I would shoot before they shot me. I crept softly in and there came upon them on their knees giving thanks to God for having let them escape from the inferno behind them. I turned and came softly out feeling rather like a murderer must when viewing his unsuspecting victim."

The French First Army under General Blanchard and the Cavalry Corps under General René Prioux were south of the B.E.F., holding a line between Warve and the valley of the Meuse (Maas) river. Today the autoroute takes advantage of the flat, easy, rather dull country alongside the deep river valley and the hills of the western Ardennes. In 1940 the French saw it as the perfect route for the Panzers to sweep forward and the Germans were eager to behave in such a way as to convince the Allies that this was, indeed, their principal axis of attack. The vehicles of the 3rd and 4th Panzer Divisions poured over the Maastricht crossings and gathered themselves for action.

On the afternoon of May 12th, Lieutenant Robert Le Bel, 3rd D.L.M. looked from the turret of his Hotkiss H-39, near Jandrain, to see:

> "... an extraordinary show which was played out about three kilometres [2 miles] away: a panzer division shaping itself for battle. The massive gathering of this armoured armada was an unforgettable sight, the more so that it appeared even more terrifying through the glasses ... Some men, probably officers, walked to and fro gesticulating in front of the tanks. They were probably giving last-minute orders ... Suddenly, as if swept away by a magic wand, they all disappeared ... A dust cloud soon appeared on the skyline, disclosing the enemy move. I got down into the tank, closed the hatch and peered through the episcopes."

The battle continued through the afternoon and into the evening. By nightfall the French had lost 24 Hotchkiss H-39s and four Somua S-35s, and the German losses caused them to remark on the bitter resistance offered by the French, necessitating repeated attacks by VIII Air Corps. The town of Hannut had fallen to the Germans. The next day, late in the morning, the Stukas were present in yet greater force. The French prevented the Germans from taking Merderop and, when the attackers tried to circumvent it, themselves launched an attack on the enemy's supporting infantry. In the close action that followed 4th

BELOW *The British blow the bridge at Louvain, west of Brussels. They withdrew from the town on May 16th as the German 6th Army broke through the Dyle line. By the next day Brussels, Louvain, and Maline had all fallen, and on the next, Antwerp. The regions ceded to Belgium by the Treaty of Versailles (1919) had been re-incorporated into Germany.*

Panzer no doubt enjoyed the advantages of their better radio communications. By the end of the day they were in Ramilles, 6 miles (10 kilometers) west of Hannut and had inflicted the loss of 11 S-35s and four H-39s upon the 2nd Cuirassiers. To their north 3rd Panzer had pushed 1st Cuirassier back beyond Jauche and destroyed 25 of their tanks. Prioux was forced to order a withdrawal to a line east of the Wavre-Namur position.

The French infantry were now well-established, so that when elements of 3rd Panzer slipped through the line in pursuit of 3rd D.L.M., they were comprehensively shot up by the 1st Moroccan Division. The battle resumed on May 15th between Gembloux and Perbaix, on both sides of Ernage, and in confused fighting neither side had the upper hand, tanks and infantry attacking and counter-attacking. But now it was becoming clear to the Allies that the major stroke was falling elsewhere. The crossings of the Meuse to the south of the 1st Army, at Sedan on May 13th and Dinant on May 14th threatened to outflank the Allies' Belgian positions. General Bilotte ordered a retreat, first to a line on Waterloo and Charleroi, and then, on Thursday May 16th, to the river Escault from where they had started only six days earlier.

To the troops of the BEF, who had seen some shelling but little else, the orders came as a terrible shock and to the Belgians it was even worse. Sysonby, who was in reserve on the Escault near Audenarde, wrote in a letter:

> "The day before yesterday [Friday, May 17th] I was told at quarter to six p.m. to start a traffic control post at a crossroads five miles [8 kilometers] away … By the time I got there portions of the Army had started pouring through. I can never describe to you the amazing scenes which took place. The inhabitants of the small village we were in were quite unprepared for this withdrawal and were completely stunned at the news that we were not advancing or even holding our ground … these wretched people had to leave, carrying everything in one suitcase and leaving their life's work and possessions behind them. All that day, all the next day and all last night the traffic never ceased pouring through."

From the Dyle to the Senne, from the Senne to the Dendre, and from the Dendre to the Escault, the British fell back. The bridges were blown as they went. The artillery was in a constant state of redeployment as they, too, withdrew and took up positions to cover their comrades' withdrawal, before moving to the rear again themselves.

By Tuesday, May 21st the British had taken position on the Escault between Audenarde in the north, through Tournai to Maulde, halfway between Tournai and Valenciennes, there to stand against the Germans with the French First Army to their right and the Belgians to the north. Events elsewhere had already undermined this plan, as Lord Gort was becoming only too well aware. Guderian's panzers were already at the Channel coast. In a matter of days, the Allies would be on the beaches of Dunkirk.

RIGHT *British troops had entered Belgium to a wildy enthusiastic reception. By May 20th, the 19th Panzer Korps had completed its advance to the Channel coast by capturing Abbeville and Noyelles, thus dividing the BEF, the French 1st Army, and the Belgian Army from the rest of the French forces to the south of the Somme River. The little boats were beginning to muster at Dunkirk.*

Hitler's Halt Order:

Dunkirk

"At the Führer's orders the attack to the east of Arras with VIII and II Corps, in co-operation with the left wing of Army Group B, is to be continued towards the northwest. On the other hand, forces advancing to the northwest of Arras are not to go beyond the general line Lens-Béthune-Aire-St Omer-Gravelines (Canal Line). On the west wing, all mobile units are to close up and let the enemy throw himself against the above-mentioned favorable defensive line."

The famous halt order, (above) made by telephone on May 24th, 1940, has been the subject of endless speculation; and the key to the puzzle is mapping. As the French and British stumble back across the fields of Flanders and the Panzer divisions push up from the south, it appears that a German victory is assured, but that the Allies still have considerable strength. So why did the Panzers that would surely have driven the Allies into the sea not press on? It has been suggested that Göring pressed Hitler to give the Luftwaffe the honor of destroying the British. The "Golden Bridge" idea, the possibility that Hitler deliberately gave the British the chance to escape, has also been mooted. Finally, it is said that Hitler over-ruled everyone and quite simply failed to appreciate the opportunity before him. None of these explanations holds the answer to the "miracle" of Dunkirk.

The impact of *Frankforce*, the BEF counter-attack at Arras went all the way to the top, Hitler himself. Rundstedt, and Kleist had both been worried about the exposure of their flanks, and Hitler shared their nervousness. The success they had enjoyed so far was almost too good to be true, and the blow delivered by a perceived five divisions on the previous Tuesday confirmed their fears. A French attack towards Cambrai on Wednesday, although turned back, reinforced their apprehensions. Moreover, the Luftwaffe's VIII Air Corps was now admitting to heavy Stuka losses inflicted by England-based R.A.F. fighters. The conclusion of the business in Flanders had not been planned in detail beforehand. Indeed, that they were here so soon was as much of a surprise to the Germans as it was to the Allies, and now the situation was that Army Group B was pressing from the north and east while Army Group A was applying pressure from the south and west on the remaining pocket of Allied troops. To whom should the control of the final phase be given? And which of the groups was best equipped for the task? The panzer-heavy Army Group A or the infantry-rich Army Group B? In the event the infantry of the 4th Army was taken from Runstedt and placed under Bock's command in Army Group B. The concept then became one of infantry, artillery, and air power breaking the remainder of the Allies against the anvil of Army Group A holding a line from Gravelines through St Omer (i.e. along the river Aa) and on by way of Aire and

RIGHT AND INSET *The routes to the English Channel suitable for armored fighting vehicles and mechanized infantry support become evident on the German map produced in February 1939. Rommel's path from Dinant (above square 15) was by Cambrai and left around Arras and up to Lille (5). The hatched areas are below sea level. Hitler is photographed (above) on June 23rd 1940.*

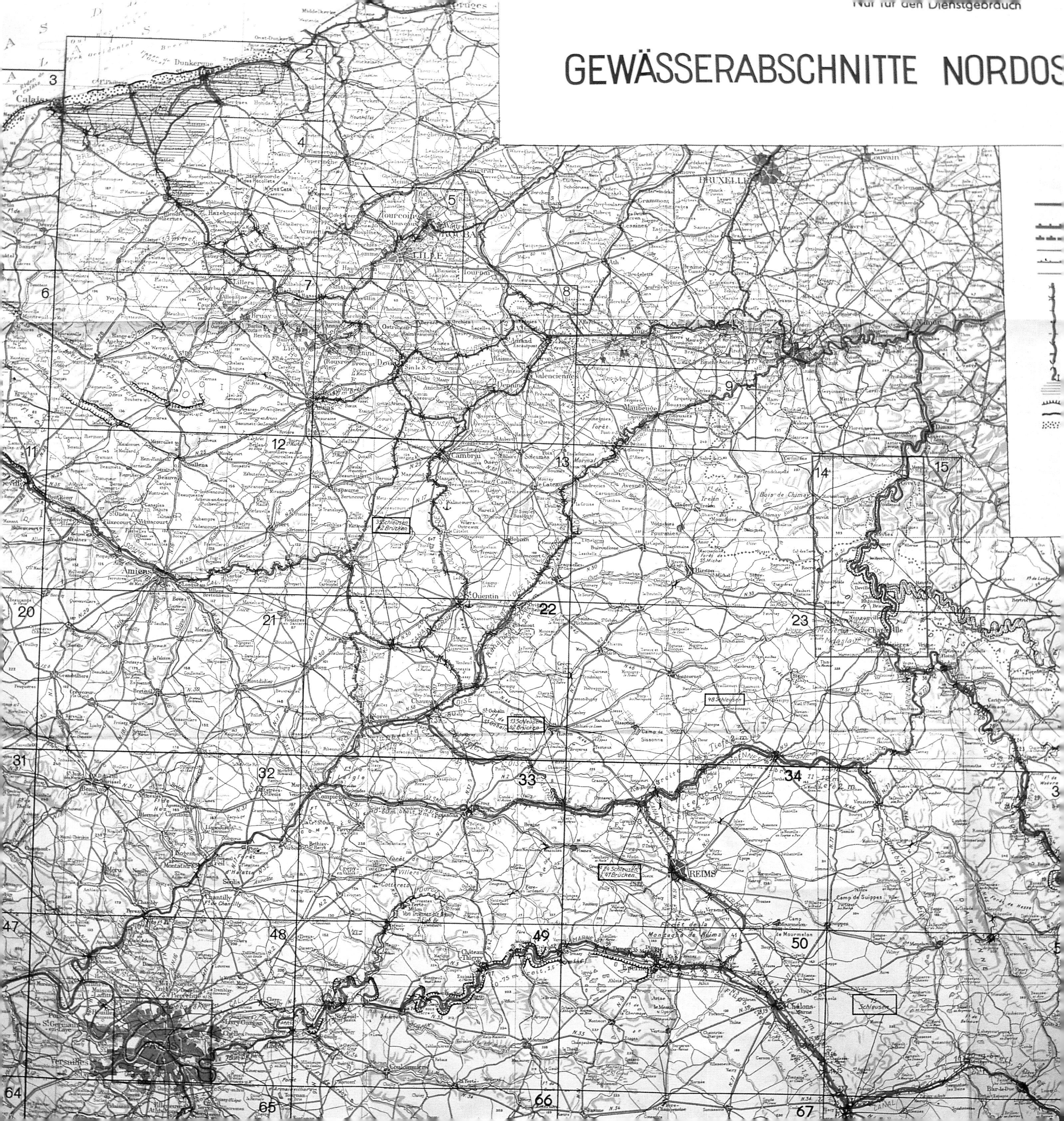
Nur für den Dienstgebrauch
GEWÄSSERABSCHNITTE NORDOS
35 Schleusen 42 Brücken
13 Schleusen 42 Brücken
48 Schleusen
24 Schleusen 41 Brücken
Schleusen
Dunkerque
Calais
BRUXELLES
LILLE
ROUBAIX
Tourcoing
Hazebrouck
Armentières
Béthune
Douai
Arras
Cambrai
Valenciennes
Mons
Maubeuge
Amiens
St Quentin
Laon
Soissons
Compiègne
Noyon
Chauny
REIMS
Epernay
Châlons
Château Thierry
Beauvais
Senlis
Chantilly
Versailles
Mézières
Charleville
Namur
Dinant
Camp de Sissonne
Camp de Suippes
Montagne de Reims
Forêt de Villers-Cotterets
Coulommiers
Bar-le-Duc

Béthune to Lens (the Canal d'Aire or La Bassée line). The nature of the terrain between Calais and Nieuport was well known to the Germans, and their maps and landscape assessments were consistent. The opinions of the generals at the front have been used to suggest that the terrain presented no problems for the panzers, but these generals themselves were to the rear. Actual valuations were being made and the decisions taken by men far away, and they were surely influenced by their official reference manuals. On 29 February 1940 a collection of maps and manuals was published in Berlin for the use of services only. Volume I of *Militärgeographische Beschreibung von Frankreich* states:

> "In wet weather wide areas become boggy and impassable on foot. Vehicles can in general only move on the roads available which are very numerous and mostly fortified. These and the little railways run throughout on dykes; these form with the numerous, in general not very wide, waterways, canals, and ditches a dense mesh of sections suitable for rearguard defense."

The map that accompanies the manual, *Wehrgeolische Übersichtskarte*, comments on the terrain around Dunkirk (see picture, above):

> "Predominant soil type: peat, groundwater near surface. Passability by traffic and on foot: At all times passable with difficulty. Accessible to infantry in dry season. Obstacles: Soft ground, criss-crossed with many ditches, shallow ground water, can be dug out to form water obstacles. Artillery firing positions: wet, ground not able to take a load … Artillery observation opportunites: flat, low lying, without rises in ground."

The land not actually shown as being below sea-level gets a rating only marginally better. This was the official assessment and it was relied upon. To men who had seen the torn landscapes of the previous war the message was obvious and inescapable; this land would be death to tanks with the first drop of rain.

Guderian was shocked. He wrote:

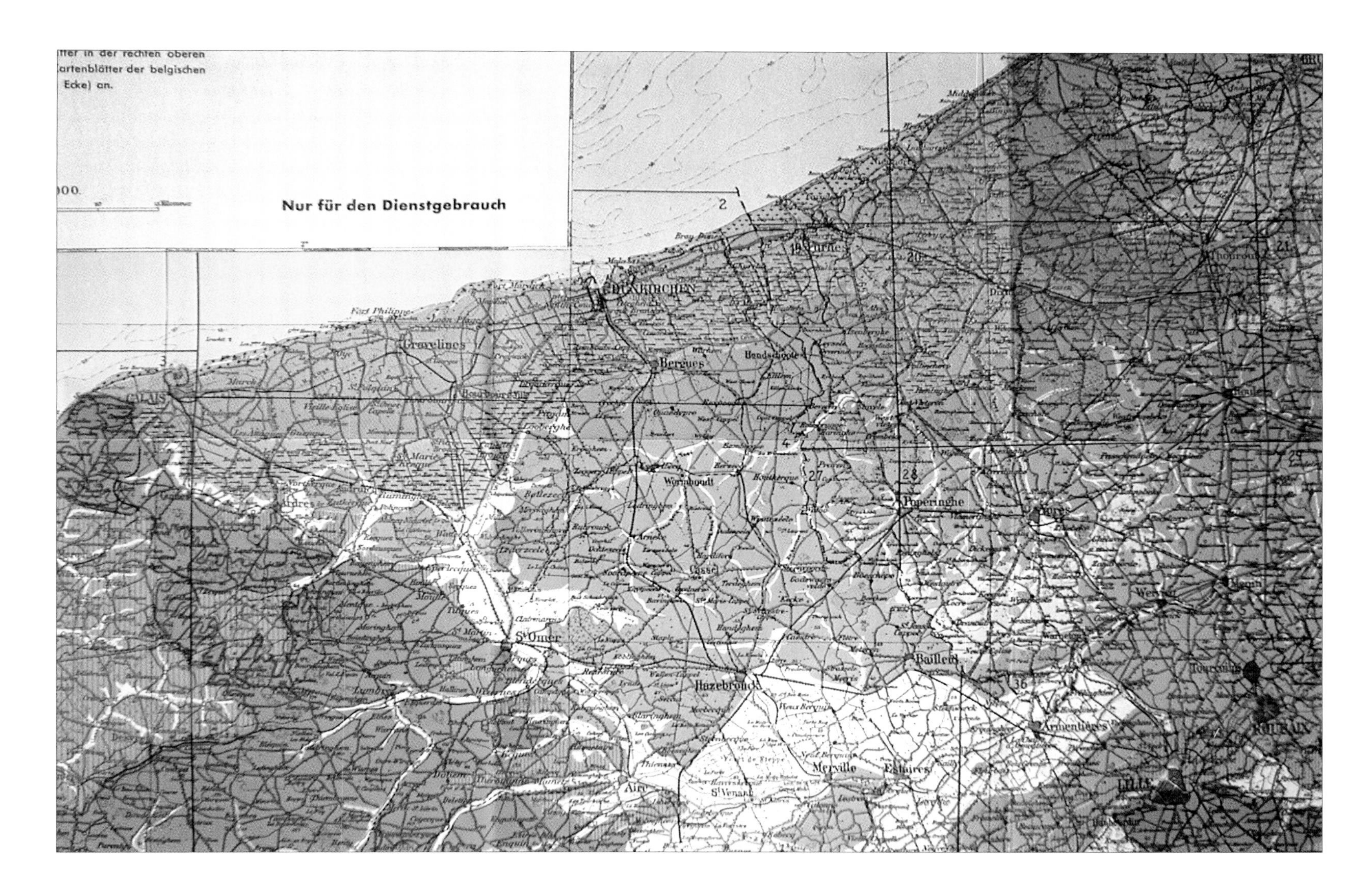

LEFT AND ABOVE *This 1940 German Army geological map is what stopped the panzers. The blue surface is described by geologists as Ypres Clay. Ypres Clay is impermeable, so water does not sink in but tries to flow off: bad tank country.*

> "On this day (the 24th) the Supreme Command intervened in the operations in progress, with results which were to have a most disastrous influence in the whole future course of the war. Hitler ordered the left wing to stop on the Aa. It was forbidden to cross that stream."

He went on to say that the order stated that Dunkirk was to be left to the Luftwaffe, as was Calais if it proved difficult to take.

> "We were utterly speechless. But since we were not informed of the reasons for the order, it was difficult to argue against it. The panzer divisions were therefore instructed: "Hold the line of the canal. Make use of the period of rest for general recuperation."

The SS Division *Leibstandarte Adolf Hitler* had been placed under Guderian's command and he ordered it to advance on Watten, six miles (10 kilometers) north of St Omer on the Aa. Early on the Saturday morning he went to check on them and discovered that the commander Sepp Dietrich had disobeyed the order, crossed the canalized river, and taken position on the top of a hill among the ruins of a monastery. When Dietrich explained that, from the top of that hill, an enemy could look down the throat of his men on the western bank, Guderian agreed the position should be held.

By the time, on Sunday 26 May, Hitler gave the order to proceed once more, Boulogne had fallen, 2nd Panzer having finally breached the ancient walls with an 88mm flak gun. The 20th Guards Brigade had been ordered to evacuate the town by sea at 6.30 p.m. on Thursday 23 May and proceeded to do so without liaising with the French, who had held on for another two days. *Operation Dynamo*, the evacuation of Dunkirk, was made possible not by an error, but by "anti" tank terrain. It seems that Hitler was right to withold the panzers.

A False Assumption:

Defenses of Dover

The ease with which the Germans had defeated France and ejected the British from mainland Europe in 1940 surprised them. No detailed plans existed to pursue the war by invading England, but preparations were soon put in hand. The pressing need was to secure a port for re-supply of the conquering armies, in the same way as the victorious campaign in Norway had been sustained. The port of Dover was an obvious target.

The information available was impressively detailed. Aerial photographs, holiday snapshots, old postcards, and British Ordnance Survey maps purchased quite openly before hostilities began contributed to the enterprise of creating maps suitable for the attacking forces. They gave information on the positions and types of defensive weapons, radio masts, and areas rendered unsuitable for landing aircraft. The British had raided the junk yards and wrecked automobiles, trucks, and buses littered the open spaces.

Dover harbor was to be captured by the German 16th Army as a priority as soon as the invasion, "Operation Sealion," was launched. A special combat formation called *Hoffmeister* was formed from 7th Airborne and 17th and 35th Infantry divisions. It was to be given the support of the specialist assault unit, 1 Company, I Brandenburg, and Underwater Tank Detachment D, landing on the coast to the west. This would consist of two platoons commanded by Oberleutnant Dr. Hartmann and would total two officers, fifteen NCOs, and 114 men with 50 motorcycles.

A direct assault on Dover was also planned to take place under Hauptmann Hollmann commanding the greater part of 4 Company. Their task was to prevent blockships being sunk in the harbor and then to neutralize the gun batteries on the cliffs. The original idea was to land them from gliders, but the study of the defenses shown by cross-hatching on the map (above right) showed that everything that was not a steep hillside was blocked. The bulk of the invasion force was to approach the English coast in barges towed by fishing boats, but they would be slow and vulnerable to the artillery mounted with such a good view on the white cliffs.

What alternative method was to be used is not clear. The scholar Dr. Peter Schenk has found a possible answer in a secret order to Korvettenkapitän Strempel, Head of the Motorboat Section of the German Navy. He was instructed to pick 25 of the fastest vessels from police and customs units and assemble them at Dordrecht in order to carry a commando force in connection with "Sealion." Strempel was never told what the target was, and the invasion never took place.

A vital gap in German intelligence information emerges from the defense maps. On February 26th 1935 a team led by Robert Watson-Watt carried out an experiment. A Morris car towing a caravan drew into a field west off the A5 road, Watling Street, south of Weedon on the road to Lichborough, (close to the point at which a small memorial stands today). A Handley Page Heyford biplane bomber from Farnborough then flew up and down a pre-determined course and the signals broadcast from the BBC's 10-kilowatt transmitter at Daventry were reflected from it and observed on a cathode ray oscillograph in the van. Radar had been demonstrated for the first time. By August 1940, 51 fixed radar stations had been built around the British coast. Chain Home stations, of which there were 21, could detect aircraft at a range of up to 120 miles (193 kilometers) and the coastal defense system, Chain Home Low, could spot low-flying aircraft 50 miles (80 kilometers) away.

The radar stations appear on the German maps, but they are called *Funkstation*, radio stations. That they were radar installations was not even suspected. Victory in the Battle of Britain would depend on these radar stations and the men and women who served in them.

RIGHT Befestifungskarte Grossbritannien, scale 1:50,000 *(above right); the cross-hatched areas on the map represent blocked areas, which would make an airborne invasion, if not impossible, then extremely costly. The* stellungskarte *(scale 1:25,000), right, indicates* funkstaton *(radio station) for radar stations.*

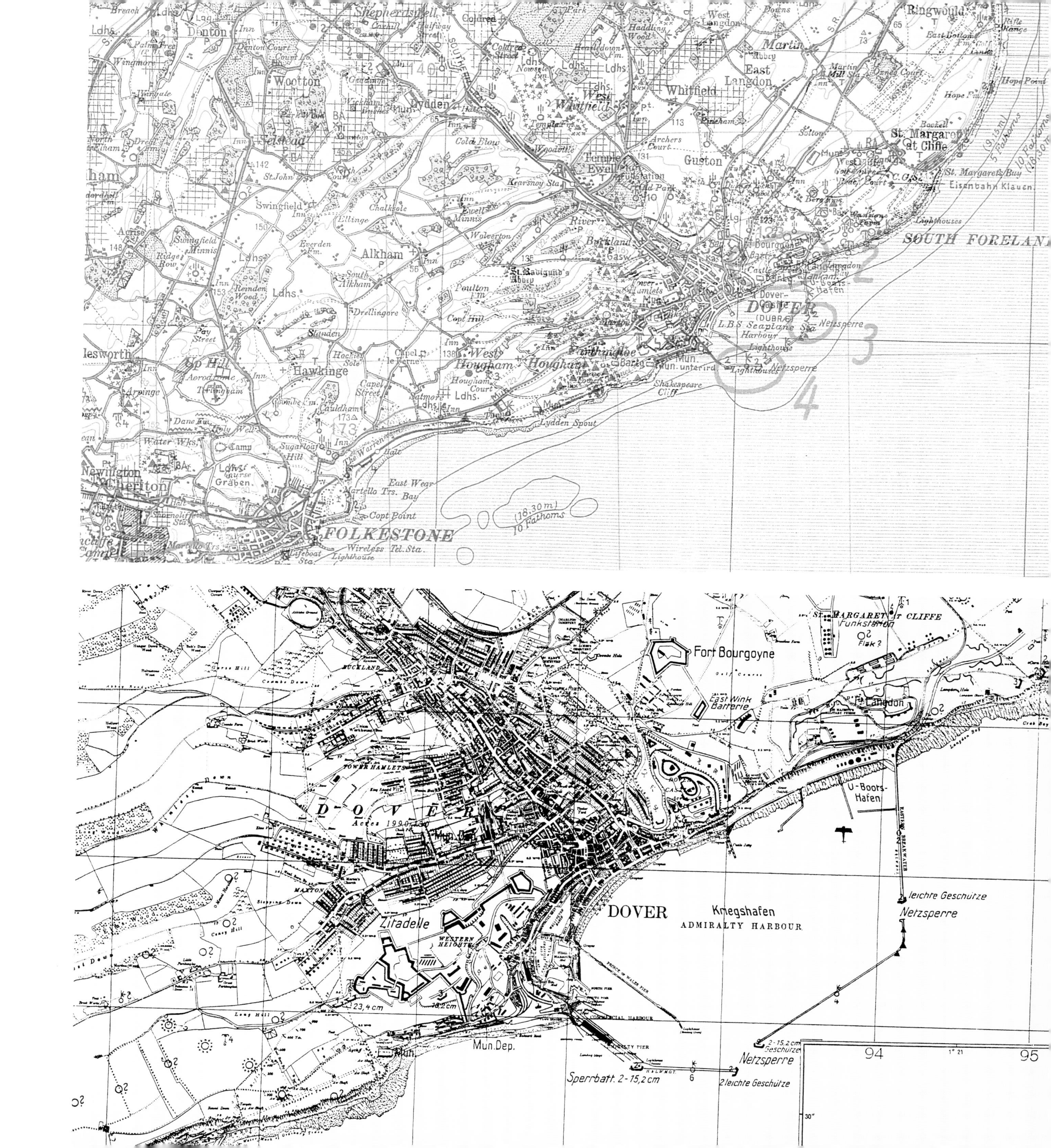

SOUTH FORELAND
DOVER
L.B.S. Seaplane Sta.
Harbour
Netzsperre
U-Boots-Hafen
FOLKESTONE
Shakespeare Cliff
Lydden Spout
Hawkinge
Alkham
Guston
Whitfield
St. Margaret's at Cliffe
St. Margaret's Bay
Eisenbahn Klauen
Fort Bourgoyne
East Wink Batterie
Funkstation
Flak ?
Langdon
U-Boots-Hafen
D O V E R
Zitadelle
DOVER
Kriegshafen
ADMIRALTY HARBOUR
leichte Geschütze
Netzsperre
23,4 cm
18,2 cm
Mun. Dep.
Sperrbatt. 2-15,2 cm
2 leichte Geschütze
2-15,2 cm Geschütze
Netzsperre
94
95

D-Day on Dog:

Omaha Beach

On June 6th 1944 the Allies invaded Normandy. The armada that brought them was huge; never before had so vast a fleet been assembled and never had an opposed landing on enemy territory been attempted on this scale. The experience of the men assigned to different points of attack were as different as they possibly could be. Some troops walked off the beaches unopposed. For those on Omaha, it was different.

The landing beaches were allocated to American and British commands, the British in the east and the Americans in the west where their landing beaches were code-named Utah, on the south of the Cotentin Peninsula, and Omaha, north-west of Bayeaux. The work of planning the invasion had begun as long ago as October 1941 and two years later the detailed military planning started under the command of Lieutenant General Frederick Morgan. By then a mass of local information from numerous sources had been collected. The French resistance sent information, clandestine landings yielded data, and overflying aircraft secured photographs. The maps prepared showed the defense works built under the direction of Field Marshal Erwin Rommel. There were welded steel hedgehogs scattered as if a giant had being playing at jacks, posts capped with anti-tank mines and various other devices to bar the way of landing craft. On the cliffs the trenches and barbed wire are shown and the gun emplacements are marked. The document must have inspired some confidence, but the hostility of the terrain is evident. High cliffs with a few narrow exits (called "draws'"by the US forces), such as the one west of Vierville-sur-Mer, face the narrow beach. At the western extreme the cliffs are washed by the sea, leaving no beach at all. Landing here would depend on the success of the naval shellfire and the use of supporting arms such as the D-D tanks, amphibious machines, and the DUKW-carried field artillery.

The western end of Omaha Beach was allocated to the US 29th Division under Major General Charles H Gerhardt. The area was given the code-name "Dog." On their left, on "Easy," the US 1st Division, the Big Red One, would be alongside them.

At 6.30 a.m. on June 6th the 16th Infantry regiment approached Dog Sector and stepped into a torrent of gunfire. They had already been in their landing craft for more than two hours and although the storm of the previous couple of days had abated, the sea was rough. The D-D tanks had been launched at 5.40 a.m. 6,000 yards (5,490m) out to sea. The waves were higher than the trivial eighteen inches (46 centimeters) freeboard of the canvas sides and of the 32 machines only five made it to shore, the rest taking their crews with them to the bottom of the sea. The field guns rendered the amphibious trucks, the DUKWs, top-heavy and they capsized, spilling their precious cargo into the water. When the Americans hit the beach they lacked vital supporting arms.

What was more, the naval bombardment was not doing its job, for the Germans had built bunkers into the cliffs facing east, along the beach, as well as on the heights above the strand. The cliff-based guns were proof against shelling from the sea and laid down a deadly fire on the soldiers struggling to gain the beach. The photographer Robert Capa was with the 1st Division landing on Easy Red. He recalled dropping into the water a hundred yards off the beach.

> "The bullets tore holes in the water around me, and I made for the nearest obstacle. A soldier got there at the same time, and for a few minutes we shared its cover. He took the waterproofing off his rifle and began to shoot without much aiming. The sound of his rifle gave him the courage to move forward, and he left the obstacle to me. It was a foot larger now, and I felt safe enough to take pictures of the other guys hiding just like I was."

RIGHT *Omaha Beach, east and west. The legends include a symbol for "cliffs 40-90°" as well as those for pillboxes, gun emplacements, "hedgehogs," and land mines, based on aerial photographs up to May 22nd 1944. The deadly guns in bunkers in the east-facing cliffs overlooking "Dog" are not marked. Coxwains are advised to rely on the panoramic sketch for visual reference, rather than the map.*

VIERVILLE-SUR-MER
Gruchy
le Haut Chemin
Hamel au Prêtre
BEACH OBSTACLES AND DEFENSES
BASED ON AERIAL PHOTOS UP 22 MAY 1944
RANDOM STAKES OCCUR THROUGHOUT
Pointe et Raz de la Percée
B A I E D E L A S E I N E
DOG RED BEACH
DOG WHITE BEACH
DOG GREEN BEACH
CHARLIE BEACH
Revised 15 June, Ground information
OMAHA BEACH-WEST (Vierville-sur-Mer)
NOTE to COXSWAIN or NAVIGATOR
LEGEND
6047
TOP SECRET - BIGOT

COLLEVILLE-SUR-MER
Cabourg
Les Moulins
BEACH OBSTACLES AND DEFENSES
BASED ON AERIAL PHOTOS UP TO 22 MAY 1944
RANDOM STAKES OCCUR THROUGHOUT
B A I E D E L A S E I N E
FOX RED BEACH
FOX WHITE BEACH
FOX GREEN BEACH
EASY RED BEACH
EASY WHITE BEACH
EASY GREEN BEACH
DOG RED BEACH
Revised 15 June, Ground information
OMAHA BEACH-EAST (Colleville-sur-Mer)
NOTE to COXSWAIN or NAVIGATOR
LEGEND

1ST U.S. DIV.
1 COY. OF RANGERS
29TH U.S. DIV.
116 RCT.

ABOVE *US Rangers rest at the foot of the cliffs at Pointe du Hoc. Amazingly, they scaled the cliffs—boobytrapped with 240mm shells hooked at intervals on trip-wires—with the loss of only 40 casualties, aided by a terrific naval bombardment.*

ABOVE LEFT *High level reconnaissance photograph of Vierville Sur Mer, indicating the landing areas for the 1st and 29th US Divisions on Omaha.*

LEFT *Omaha Beach from an unmapped German artillery position concealed in the cliff. The consequences of such a field of fire are obvious. The author Ernest Hemingway described how "The first, second, third, fourth, and fifth waves lay where they had fallen, looking like so many heavily-laden bundles on the flat pebbly stretch between the sea and the first cover."*

A company of US Rangers which had strayed from its intended landing at the Pointe du Hoc to the west managed to establish itself under the cliffs on the right. More landing craft poured in, more men floundered dying in the water but even more made it to the beach. The Rangers took out German bunkers one after another and as they day went on German resistance faltered. Their ammunition ran low. No reinforcements came to their aid. By evening the Americans had gained the cliff tops. It had cost them more than 3,000 killed, wounded or missing. If the map had told the whole truth would the outcome have been different? At 0900, Lieutenant General Omar Bradley had asked permission to abandon the beachhead. Eisenhower did not get the message until late in the day, by which time the situation had changed.

Andy Warhol:

Cold War Subversion

It is 1986; in the US, Republican President Ronald Reagan has just embarked on his second term of office, while in the USSR, the new Soviet leader Mikhail Gorbachev is beginning to institute his reformist policies of glasnost and perestroika. Yet for all the incipient change in world affairs, the Cold War is still in a dangerous state of flux. It is in this political climate that New York Pop artist Andy Warhol (1926-87) paints a picture of a map showing Soviet missile bases. What is its purpose, and what (if anything) does it say about its times?

Warhol's enigmatically simple USSR map (acrylic and serigraph on canvas; Andy Warhol Museum, Pittsburgh) turns the conventional relationship between cartography and art on its head. Maps such as those by Mercator, or the Blaeus—for all the craftsmanship their makers invested in them—were intended, first and foremost, as functional tools, and have become ever more highly prized as works of art over time. By contrast, Warhol's map was created from the outset as a work of art.

Does it have a utilitarian purpose? What hard-and-fast information (one defining characteristic of a map) does it convey? On a rough outline map covering the Soviet Union east of the Urals, northern China, and Japan, the sites of nuclear missile bases are marked. Some, but not all of them, are labeled either "SS4," "SS11," "SS17," or "SS20"—designations of different types of missiles deployed by the Soviets. Few geographical labels are supplied, the only other guide being a key distinguishing two categories of weapon, couched in the "alphabet-soup" acronyms favored by military strategists: ICBM (Inter-Continental Ballistic Missile), IRBM/MRBM (Intermediate-range Ballistic Missile/ Medium-Range Ballistic Missile), plus SLBM (Submarine-launched Ballistic Missile) on the Kamchatka Peninsula.

Even at a glance, however, the information value of Warhol's map becomes suspect. First, the key turns out to be useless; so crudely are the types of weapon executed that we can hardly differentiate them, let alone tally them with the equally sketchily drawn missile sites on the map. No color coding comes to our aid—the map is relentlessly monochromatic. Second, the very nature of nuclear strategy precludes missile bases being identified and pinpointed in this way—at least by the general public, or artists for that matter. Satellite surveillance would tell the war strategists of each superpower where the enemy's major missile silo complexes were located, but they would not disseminate this information, neither in a secretive and autocratic society like the Soviet Union (which Reagan, three years before, had dubbed the "Evil Empire") nor in the US, unless there was a strategic advantage in doing so. Finally, a key point about the 1980s arms race was that SS20s, along with their US counterparts Cruise and Pershing, were mounted on mobile launchers to evade detection. They had no fixed location. In this context of vanishing factuality, the bona fide city names (e.g. Olovyannaya, Svobodnyy) become as exotic as those of ancient towns on the Silk Route.

What we are faced with is a deliberate act of subversion by the artist. Its focus could simply be the traditional artistry and usefulness of maps; alternatively, it might be directed at military planners' pretensions to sovereign knowledge about the whereabouts of their opponents' WMDs. Yet Warhol's stock-in-trade was not political comment or protest, but instead (even from his early "repetitive" silkscreen prints like his 1965 "Atomic Bomb" onward) the ephemeral nature of information and the banality of consumer culture. It is tempting to see his target here, then, as being the type of instant graphics so beloved of the editors of newspapers and current affairs magazines in times of conflict, purporting to show weapons deployments and possible war tactics. Like Warhol's map, these exude an air of authority that turns out, in fact, to be a will o' the wisp, a chimera. Warhol's map is scornfully ironic and playful—an "anti-map" offering specious data for a "soundbite" mass media age.

RIGHT *Warhol's stripped-down map of the USSR, with its sketchy and spurious data, deliberately plays a game of "blind man's buff" with the viewer—in much the same way that nuclear strategists of the two superpowers constantly strived to make their ballistic missiles elusive to the other side!*

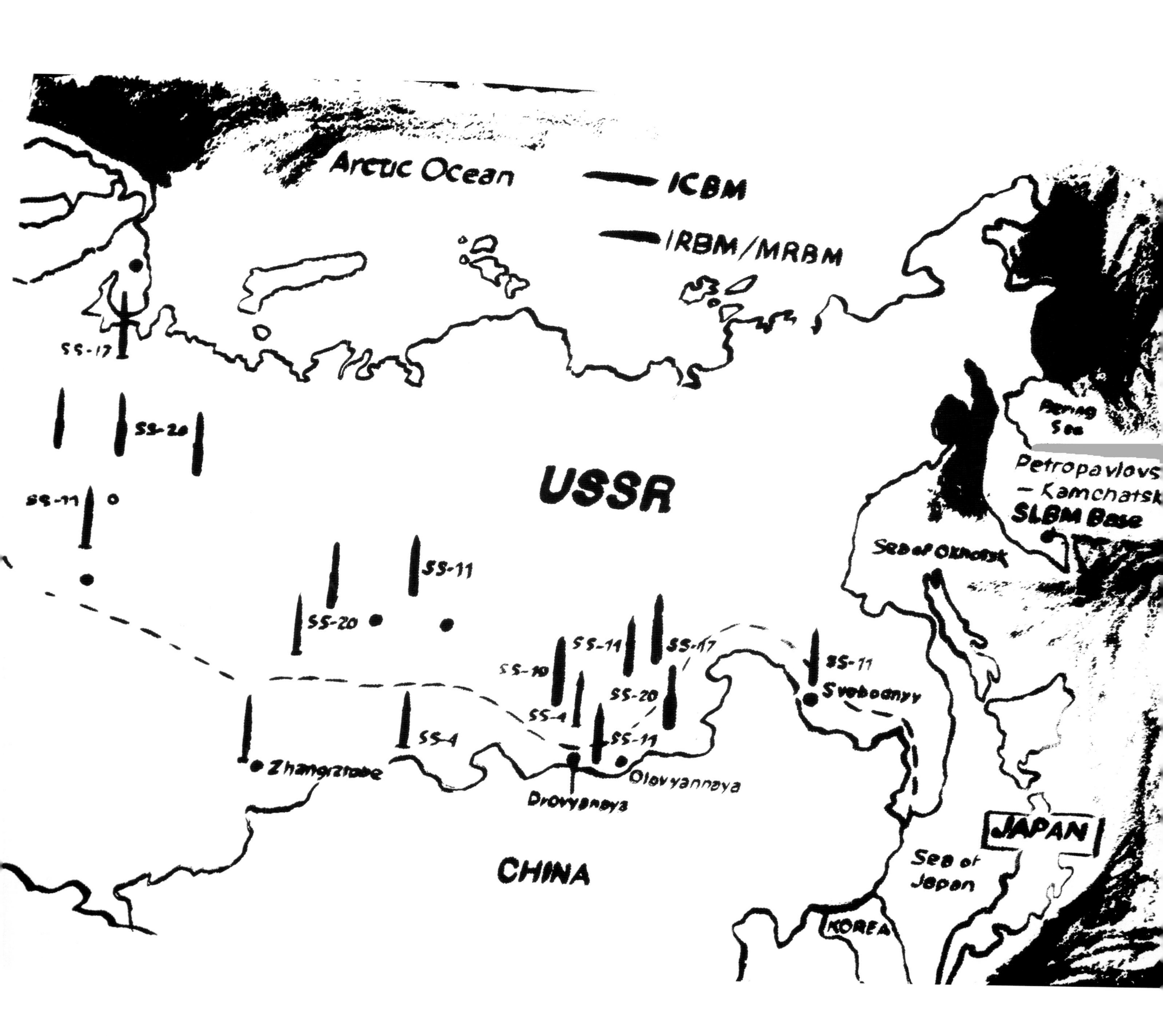
Arctic Ocean
ICBM
IRBM/MRBM
SS-17
SS-20
SS-11
USSR
Petropavlovsk
– Kamchatsk
SLBM Base
SS-11
SS-20
SS-11
SS-17
SS-20
SS-4
SS-4
SS-11
SS-11
Svobodnyy
Zhangiztobe
Olovyannaya
Drovyanaya
CHINA
JAPAN
Sea of
Japan
KOREA

MAP
of the
WEST COAST OF AFRICA,
from
SIERRA LEONE to CAPE PALMAS;
including the Colony
OF
LIBERIA:
Compiled chiefly
from the
SURVEYS and OBSERVATIONS
OF THE
LATE REVᴰ J. ASHMUN.
J.H. Young Sc.
Published by A. Finley Philadᵃ
1830
REMARKS.
The Colony of Liberia extends from Gallinas river to the Territory of Kroo Settra, a distance of about 280 miles in length, along the Coast,* & from 20 to 30 miles inland, (in some places much more,) it includes within its Jurisdiction, the territories of several native tribes, the names of which are as follows; the Feys or Veys occupy the country from Gallinas R. to Little Cape Mount, a distance of about 50 miles along the coast, and 25 to 30 miles inland; they are an active warlike and proud people. Population 12,000 to 15,000. The Dey Tribe extends from Little Cape Mount to Mesurado river about 30 miles in length, and 12 to 16 miles inland; an indolent & inoffensive people. Pop. 6000 to 8,000
The Bassa Tribes extend from Mesurado River, southward; they are generally domestic, industrious and averse to war; and supposed to be in numbers about 125,000 souls. The country abounds in rice, oil and cattle, and rivals in fertility any part of the African coast.
At a distance of from 30 to 50 miles inland, a belt of dense, and almost impassable forest occurs along the whole of this coast, of from one to two days journey in breadth, which nearly prevents all intercourse between the maritime and interior tribes, and is one of the principal causes why the inland parts of this section of Africa are so entirely unknown to the civilised world.
* The territory, at present under the actual jurisdiction of the Colony, extends from Grand Cape Mount, to Trade Town, a distance of about 150 miles.
Longitude East from Washington
Port Logo
Timanees
Rokelle or Mitambo R.
Mayosa
Maburg
Rochetkik
Kamaranka R.
Leopards I.
Sierra Leone R.
Cape Sierra Leone
FREE TOWN
COLONY OF SIERRA LEONE
Boolams
False Cape
Kates R.
Cape Shilling
Turtle Rocks
Banana Isles
Kamaranka R.
Oyster Banks
BOLM
Cockboro R.
Tassa Pᵗ
Yaltooka R.
Bengal Rocks
Plantain Isles
Mindo Trees
Shebar River
Country of the Kittims
Turtle or Bashaw Iᵈ
Jenkin I.
Gallinas R.
Bendo
Condima
Juning
Femne
Madima
Soulima
Sherbro River
C. Mount Landing
Mountain from 1000 to 1000 Fᵗ above the Sea
Grand Cape Mount
Gambia
Pissou R.
The Pissou River has been traced 100 miles from the sea, and affords a safe navigation of 1½ days sail.
Country of the Feys or Veys
LIBERIA
Little C. Mount
Poor R.
Country of the Deys
Sᵗ Pauls R.
The Sᵗ Pauls River is supposed to have a course of from 250 to 300 miles.
Millsburg
Caldwell
Stockton
Mesurado Terʸ
CAPE MESURADO
MONROVIA
Mesurado R.
False Cape
Junk River District
Junk Territory
Colonial Factory
Saddle Land
Little Bassa
Bob Grays Colonial Factory
Factory I.
Sᵗ Johns R.
Bullock Town
Colonial Factory
Grand Bassa
Bassa Cove
Tabocannee Terʸ
Proposed Town
Colonial Factory
Poor R.
Young Sesters Terʸ
K. Wests old domʳ
Trade Town
Grand Colo
Teembo
The Kings Vill
Manna
R. Sesters or Cestos
Rock Sesters
ATLANTIC OCEAN
WINDWARD COAST
NEGRO
Country of the GRAIN
Scale of Miles.
5 10 20 30 40 50 60 70 80 90 100

Drawing The Line

"Map of the West Coast of Africa from Sierra Leone to Cape Palmas, including the colony of Liberia; compiled chiefly from the surveys and observations of the late Rev. J. Ashmun," published in Philadelphia in 1830. The local tribes are described in the "Remarks"—"proud and warlike," "indolent and inoffensive." In 1816, a group of white Americans founded the American Colonization Society to deal with the "problem" of the growing number of free blacks by resettling them in Africa. Land was bought—probably through coercion—from local tribes in west Africa, a line was drawn; and in 1847, the free state of Liberia declared independence. The US did not recognize the new nation that it had itself created until 1862, for fear of the impact on the domestic issue of slavery.

A Very Incorrect Map:

The Anglo-French Map Wars

During the 17th and 18th centuries, maps became the primary political weapons in a series of "Map Wars" between France and England. Instead of relying on violence in the dispute of territory, each country's politicians turned to their cartographers and commissioned maps that would carve up North America in their favor, marginalizing their enemy's lands and pushing their own borders outward.

Despite the politically motivated inaccuracies that characterized these portrayals of America, the period nevertheless saw the production of the most geographically accurate maps of North America yet produced. The Frenchman Guillaume Delisle was the first truly outstanding mapmaker of the period. His 1718 map of Louisiana and the Mississippi was unsurpassed at the time of its production in its quality, accuracy, and topographical detail, but it undeniably compressed the English colonies along the Eastern seaboard and exaggerated French claims to land beyond the Appalachians.

English mapmaker Henry Popple was engaged by the Lord Commissioners to produce an alternative map to challenge Delisle's, and his *America Septentrionalis*, a giant 20-sheet map, was the first large-scale printed map of North America. Printed in 1733, Popple's map proved just as controversial as Delisle's before it. George Washington, himself a mapmaker, owned a copy of it; Benjamin Franklin is on record as having ordered two copies for the Pennsylvania Assembly; and John Adams, second American President, declared the map "the largest I ever saw, and the most distinct." Yet the English Lord Commissioners were not pleased by the map and disowned it, finding its territorial claims insufficiently aggressive.

Perhaps the most extreme map drawn during the Wars was that of William Herbert and Robert Sayer, issued by a "Society of Anti-Gallicans" in 1755, and shamelessly laying claim to swathes of French-owned territory. The disputes were ultimately resolved with the help of another 1755 map, produced by John Mitchell. This more universally acceptable map was used to settle English boundary disputes with France in 1763 and with North America in 1783, and later served to set the boundaries between the newly formed United States and her neighbors during the Treaty of Paris.

Consider the genesis of each of the maps reproduced here in turn.

Guillaume Delisle, himself the son of a cartographer, produced some of the most outstanding and accurate maps of his day. He was prolific, producing more than 100 maps, and was consulted by scientists, politicians, and kings. His admirers included Peter the Great and Louis XIV, and he was finally named "Premier Geographe du Roi"—Chief Geographer to the King.

Delisle's "Carte de la Louisiane et du Cours du Mississippi: Dressée sur un grand nombre de Memoires" (Map of Louisiana and the Course of the Mississippi: Drawn from a large number of Accounts) brought a number of "firsts" to the discipline of American mapping. In producing it, Delisle went back to the drawing board: while his contemporaries produced maps that were largely derivative of one another, trusting existing work to be accurate and perpetuating old errors and misconceptions, he sought out fresh, documented evidence. Using reports from fur traders, explorers, and priests, Delisle set out to discover the true lie of the North American land. Where no evidence was available, Delisle simply left the area blank—a significant departure from the norm, whereby unknown areas were usually filled with fabricated and hypothetical geographical features.

Delisle's was one of the first maps to focus on the interior of the American continent, and the first to represent accurately the Ohio, Missouri, and Mississippi Valleys. It was the first map in which California did not appear as an island, and, correcting the longitudes of North America, Delisle introduced many previously unrecorded place names including that of "Texas." The routes taken by early explorers are also shown in some detail on the 1718 map.

The remarkable topographical and geographical accuracy of Delisle's work made it the template for 50 years of American mapping,

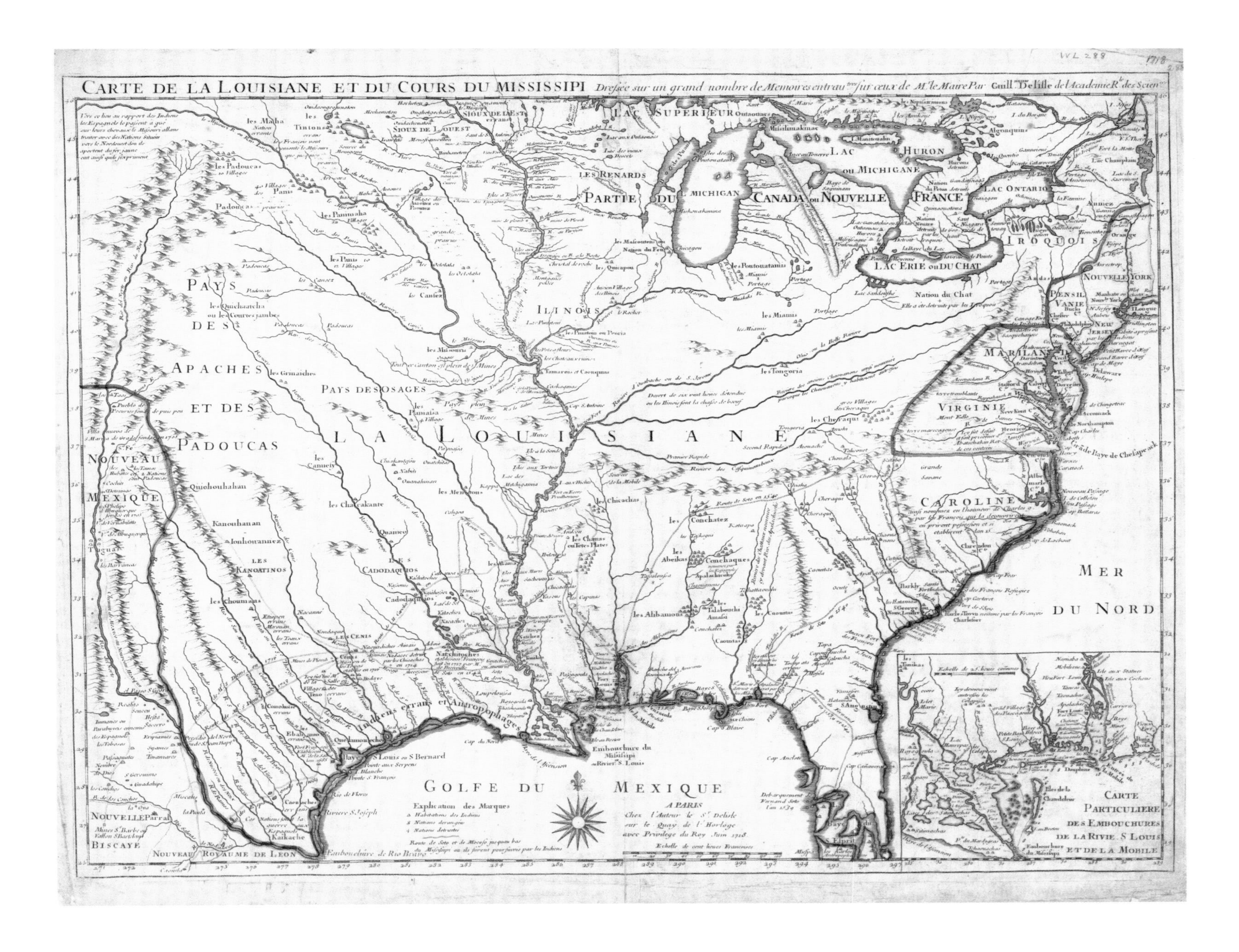

ABOVE Carte de la Louisiane et du Cours Mississippi: Dressée sur un grand nombre de memoires *(1718, Delisle). This map, drawing on the records of French fur traders, priests and explorers, influenced European notions of American geography for over 50 years. As a young man, Delisle studied maps and globes with his father, as well as studying astronomy with Jean-Dominique Cassini. His expert understanding of latitude and longitude was owed in part to Cassini's astronomical teachings.*

including the map produced by the Dutch emigré to England, Hermann Moll, "A New Map of the North Parts of America Claimed by France Under Ye Names of Louisiana, Mississippi, Canada and New France" of 1720 (see page 54). The content of Delisle's map, however, remained contentious, as he deliberately reduced England's Eastern colonies and failed to acknowledge Spanish claims above the Rio Grande.

The "Map of the British Empire in North America" produced by Henry Popple in 1733 represented the first serious attempt to survey America following the publication of Delisle's work in 1718. Popple worked for the English Board of Trade and Plantations in America, where he was involved in the resolution of boundary disputes such as the one that erupted between New Hampshire and Massachusetts Bay. A keen awareness of the degree to which boundaries were indeed disputable, and the difficulty involved in fairly resolving such disputes, brought home to Popple the political significance of accurate—or strategically inaccurate—mapping.

RIGHT *A section of "America Septentrionalis: A Map of the British Empire in America with the French and Spanish Settlements Adjacent Thereto" (1733, Henry Popple). Henry Popple's highly influential map was the first large-scale map of North America ever produced, and was one of the largest maps drawn in the 18th century. The work is made up of twenty regional maps, each appearing on a separate sheet, plus one index map that shows the whole continent including the Caribbean. Joined up, the map measures 94.5 x 92 inches (240 x 234 cm)—a truly impressive creation. (When the keepers of the UK Public Records Office were asked to photograph their copy specifically for this book, they were aghast at the technical difficulty of doing so). It was originally available as a single, bound work, or in loose sheets, as we see from Benjamin Franklin's order of "two setts of Popple's Mapps of N. America one bound the other in Sheets" for the Pennsylvania Assembly in 1746, and later "another of Popple's Maps of North America, large, on Rollers" (1752). This map was the fruit of several year's work for Henry Popple, yet despite its influential nature and remarkable accuracy, "America Septentrionalis" was first rejected, and later disowned by the Lord Commissioners for Trade and Plantations who sponsored its creation. A list of the map's alleged defects was published in* The Gentleman's Magazine *in 1746, as the British government judged it insufficiently aggressive in asserting Britain's North American territorial claims.*

Popple's first map of North America, "A Map of the ENGLISH and FRENCH Possessions on the Continent of North America," was drawn up in the year of his appointment to the Board of Trade and Plantations. The existence of this earlier map allows us to see how Popple's representation of North American geography was changing, as, by the time that the 1733 map was published, his representation of rivers north of the Missouri and Mississippi had changed significantly. Both works, however, were remarkable in their detail: rivers, Indian names, forts, and topographical features were shown more clearly than ever before.

In creating his "Map of the British Empire in North America" Popple drew heavily on the first-hand information that he had accumulated during his time in America—like Delisle before him, he was not content merely to copy the work of earlier mapmakers. The boundaries illustrated on his large-scale map shifted significantly as he continued to work on it: for instance, a 1727 manuscript shows New Hampshire's borders reaching the Wenipisiocho Lake and stretching to the Merimack River in the south, yet the published engraving places Wenipisiocho Lake in the middle of the Province of Maine, under Massachusetts jurisdiction.

It was generally accepted that 18th century mapmakers in America had two choices: to map accurately, or to map politically.

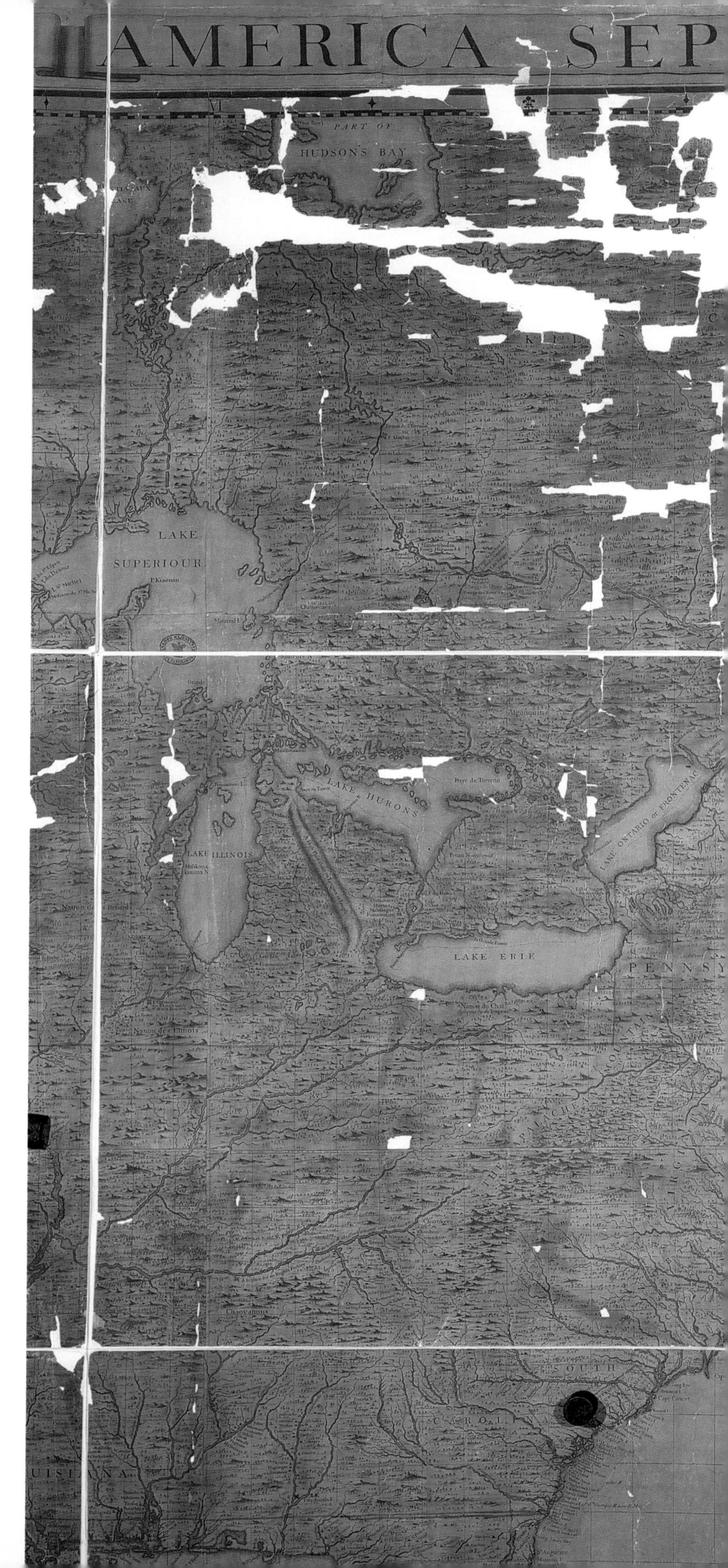

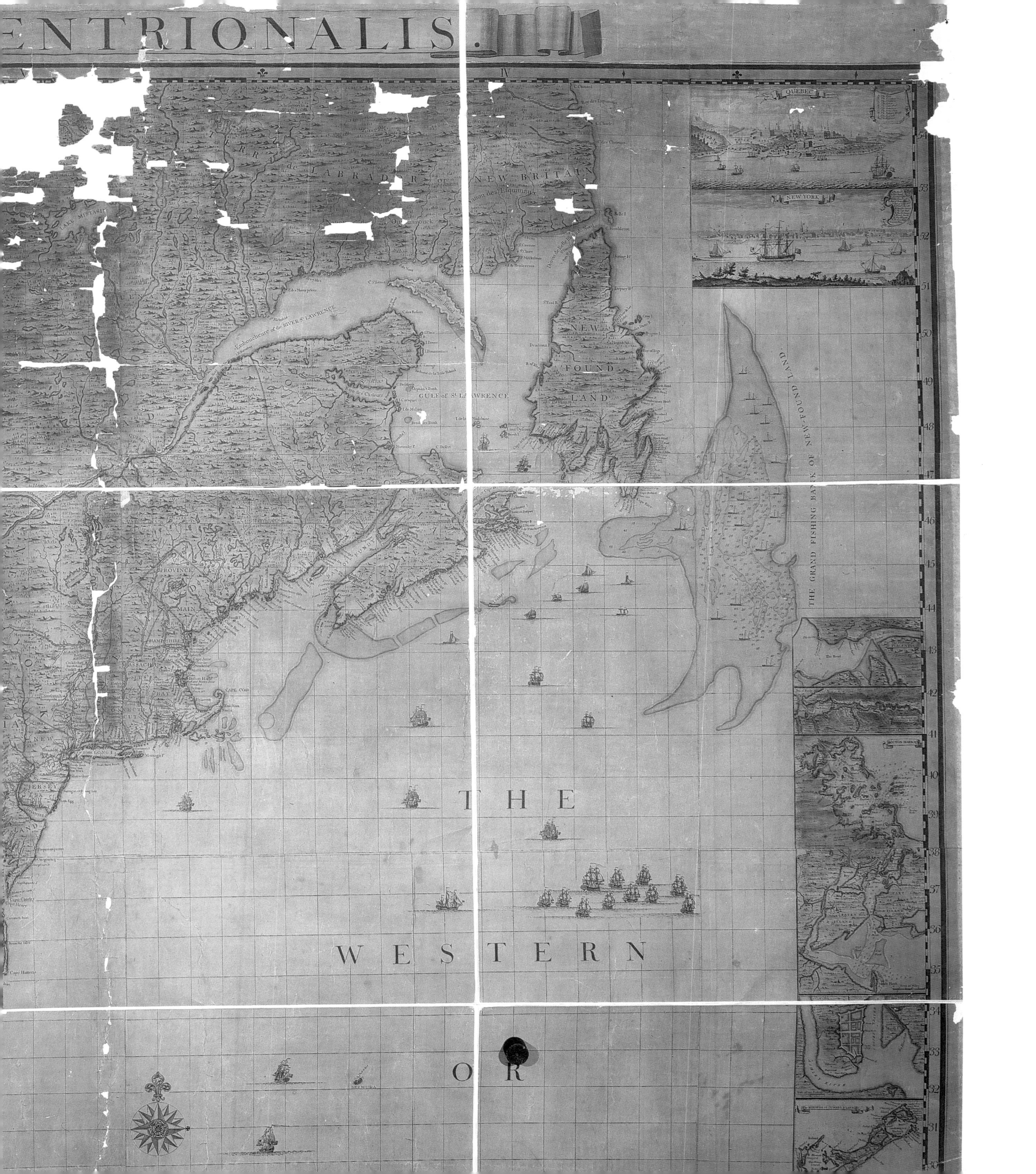
ENTRIONALIS.
NEW BRITAIN
NEW FOUND LAND
GULF of ST LAWRENCE
THE GRAND FISHING BANK OF NEW-FOUND LAND
QUEBEC
NEW YORK
CAPE COD
THE
WESTERN
OR
BERMUDA

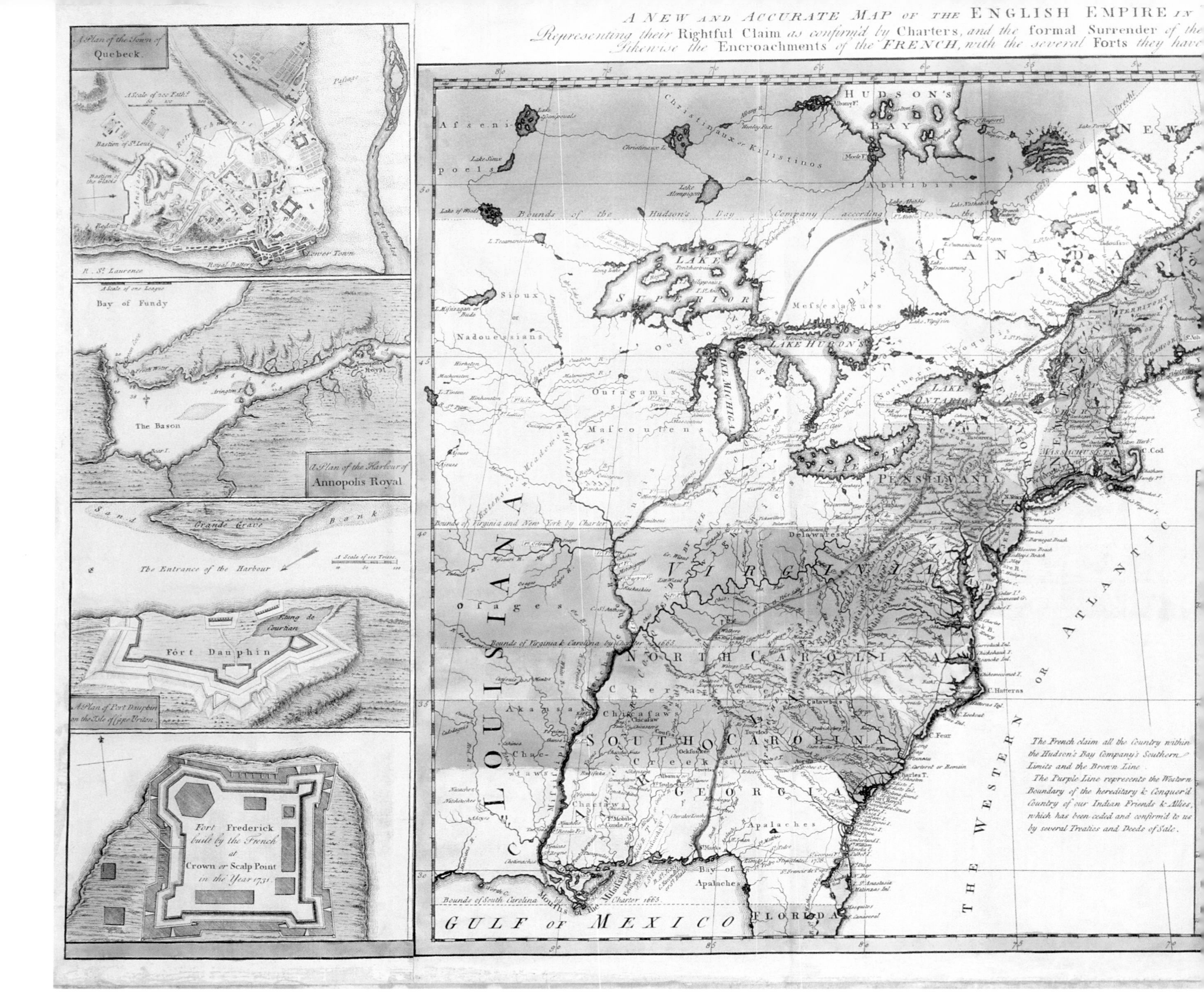

ABOVE *"A New and Accurate Map of the English Empire in North America" (1755, William Robert and Herbert Sayer). Espousing the agenda of "a Society of Anti-Gallicans," this is an overtly propagandist work. Like many maps of the period, it shows the area from Newfoundland to Florida and west to the Mississippi; Canada and the Caribbean appear on the largest of the inset maps, and the additional insets feature Crown Point and Canadian locations.*

Henry Popple's map, admirable for its accuracy and quality, proved controversial in its political content as the Lord Commissioners felt that it did not support their position in America. Unable to overtly reject the map on political grounds, they tried to discredit its accuracy and claimed:

> "He published it on his own single Authority; the Board of Trade at the Time gave it no extraordinary Sanction. It is inconsistent with the very Records it pretends to have copied; it came into the world as the Performance of a Single Person; it

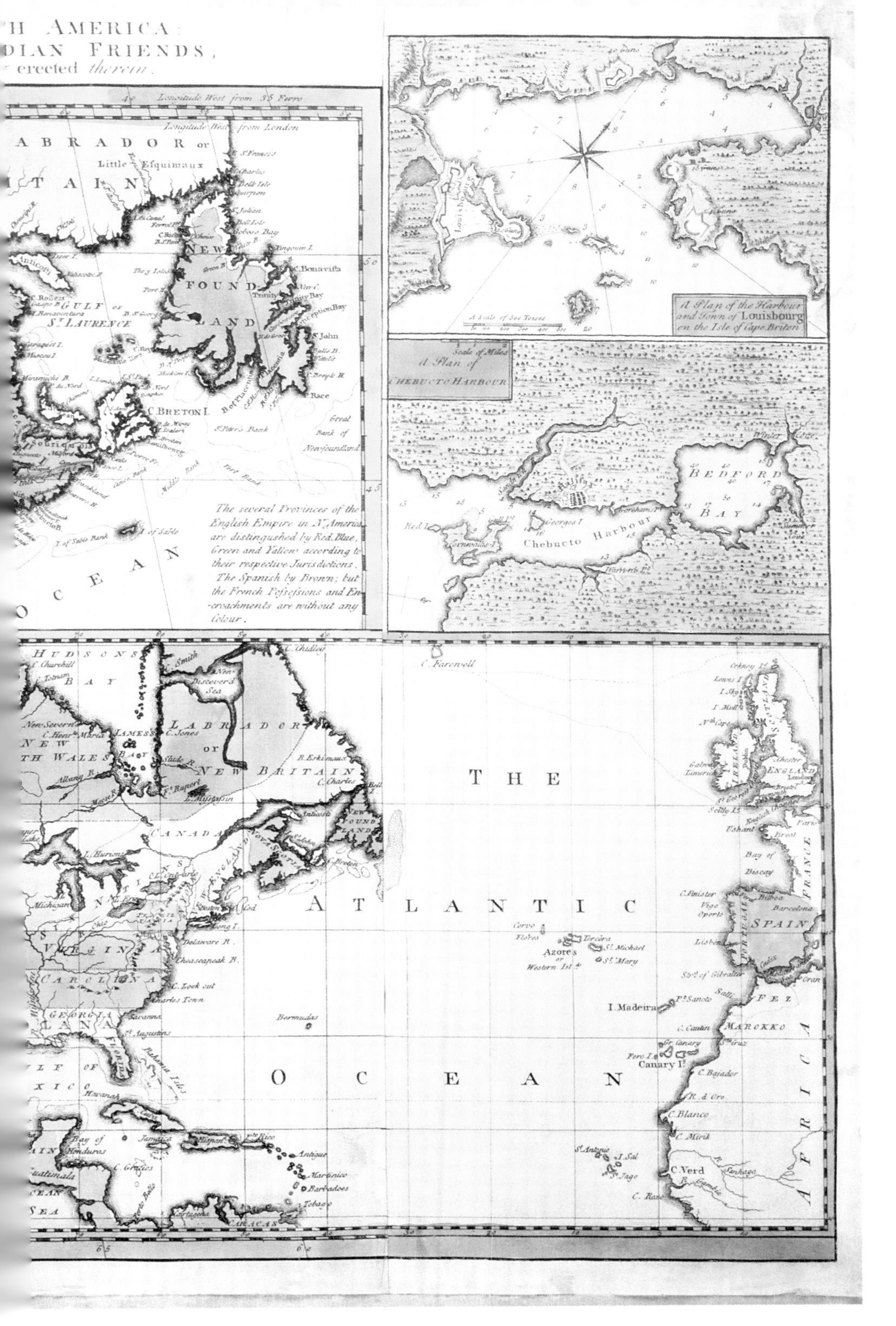

ABOVE *George Washington (left) and Benjamin Franklin (right) are shown flanking a map of North America. The tensions that grew throughout the Anglo-French "Map Wars" of the 1750s erupted with the outbreak of the French-Indian War, and Washington and Franklin emerged as two of the most influential colonial leaders. The allegiances and enmities that developed during this time would, in turn, influence the course of the American Revolution ten years later, when control of the much-disputed territories was finally wrested from both English and French.*

has ever been thought in Great Britain to be a very incorrect Map, and has never in any negotiation between the two Crowns been appealed to by Great Britain as being correct, or as a map of any authority."

Popple's map, nevertheless, remained enormously influential in the field of American mapmaking for many years to follow.

In 1755, William Herbert and Robert Sayer of a "Society of Anti-Gallicans" produced one of the most overtly political maps ever. Published after the outbreak of the French-Indian war, it shed all pretence at impartiality. The very title, "A new and accurate map of the English empire in North America; Representing their rightful claim as confirmed by charters and the formal surrender of their Indian friends; likewise the encroachments of the French, with the several forts they have unjustly erected therein," says it all. This map was the product of several years of pamphlets, propaganda, and proclamations, as English and French accused one another of illegal "encroachments" on North American land. It shows hugely exaggerated British territories, delineated by bold color. What remains of the French territories are disparagingly described as "French possessions and encroachments," and left uncolored on the map.

The Great Survey of India:

A Triumph of Trigonometry

At the beginning of the 19th century the British could proudly boast that the Sun never set on the British Empire. From Canada in the west, across parts of Africa, through India, Ceylon, and Malaya to Australia and New Zealand vast areas of the world map were colored pink. But the problem with such a far-flung empire was that it was, well, far flung.

Before the advent of the electric telegraph, the quickest means of communication was the fastest sailing ship. So Britain had a problem. How do you govern a place like India, for instance, with a population of 350 million people living in an area of 1,544,000 square miles (4 million square kilometers)?

Britain's answer was to set up separate colonies—later some of them became dominions—appoint a governor, give him an army of soldiers (though usually very limited in number), another army of civil servants, and let him get on with governing as he saw fit. The governor was the representative of the king (or later, Queen and Empress) and ruled like one. But to build roads and railways, to collect taxes, and to maintain law and order, the authorities needed accurate maps. And India did not have them, apart from a few sketchy maps of the areas around the major ports. It is said that the central part was labeled "A great expanse of country about which we have no exact knowledge." Impeccable grammar, but no way to run a railroad.

In 1763 the East India Company commissioned mapmaker James Rennell (1742-1830) to map the River Ganges and its surroundings in Bengal. Rennell's *Bengal Atlas* was published in 1779, based on a fairly rough survey, which he had to undertake under dauntingly difficult conditions. In 1766, for example, Rennell was nearly killed by saber cuts received in a skirmish with tribesmen. In 1782 he extended his coverage to the whole subcontinent and produced the *Map of Hindoostan*. This went through several editions and by 1826 included also Nepal and the Himalayas. Rennell would become a great hydrographer and his theories, regarding the Gulf Stream in particular, would furnish an essential background to Maury's *Physical Geography of the Sea*.

What the East India Company really wanted was an accurate, detailed map of the whole country. The task fell to an engineer named William Lambton (c.1756-1823), who was a colonel in the British army. In 1802 he began to carry out what was to become known as the Great Trigonometrical Survey. He imported the best surveying instruments from Britain, and spent the first two months setting out and measuring his baseline. To do this he used an apparatus called Colby's bars, which are bars 3 meters long made from two different metals and constructed in such a way that they do not change length with changes in temperature. Similar bars were later used to construct the baseline for the Ordnance Survey of Britain (also carried out by military engineers).

The whole structure of a trigonometrical survey depends on the accuracy of its baseline. The surveyor marks out a line between two fixed points, and checks and rechecks its length—Lambton made 400 separate measurements to ensure that he got the baseline correct. The surveyor then chooses a prominent object roughly opposite the center of the baseline and some distance from it, a church spire or a tower are favorite triangulation points. In open country, where there are no such objects, the surveyor has to make one. One method is to build a cairn of stones and erect a tall—say, 60-foot (18-meter)—pole on top of it; some of Lambton's original towers still stand. The surveyor then uses a theodolite to measure the angle between one end of the baseline and the tower, pole, or whatever he is using as a triangulation point. (A theodolite is a low-power telescope mounted on a horizontal, circular table calibrated in degrees with a Vernier scale; a good instrument will

RIGHT The Index Chart to the Great Trigonometrical Survey of India *shows at a glance the huge extent of the mapmaking survey conducted by William Lambton and George Everest between 1802 and 1843. This version dates from 1876 and plots the extension around the coast to Burma (modern Myanmar).*

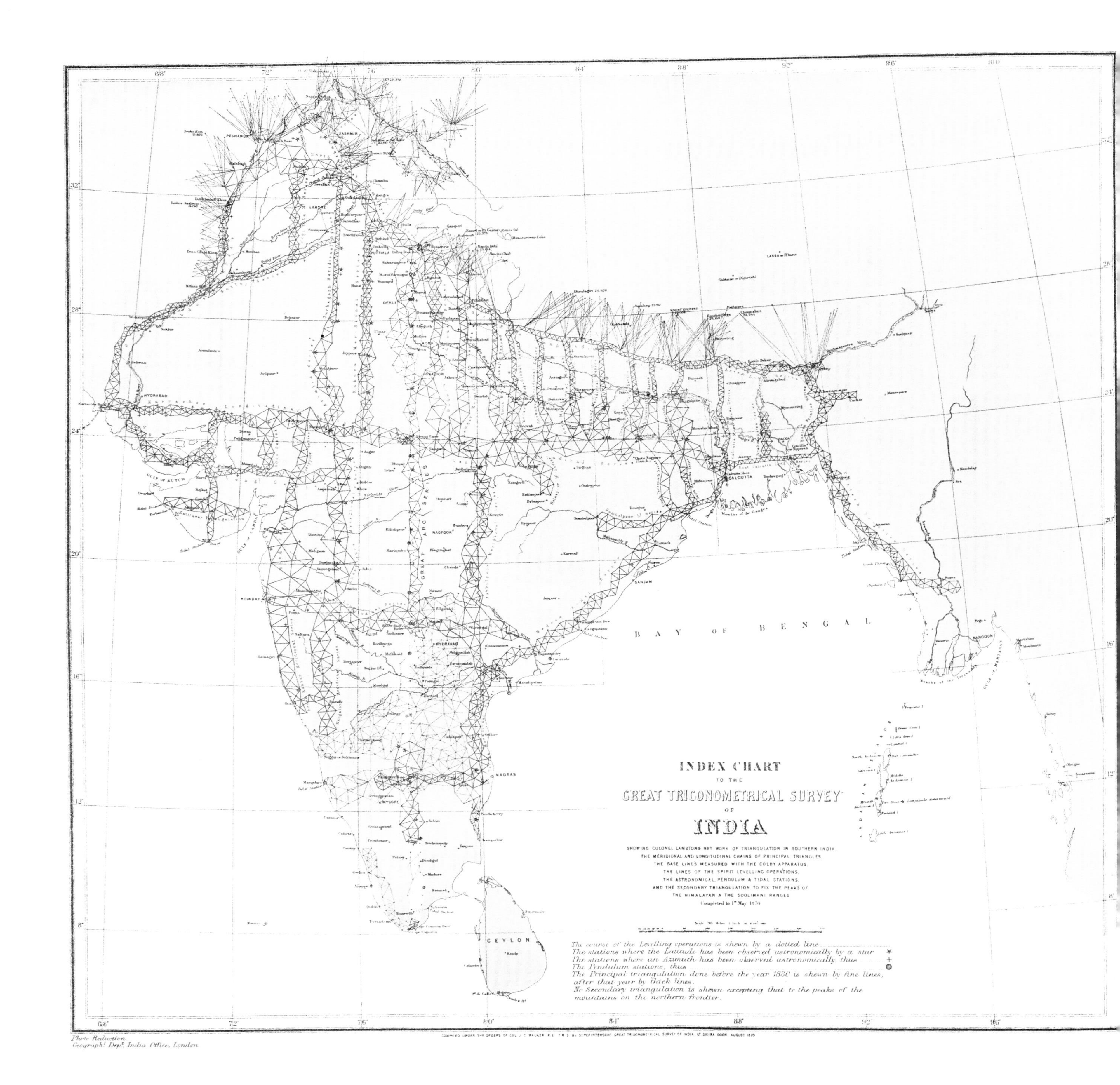
INDEX CHART
TO THE
GREAT TRIGONOMETRICAL SURVEY
OF
INDIA
SHOWING COLONEL LAMBTONS NET WORK OF TRIANGULATION IN SOUTHERN INDIA,
THE MERIDIONAL AND LONGITUDINAL CHAINS OF PRINCIPAL TRIANGLES,
THE BASE LINES MEASURED WITH THE COLBY APPARATUS,
THE LINES OF THE SPIRIT LEVELLING OPERATIONS,
THE ASTRONOMICAL, PENDULUM & TIDAL STATIONS,
AND THE SECONDARY TRIANGULATION TO FIX THE PEAKS OF
THE HIMALAYAN & THE SOOLIMANI RANGES
Completed to 1st May 1870
The course of the Levelling operations is shewn by a dotted line
The stations where the Latitude has been observed astronomically by a star
The stations where an Azimuth has been observed astronomically thus
The Pendulum stations, thus
The Principal triangulation done before the year 1830 is shewn by fine lines, after that year by thick lines.
No Secondary triangulation is shewn excepting that to the peaks of the mountains on the northern frontier.
BAY OF BENGAL
CEYLON
CALCUTTA
MADRAS
BOMBAY
HYDRABAD
MYSORE
NAGPOOR
GANJAM
RANGOON
PESHAWUR
KASHMIR
LAHORE
DEHLI
GREAT ARC SERIES
Photo Reduction
Geograph! Dept. India Office, London

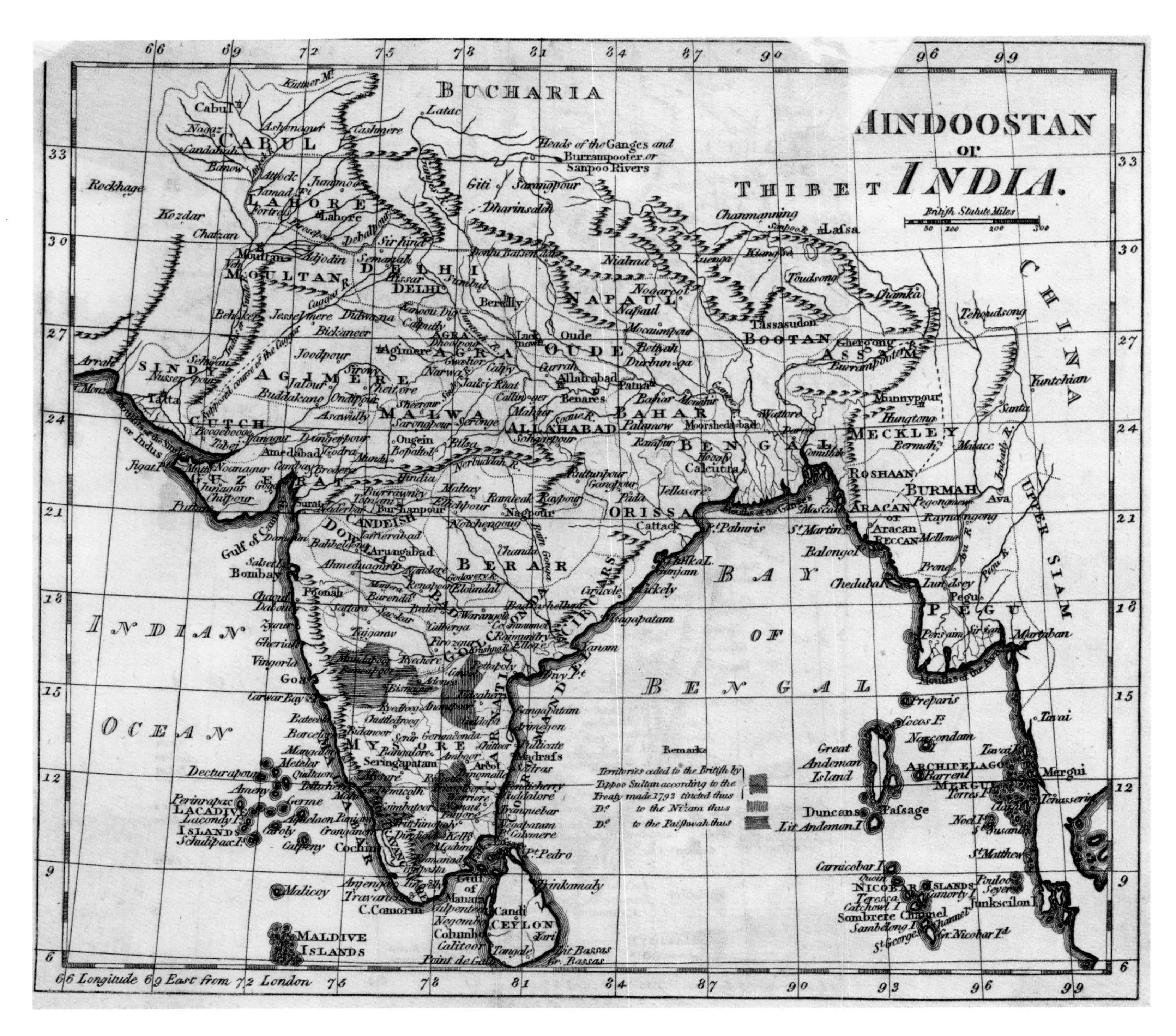

ABOVE *Even in the late 1790s maps of the Indian subcontinent still carried the alternative name Hindoostan. Parts of Afghanistan are labeled Cabul, and northwest of Thibet is Bucharia, corresponding to the present-day Sinkiang-Uighus province of China (Chinese Turkestan). Although printed in English, the map shows close similarities to that of India published in Rome in 1797 by Giovanni Cassini in his* Nuovo Geografico Universale.

measure angles to within a small fraction of a degree.) He then measures the angle to the triangulation point from the other end of the baseline. The mapmaker now has a triangle of which he knows the length of one side and two angles. Using basic trigonometry (cast your mind back to math at school, if you dare) he can calculate the lengths of the other two sides and the other angle of the triangle. Taking one of the sides as a new baseline, and choosing a new triangulation point, the surveyor repeats the whole process. In this way, he measures the

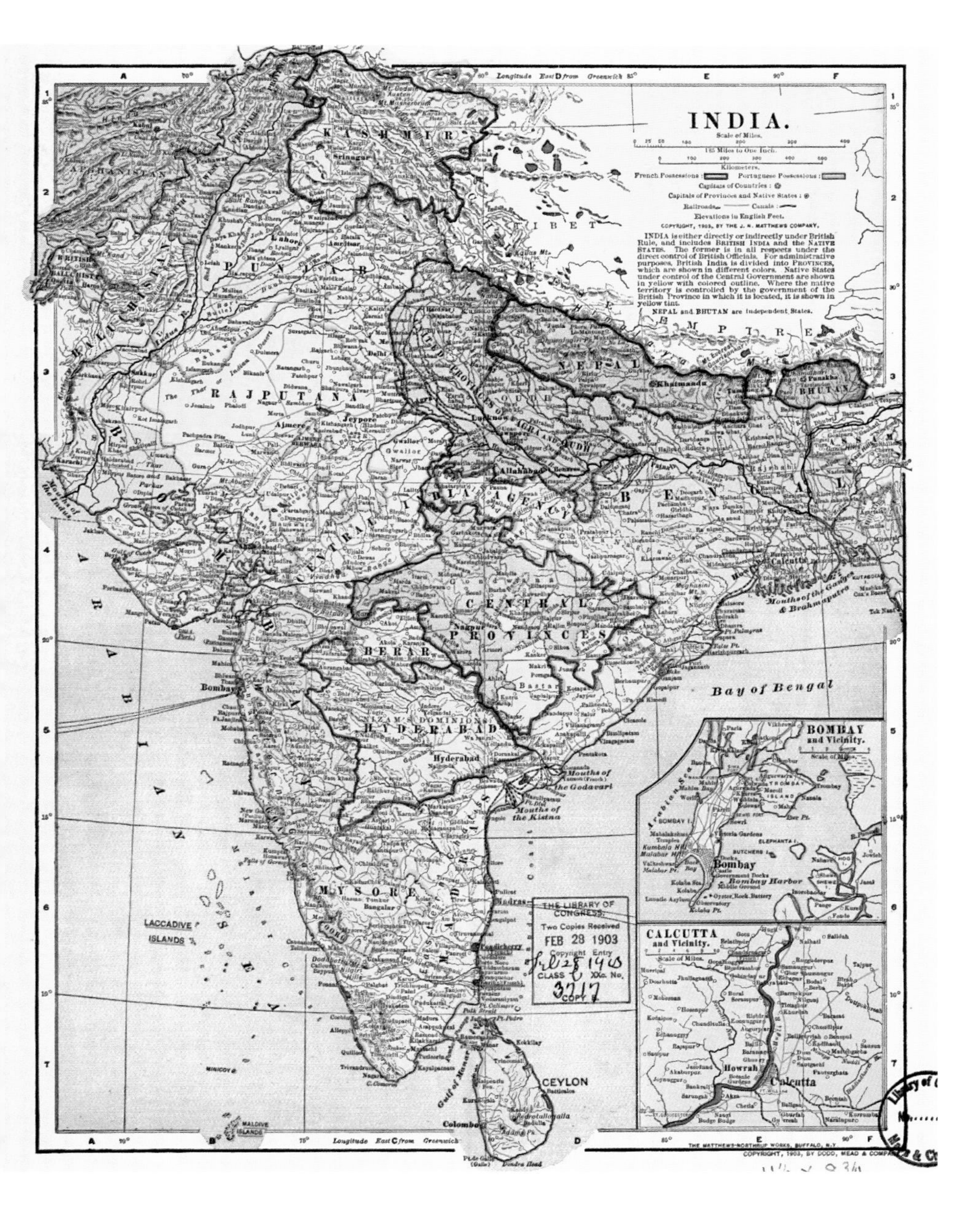

LEFT *This map of 1903, copyrighted by the J. N. Matthews Company, states categorically that "India is either directly or indirectly under British Rule. For administrative purposes, British India is divided into provinces, which are shown in different colors."*

ground as a series of zig-zagging triangles that slowly sweep across the countryside as the survey is carried out.

Obviously the bigger the triangles, the faster the work proceeds. But the consequences of any errors are magnified by big triangles, and so the surveyor has to check, recheck and check again all the angles and all the calculations.

Clearly, the ground is not flat like a piece of paper and on hilly or mountainous ground, only small triangles are possible. It took Lambton four years to complete his survey of most of the southern part of the Indian subcontinent. It was a massive job, and required the use of a suitably massive theodolite: Lambton's weighed more than half a ton and was the size of a small tractor! Then the baton passed to George Everest.

Everest was born in 1790 in Gwernvale, Breckonshire (now Powys), Wales. He trained as an engineer in various military schools in England and in 1806 joined the East India Company as a military engineer. After serving in Bengal for seven years, he went in 1814 to work for two years on a survey of the Indonesian island of Java. At the time the British were occupying what were then known as the Dutch East Indies. After some home leave to recover from illness contracted in the jungles of Java, he returned to India and in 1818 took up the trigonometrical survey, concentrating at first on completing the series of triangles up the "spine" of the subcontinent, called the Great Arc Series, a distance of 1,550 miles (2,500 kilometers); he became superintendent of the project in 1823. The Great Arc begun by Lambton was generating the most complex mathematical equations of the pre-computer age. Wherever possible, Everest worked due north along the lines of longitude, although he also had to work along the coastlines and the borders with Persia (Iran) and Afghanistan. To the east, the survey continued along the coast of Burma (Myanmar) and across the Irrawaddy Delta to Rangoon.

In 1830 Everest was made surveyor-general of India, and by 1837 he had completed the main part of the survey. He obtained the most accurate surveying instruments available, and in the course of the work measured the length of the meridional arc of 11.5 degrees from the Himalayas in the north to Cape Comorin, the southernmost point in India. He continued working northwards, surveying the Himalayan range and measuring the heights of the mountains. The Sahib paid a price—dysentery, malaria, bouts of paralysis, and even what appear to have been periods of madness. Everest retired in 1843 and was knighted by Queen Victoria in 1861.

In 1855 at the headquarters of the Great Trigonometrical Survey at Dehra Dun in the foothills north of Delhi, the Bengali Radhanath Sikdhar, head of computations, burst into the office of Superintending General, Andrew Scott Waugh (Everest's successor), and announced that he had discovered the highest mountain in the world. The computations were based of course upon triangulation. The method can measure distances upward as well as across the land. Waugh renamed "Peak XV"—he had given all the peaks Roman numeral designations— as Mount Everest in his predecessor's honor. A century later it became part of Nepal's new Sagarmatha National Park.

Terra Nullius:

Africa after the Berlin Conference 1885

The West Africa Conference, held in Berlin from November 15th 1884 to February 26th 1885, was convened by the German Chancellor Otto von Bismarck to resolve border disputes between European powers as they encroached upon Africa. Representatives of 12 nations attended, along with the Ottoman Empire and the US. Although no territorial divisions were formally settled at the meeting, it was a major catalyst to the "Scramble for Africa" which brought almost all of the continent under foreign domination by the end of the century.

The signatories' high-minded claim to be upholding David Livingstone's mission to bring Commerce, Christianity, and Civilization to the peoples of Africa rang hollow from the outset.

The Berlin Conference was popularly referred to at the time as the "Congo Conference," revealing its principal focus. The prime mover in persuading Bismarck to host the meeting was the Belgian King Leopold II, whose concern centered on the Congo Free State, which he had "founded" in 1876 under the auspices of a supposedly philanthropic body, the African International Association for the Civilizing of Africa (AIA). In the meantime (indicating the real reason for his interest), Leopold secretly increased his stake in a related commercial enterprise, the International Congo Society. In formally recognizing the sovereignty of the AIA "front" organization over the region, the conference effectively set the seal on Leopold's appropriation of the vast Congo territory 965,0000 square miles (2.5 million sq km) as a personal fiefdom. In Bismarck's closing address, he spoke of the "careful solicitude" that the signatories had shown for the welfare of the indigenous peoples in drafting the General Act that concluded the conference. Nowhere was this statement more grotesquely at variance with reality than in Leopold's Congo. His agents systematically abused the Congolese as a slave labor force; brutal treatment was meted out to those who did not meet punitive production targets for the commodities the colonialists wished to exploit. As many as 6 million are thought to have perished. When the genocide came to light in 1905-08, Leopold was forced to relinquish control of the Congo to the Belgian state.

Yet the negative impact of the conference was not confined to the despoiling of the Congo. Before 1885, some 80 percent of the African landmass was still controlled by local rulers. Despite the fact that much of the continent had already been explored by White missionaries and adventurers—most recently by Henry Morton Stanley, who had explored Central Africa in Leopold's service in 1875-77—there were as yet only tentative moves toward colonization, and European settlement was largely confined to the coast. Yet commercial interest was growing in the mineral wealth of Africa; diamonds had been discovered at Kimberley in South Africa in 1871, followed by gold in the Transvaal in 1886, while Leopold was keen to exploit the Congo's ivory, timber, and gum. Later, he was to realize the huge potential for tapping natural rubber that the Congo offered (in the same year that the conference ended, Karl Benz launched his new automobile, which ran on rubber tires, a huge future market). As well as agreeing that the mouths and basins of the Congo and Niger rivers would be neutral territory open to free trade, the Berlin conference more significantly established the principle that the hinterland of a stretch of coast occupied by a European power could be legitimately regarded as a sphere of influence. This ruling sanctioned encroachment by settlers into the interior throughout the continent, into what they wrongly regarded as "terra nullius," or empty land. The results can be plainly seen on Hertslet's map of 1909.

The firm boundaries that were drawn up from 1885 onwards, as the Berlin signatories carved up the continent into their separate spheres of influence, created fifty arbitrary countries that virtually fill the map of 1909. These new polities were superimposed upon traditional

RIGHT *Edward Hertslet's 1909 General Map of Africa, published seven years after he died, uses border colors to distinguish British Possessions and Protectorates from those of France, Italy, Germany, Portugal, Spain, the Independent States, and the Belgian Congo. Hertslet was keeper of papers for the Foreign Office.*

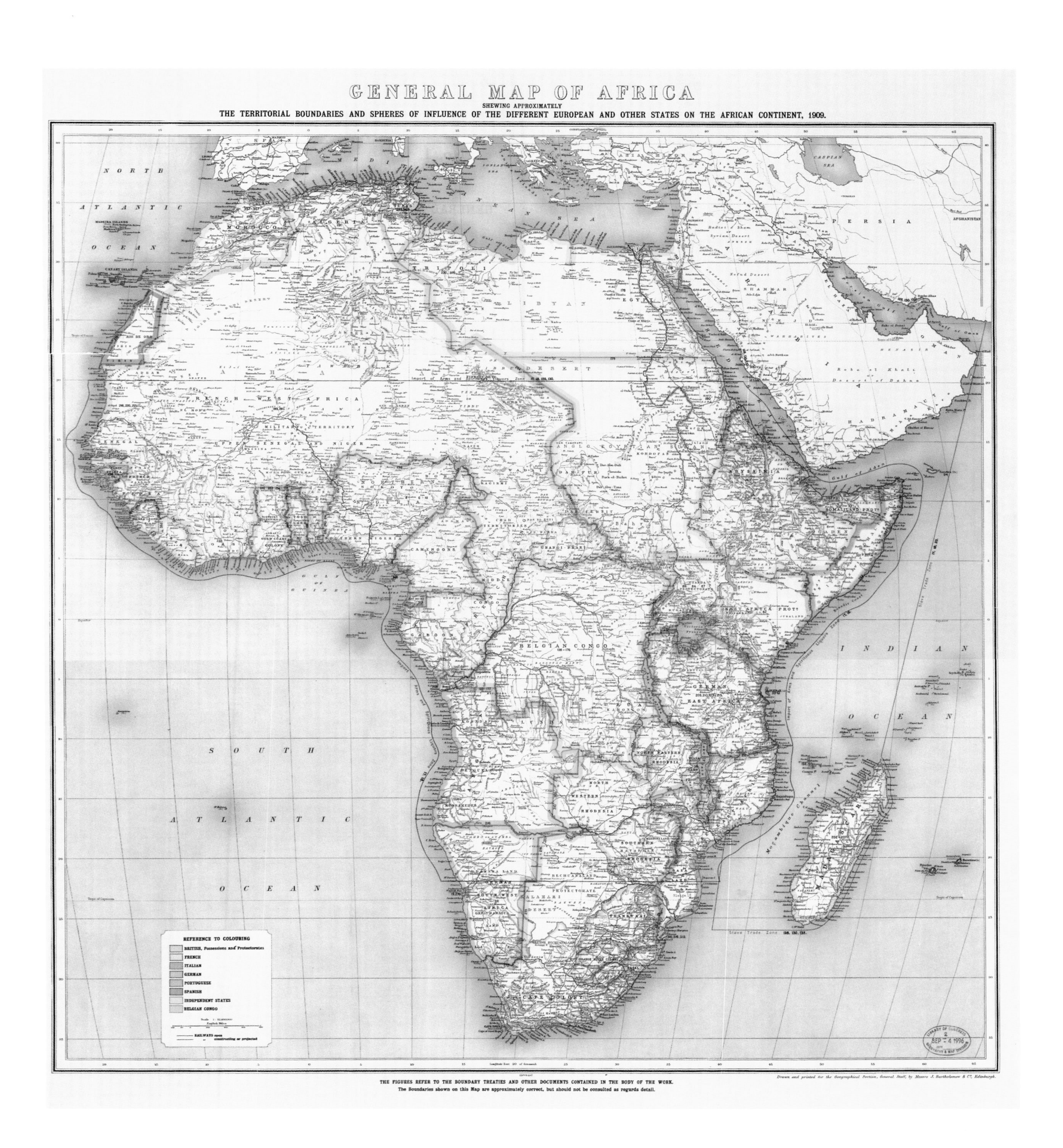
GENERAL MAP OF AFRICA
SHEWING APPROXIMATELY
THE TERRITORIAL BOUNDARIES AND SPHERES OF INFLUENCE OF THE DIFFERENT EUROPEAN AND OTHER STATES ON THE AFRICAN CONTINENT, 1909.
NORTH ATLANTIC OCEAN
MEDITERRANEAN SEA
CASPIAN SEA
PERSIA
MOROCCO
ALGERIA
TRIPOLI
LIBYAN DESERT
EGYPT
ARABIA
SAHARA
FRENCH WEST AFRICA
MILITARY TERRITORY
UPPER SENEGAL AND NIGER
SENEGAL
ANGLO EGYPTIAN SUDAN
KORDOFAN
DARFUR
ABYSSINIA
Gulf of Aden
SOMALILAND PROTᵗᵉ
GULF OF GUINEA
NIGERIA
CAMEROONS
BELGIAN CONGO
GERMAN EAST AFRICA
INDIAN OCEAN
SOUTH ATLANTIC OCEAN
NORTH EASTERN RHODESIA
NORTH WESTERN RHODESIA
SOUTHERN RHODESIA
BECHUANALAND PROTECTORATE
KALAHARI DESERT
CAPE COLONY
Mozambique Channel
REFERENCE TO COLOURING
BRITISH, Possessions and Protectorates
FRENCH
ITALIAN
GERMAN
PORTUGUESE
SPANISH
INDEPENDENT STATES
BELGIAN CONGO
RAILWAYS open
constructing or projected
THE FIGURES REFER TO THE BOUNDARY TREATIES AND OTHER DOCUMENTS CONTAINED IN THE BODY OF THE WORK.
The Boundaries shown on this Map are approximately correct, but should not be consulted as regards detail.
Drawn and printed for the Geographical Section, General Staff, by Messrs J. Bartholomew & Cº, Edinburgh.
SEP - 4 1996

"There be Elephants"

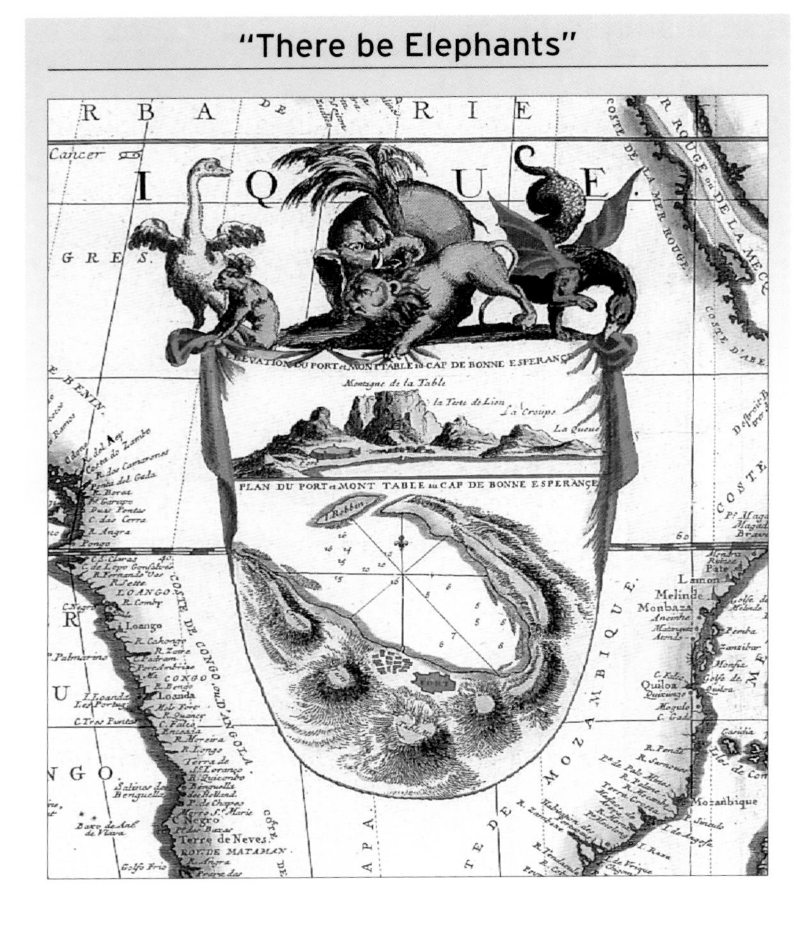

In 1733 (in **On Poetry: A Rhapsody**) Jonathan Swift had satirized mapmakers who had only the sketchiest idea of what states and peoples there were in the Dark Continent:

> 'So geographers in Afric-maps
> With savage-pictures fill their gaps
> And o'er uninhabitable downs
> Place elephants for want of towns.'

And indeed, even in maps as late as John Bartholomew's Africa map of 1885, although some indigenous states are shown, many others are omitted. The great Italian theologian and cartographer Vincenzo Coronelli simply gave up on the interior in his 1690 map (above).

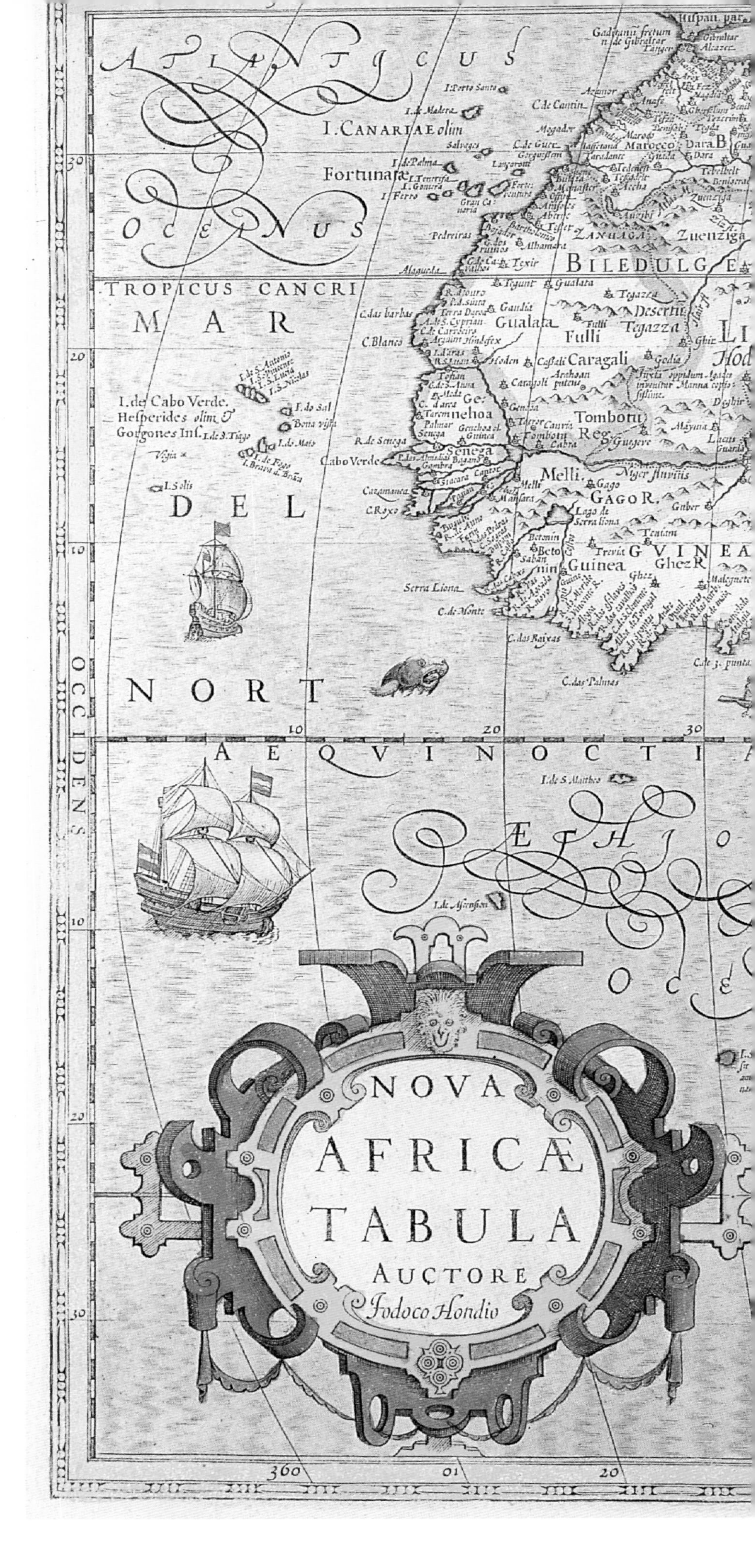

RIGHT *Joos de Hondt (1563-1611), or Jadocus Hondius as he became known, was a Flemish engraver and founder of a famous family of map-makers. His New Map of Africa of 1610 was made using the copper plates from Mercator's earlier Atlas but with some notable improvements, such as the more correct outline of Madagascar off the south-eastern coast.*

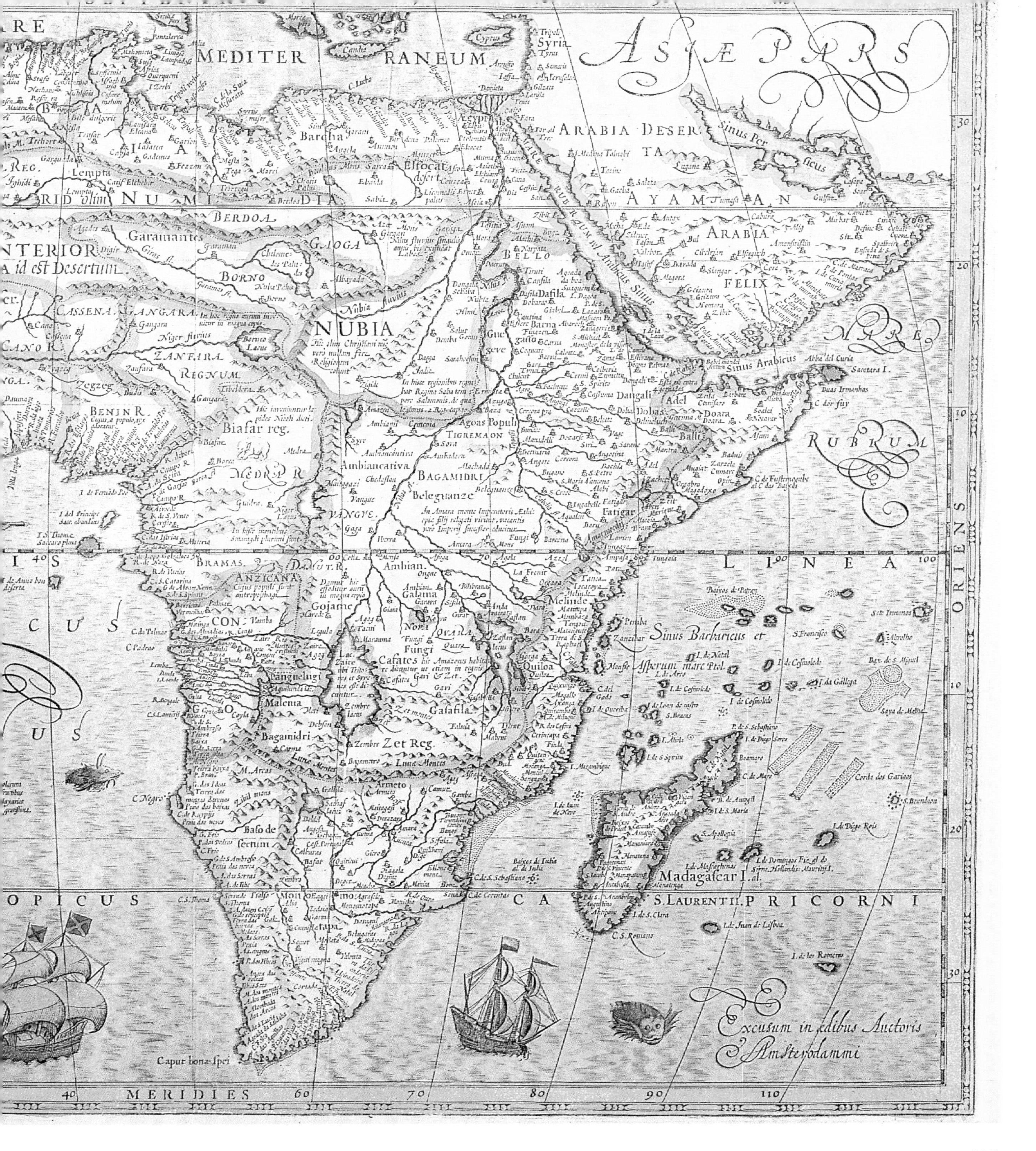
MEDITER RANEUM
ASIÆ PARS
ARABIA DESERTA
Sinus Persicus
NUBIA
ARABIA FELIX
MARE RUBRUM
Sinus Arabicus
LINEA
Sinus Barbaricus et Asperum mare Ptol.
Madagascar I. al.
S. LAURENTII
CAPRICORNI
ORIENS
MERIDIES
Capur bonæ Spei
Excusum in ædibus Auctoris Amsterodammi

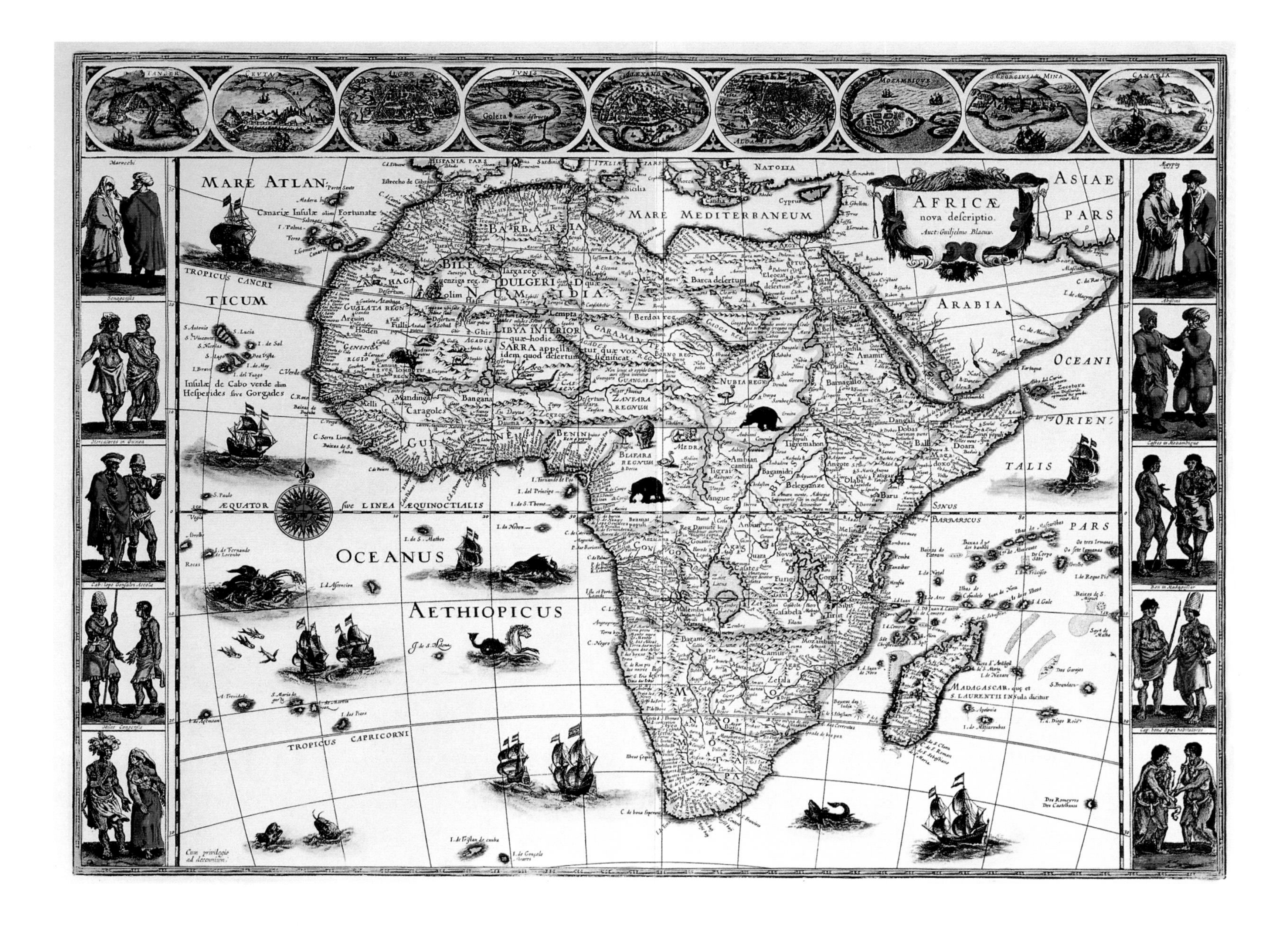

ABOVE *A so-called new map of Africa produced by Dutch mathematician and geographer Willem Blaeu (1571-1638), this version from 1640, included plans of principal cities along its upper border (see page 116). The continent's coastline is well represented, based mainly on information from Arab and Portuguese explorers, but the apparent detail in the interior is mostly fictitious. It follows a Ptolemaic lead, with the source of the Nile in Lakes Zaïre and Zaflan.*

regions, bound together by common culture and language. As a result, former tribal entities were broken up and subsumed within the new states. In addition to the Belgian acquisition of the Congo, Great Britain established a string of colonies and protectorates running from the Cape to Cairo through their control of Egypt, Sudan, Uganda, British East Africa (modern Kenya), Northern Rhodesia (Zambia), Southern Rhodesia (Zimbabwe), Bechuanaland (Botswana), and South Africa, where after two hard-fought wars they won a bitter power struggle with the Boers by 1902. Only Germany's colonization of Tanzania (German East Africa) thwarted Britain's—and specifically the colonial adventurer Cecil Rhodes'—dream of an unbroken swathe of British territory served by a north-south railway running from Cairo to the Cape of Good Hope. France annexed most of western Africa, from Mauritania to Chad (French West Africa), Gabon, and the Republic of Congo (French Equatorial Africa); Portugal took Mozambique in the east and Angola in the west; Italy overran Somalia (Italian Somaliland) and a portion of Ethiopia (where, alone among the European invaders, it suffered defeat at the hands of a native army). Where obvious anomalies do appear on the map, they came into being to accommodate the needs of the colonial powers, not the native populace; thus, the narrow Caprivi Strip, running from the northeastern corner of German Southwest Africa (modern Namibia), was created to give the German colony access to the Zambezi River (as part of a complex agreement of 1890, which saw Germany cede sovereignty over Zanzibar in exchange for control of the former British North Sea island of Heligoland). The frontiers that resulted from the conference and other imperial machinations were reluctantly recognized as *de facto* state borders by the Organization of African Unity in 1963 (which acknowledged that more bloodshed would result from any attempt to redraw boundaries).

The legacy of the Berlin Conference and the ensuing Scramble has been far-reaching and enduring. Most obviously, after decolonization, bitter ethnic rivals found themselves yoked together within newly independent states, and violent power struggles exploded: for example, the Yoruba–Igbo tension that led to the devastating Biafran War in Nigeria (1967-70) and the genocide of the Hutu by the Tutsi in Rwanda and Burundi in 1994. Perhaps even more damaging in the long term has been the economic and environmental impact. The uprooting of entire communities in the Congo to tap rubber is just the most glaring instance of how colonialism could upset the socio-economic balance of a region. Single crops were exploited or deliberately introduced into plantations for the export market (e.g. coffee, cocoa, and cotton) at the expense of diverse indigenous staples such as cassava, maize, or millet. Such monoculture farming would often destroy the fragile but practicable subsistence economy that had long sustained a particular area. In addition, compulsory resettlement of people in regions that had formerly supported only nomads brought overgrazing and drought. Much of Africa's sad history of political instability, debt, and famine can be traced to decisions taken round a table in Berlin in 1885.

BELOW *Rwandan refugee camp, 1994; between April and June that year, about 800,000 Rwandans, mostly Tutsi, were killed in 100 days. The roots of the tension between the Hutu and Tutsi lay in the decision of the Belgian occupiers to favor the Tutsi above the Hutu in 1916. When Belgium withdrew in 1962, the Hutu gained power and the Tutsi became scapegoats for every woe. There was never much difference between the two groups: they even shared the same language.*

The Transcontinental Railroad:

"Done"

This was the telegraph message sent on May 10th 1869 from Promontory Point, Utah, when the transcontinental railroad across the Unites States was finally completed. At a cost of more than $115 million, two gangs totalling at times 25,000 men forged a 2,000-mile transport link across some of the world's worst terrain for railroad building. The story of this epic achievement begins in 1849 as a clipper ship struggles into San Francisco harbor with an unusual deck cargo: a 20-ton steam locomotive appropriately nicknamed the Elephant.

The Elephant had been built in the Baldwin works in Philadelphia and shipped down the eastern Atlantic, hugging the shore of South America, through the turbulent waters of Cape Horn and up the Pacific coast to San Francisco. The first locomotive on the west coast, it went to work on the Sacramento Valley Railroad (SVRR) between San Francisco and Sacramento. Completed in 1856 and the first railroad west of the Mississippi, the SVRR carried the mail of the Pony Express from its western terminus at Folsom 22 miles (35 kilometers) to Sacramento (after 1861 the Pony Express ran to Placerville). The Elephant was later renamed "Garrison," after the railroad's president, and was finally called "Pioneer," always a popular locomotive name in the US.

In the more densely populated east there were already several hundred miles of railroads, with a line from Chicago reaching as far west as Omaha in Nebraska. Then in 1862, spurred on by the secession of the South, President Abraham Lincoln signed an enabling act (the Pacific Railway Act 1862) authorizing the construction of a new length of standard-gauge track to link the railheads at Sacramento and Omaha. Lincoln wanted to guarantee the Union's contacts with the western states, particularly California. To finance the construction, the Act authorized the issue of loan bonds repayable in 30 years of $16,000 to $48,000 for each mile of track (the price depended on the difficulty of the terrain: $16,000 over the plains, $32,000 in the Great Basin and $48,000 in the most difficult mountainous areas). To raise yet more capital, another Pacific Railway Act of 1864 doubled the land grants and permitted the railroad companies to issue their own bonds. The actual construction cost was $10,600 per mile, and later the railroad directors were accused of profiteering, if not outright corruption, in their financial dealings. In all, 33 million acres (134,000 square kilometers) of land were acquired free by the transcontinental railroad companies.

The war department's surveyors had already completed their task by 1853 aided by a Congressional grant of $150,000 and the cartographers immediately went to work, producing detailed maps of the tortuous route across the plains and through the mountains. Both surveyors and mapmakers included alternative routes where the terrain presented particularly difficult civil engineering problems. But the Civil War and lack of investors delayed the start of the undertaking until 1865. In July of that year the Central Pacific Railroad, using mainly Chinese workers, started laying track eastwards from Sacramento. They used rails that had been shipped by sea from the east coast. Their construction locomotive Number 1, named Leland Stanford after the Governor of California, had also arrived by sea and had been working on the line since 1863. Stanford was one of the prime movers in getting the western end of the enterprise under way.

At the same time, the Union Pacific Railroad—newly formed for the project—began work at Omaha with a labor force consisting mainly of Irish immigrants and Civil War veterans, building the line westwards up the Platte River valley and on across Nebraska toward Grand Island. Throughout the freezing winters and sweltering summers, the Union Pacific workmen labored on until in three years they had connected the towns of Cheyenne, Laramie and Rawlins in Wyoming, often having to beat off attacks by local Native Americans. The Plains Indians were attacking in part because the railroad workers were killing off the buffalo herds for meat. Meanwhile the Central Pacific, suffering the same harsh weather conditions, forged their way through the Sierra

RIGHT *New York publisher Gaylord Watson's* New Rail-Road and Distance Map of the United States and Canada *of 1871 includes the route of the eastern end of the Transcontinental Railroad as it was two years after its completion.*

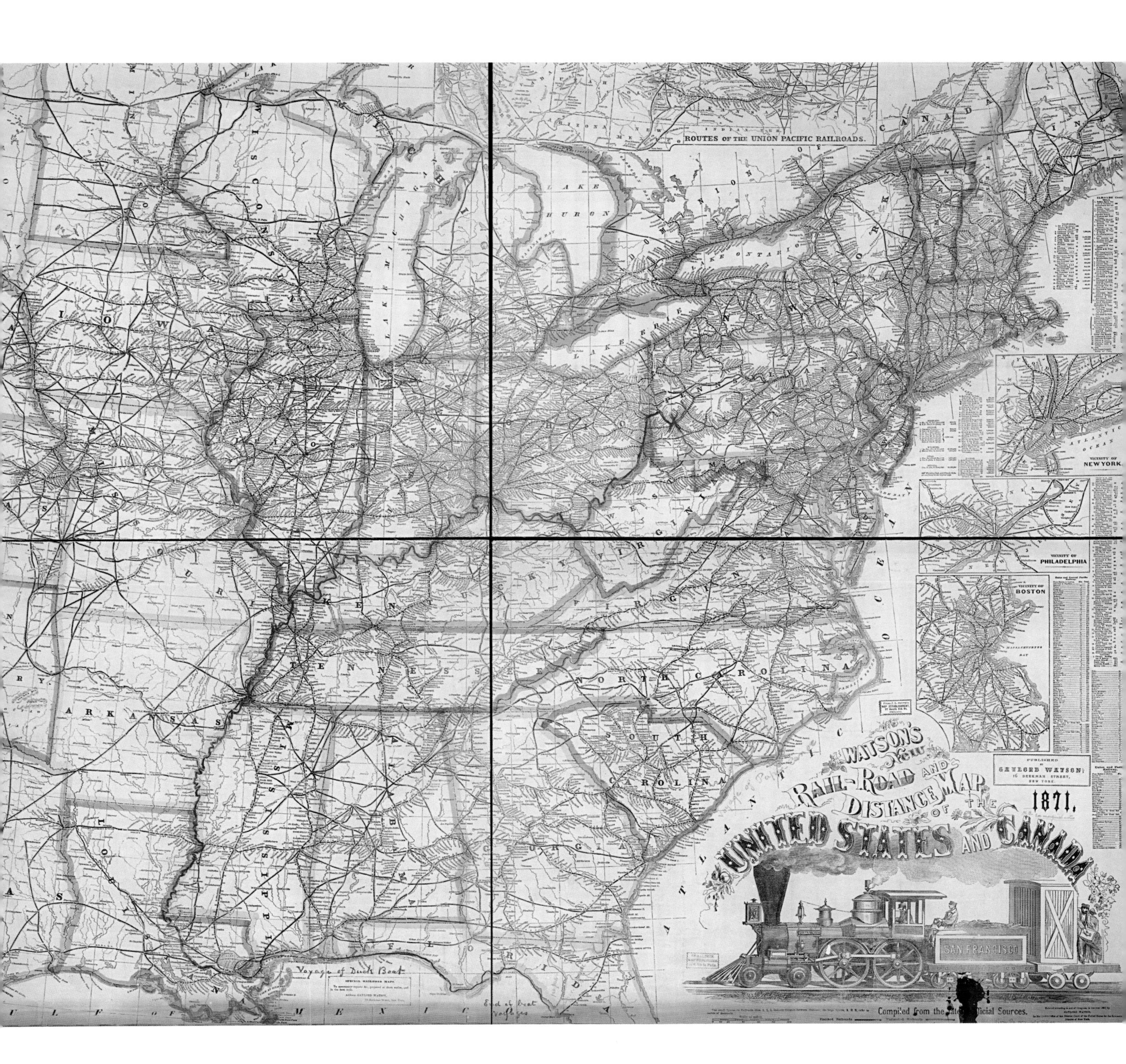
ROUTES OF THE UNION PACIFIC RAILROADS.
WATSON'S New RAIL-ROAD AND DISTANCE MAP OF THE UNITED STATES AND CANADA
1871.
PUBLISHED BY GAYLORD WATSON, 16 BEEKMAN STREET, NEW YORK.
VICINITY OF NEW YORK
VICINITY OF PHILADELPHIA
VICINITY OF BOSTON
SAN FRANCISCO
Compiled from the latest Official Sources.
Voyage of Duck Boat
End of boat voyages

Nevada toward Promontory Summit at the north of the Great Salt Lake. They had to build timber trestle bridges across dozens of gorges and miles of snowsheds to protect the line against mountain avalanches. Three gangs of men worked in succession. The first gang cleared the ground of trees and large rocks, using gunpowder to blast away major obstructions. Then came the graders, who leveled the trackbed and graded the route up and down hills. Finally the tracklayers set out rows of crossties which, for speed of construction, were made mainly from suitable branches shipped in from Wisconsin rather than from more substantial balks sawn from tree trunks. The track itself used flatbottomed rail pinned down directly onto the crossties using iron spikes. The men lived rough alongside the railroad workings and with the overriding demand for speed of progress, accidents were common.

The two gangs of workmen finally met up at Promontory Point in July 1869, with a golden spike ceremoniously driven in to fix the last length of single track; the line was laid to single track along its entire length. The Central Pacific's locomotive Jupiter stood nose-to-nose with Number 119 of the Union Pacific. Some moralistic doggerel of the time concluded with the description of:

How two engines, in their vision,
Once have met without collision.

The workmen opened bottles and drank copious toasts, while in Sacramento thirty locomotive whistles "screeched out a concert of joy". Public railroad services began five days later from Omaha, with passenger fares of $111 first class and $80 second class ("immigrant class", with no amenities, cost only $40). The trip was scheduled to take 4 days, 4 hours and 40 minutes, although straying buffaloes, train robbers and raids by Native Americans often added to this time. Occasionally floods washed out the whole of the trackbed and nothing moved until it was repaired. In 1876 the centennial Transcontinental Express completed the 3,500-mile journey from New York City to San Francisco in a record time of 83 hours and 39 minutes, establishing an average speed—42 miles per hour (68 kilometers per hour)—that was not to be matched on a regular basis for 50 years.

The immediate effects of the completion of this first transcontinental railroad (four more were to be built later) were social and psychological rather than economic or commercial. The country began to see itself as a single nation rather than a collection of semi-independent states and territories, despite the terrible recent depredations of the Civil War. Though the iron horses did fairly immediately replace the wagon trains for carrying settlers and manufactured goods, and did away with the need for the hazardous

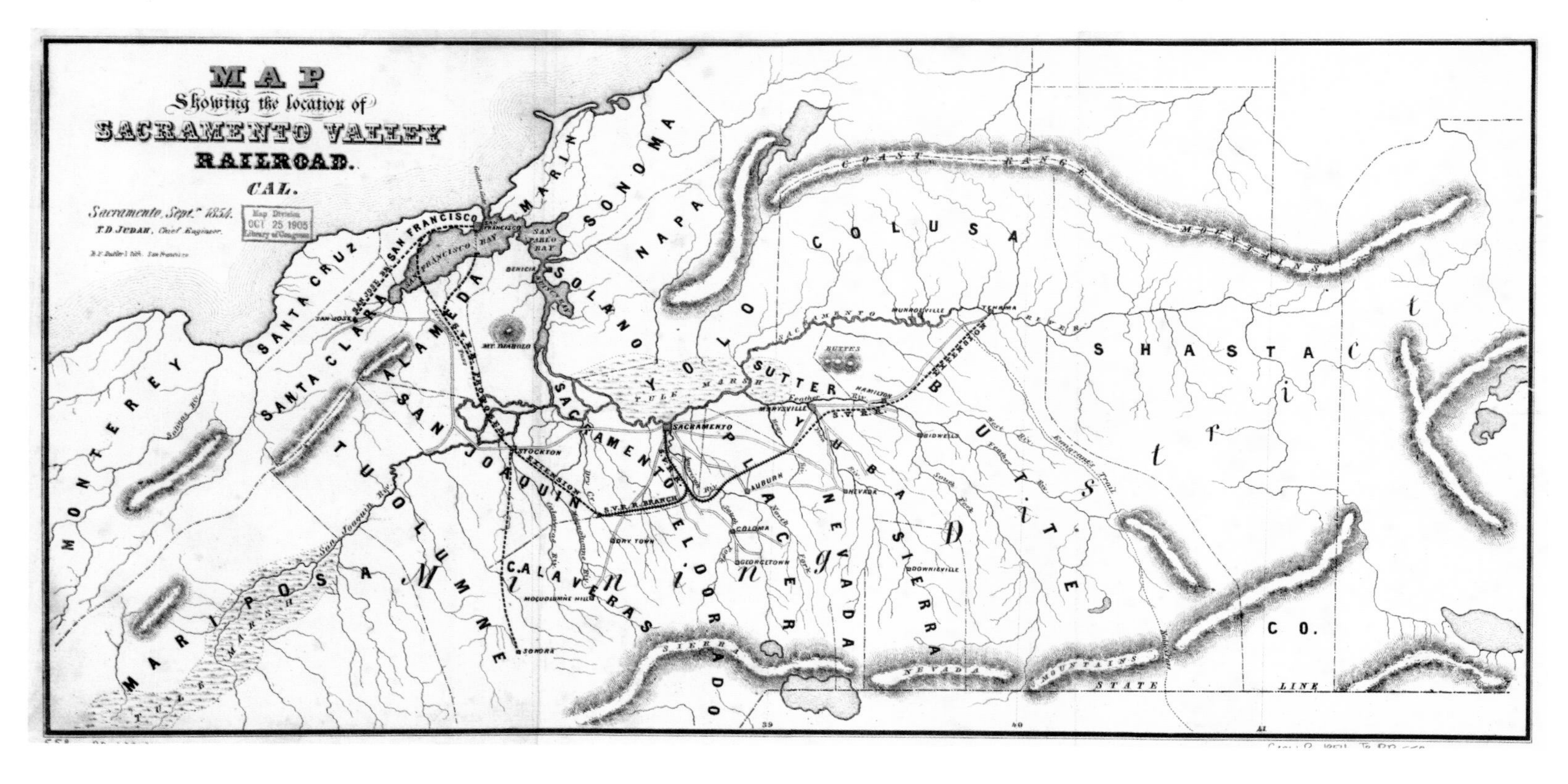

BELOW The Sacramento Valley Railroad was the first to be built in California. On his map of 1854, chief engineer Theodore Judah (1826-63) included plans for extensions eastward to San Francisco and westward to Sonora. It became part of the Central Pacific Railroad and the western end of the Transcontinental.

ABOVE *By 1876, when this map was published, the New York Central and Hudson River Railroad provided New Yorkers with their main route to Buffalo and Chicago, and then on via Omaha for access to the Transcontinental Railroad to take them all the way to San Francisco in just over four days.*

voyage by sea around Cape Horn. The four- to six-month overland journey was cut back to a mere five or six days.

The maps included in the several railway guides available in this period are of course designed to inform the traveler, but they are also celebrations of this new national spirit. *Crofutt's Trans-Continental Tourist's Guide* for example is an exuberant publication, "describing 500 Cities, Towns, Villages, Forts and Camps, Mountains, Lakes, Rivers … Where to Look for and Hunt Buffalo, Antelope, Deer, and other Game; Trout Fishing, etc., etc. ... Where to Go, How to Go … While passing over the Union Pacific Railroad, Central Pacific Railroad of Cal., their Branches and Connections by Stage and Water, from the Atlantic to the Pacific Ocean." The 1872 edition included three foldout maps, one a three-foot hand-colored map of the world, showing distances and fares to and from various world destinations. Crofutt was thinking big, as was the nation. Alfred A. Hart's *The Traveler's Own Book* of 1870 provided "by a system new and comprehensive, all the minutiae of railway travel, noting all interesting points and … illustrated by fine photo-chromo views." The maps were rather like strip maps with text running along the bottom, indicating—in California for example—such sights to look out for from the train window as: "Yosemite Valley—vineyards—insane asylum—copper mines—big trees."

JFK's First Domino:

John F. Kennedy and Laos

When John F. Kennedy was elected President of the United States of America in January 1961 he was confronted with an immediate communist threat in Southeast Asia. Not the one that would later dominate his and his successor Lyndon B. Johnson's foreign policy, that of Vietnam, but one in a neighboring country, Laos.

He had been warned about the impending communist threat by President Eisenhower and was aware of the perilous state of the country. In a press conference on March 23rd 1961 he addressed the watching journalists, emphasizing the serious nature of the matter:

"I want to make a brief statement about Laos. It is, I think, important for all Americans to understand this difficult and potentially dangerous problem. In my last conversation with General Eisenhower, the day before the Inauguration, on January 19, we spent more time on this hard matter than on any other thing; and since then it has been steadily before the Administration as the most immediate of the problems that we found upon taking office."

The difficult situation in Laos was caused by the successes of the Pathet Lao, a communist guerrilla organization that was being supplied by both the North Vietnamese and Soviet governments. The Pathet Lao controlled much of the north of the country and was launching attacks southward across the Plain of Jarres towards the Mekong River. Kennedy illustrated their growing successes with a series of maps:

"These three maps show the area of effective Communist domination as it was last August, with the colored portions up on the right-hand corner being the areas held and dominated by the Communists at that time. And now next, in December of 1960, three months ago, the red area having expanded—and now from December 20 now to the present date, near the end of March, the Communists control a much wider section of the country."

Should he intervene in the situation to prevent the collapse of the Laotian government? He was overstretched militarily and only had a few troops available, he was also wary of provoking Khrushchev. However, he was well aware of the "domino theory," and it is clear from the same press conference that he was prepared to take a stand.

"Our response will be made in close cooperation with our allies and the wishes of the Laotian government. We will not be provoked, trapped, or drawn into this or any other situation but I know that every American will want his country to honor its obligations to the point that freedom and security of the free world and ourselves may be achieved."

He decided to deploy pretty much the only real operational reserve he had—10,000 Marines based in Okinawa. The threat of escalating military action worked and Khrushchev backed down, with the flow of men and materiel across the Plain of Jarres slowing to a trickle. Later in 1961, Great Britain and the Soviet Union initiated talks in Geneva for a peaceful solution and, on January 19th 1962, a ceasefire agreement. The whole situation had "sensitized" Kennedy to communist insurrections in Southeast Asia and, now that the situation in Laos had been reasonably settled, he turned his attention to South Vietnam. In fact, Kennedy said about Vietnam that it "was the worst one we've got." In 1962 he oversaw the formation of the Military Assistance Command of Vietnam (MACV), which provided American assistance in the training of the Vietnamese Army, and was the beginning of a large-scale troop commitment that would escalate into the Vietnam War. This moment, this press conference, and these three maps would have a profound effect on the second half of the 20th century.

RIGHT *In this map (or rather series of three maps) is a classic example of the use of cartography for political effect. The successes of the communists grow, the red tide comes in. And the juxtaposition of the political leader with a map was an image as familiar to the Russians or the Chinese as it was (and is) to the American people. It denotes power and understanding, and implies control.*

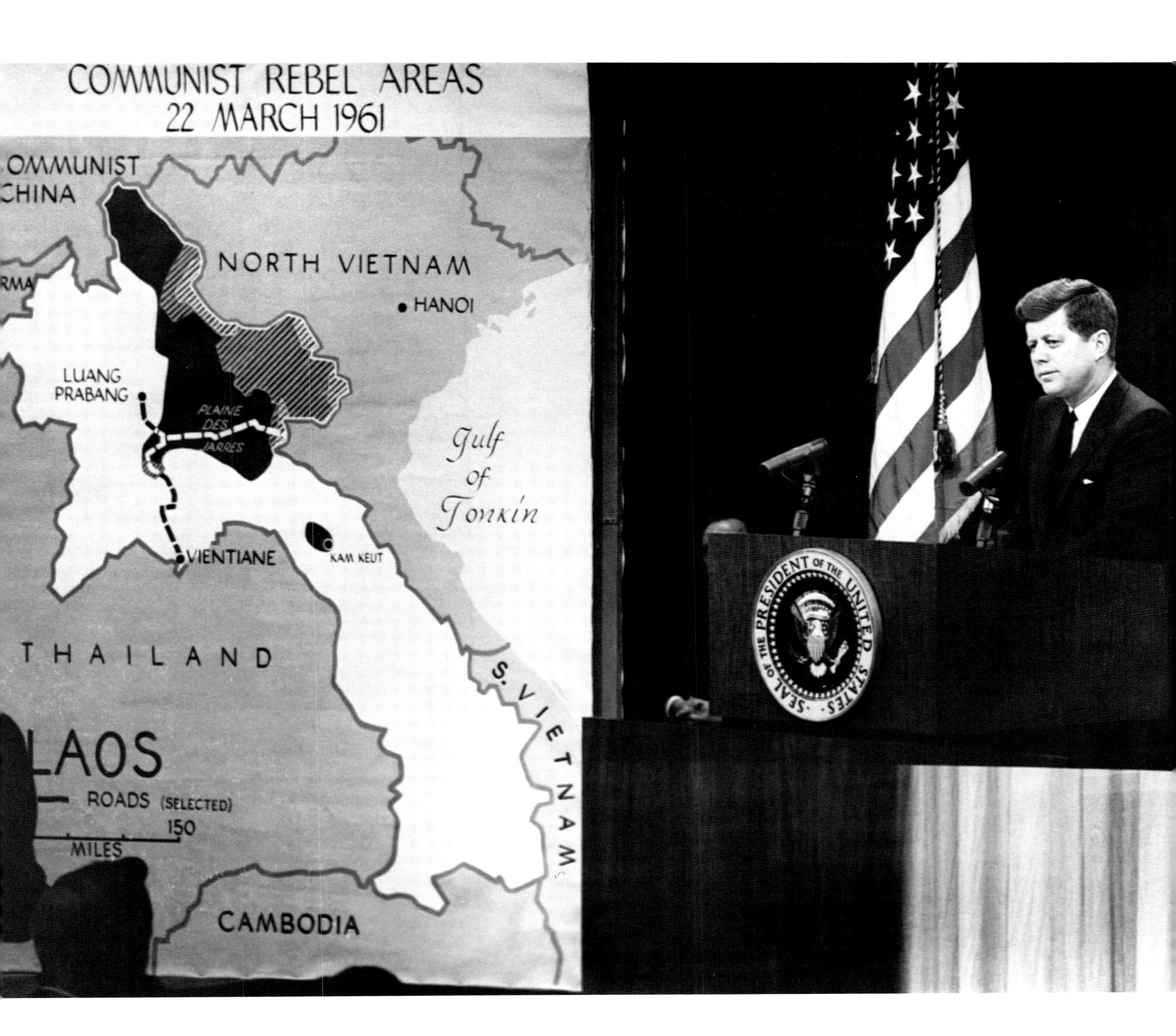

COMMUNIST REBEL AREAS
22 MARCH 1961
NORTH VIETNAM
HANOI
LUANG
PRABANG
PLAINE
DES
JARRES
Gulf
of
Tonkin
VIENTIANE
KAM KEUT
THAILAND
S. VIETNAM
LAOS
ROADS (SELECTED)
150
MILES
CAMBODIA
SEAL OF THE PRESIDENT OF THE UNITED STATES

"Provision of Space for Pursuit after Suspect":

Maps of Israel

In recent times, no nation's map has changed as often or as radically as that of Israel. This reflects the frequent conflict in which the country has been embroiled, a situation that still pertains. In addition to maps showing real territorial changes, Israel has also been the subject of many speculative maps expressing plans for peaceful partition between Jews and Arabs that never came to fruition.

Since the late 19th century, Zionist emigrants from Europe had pioneered the concept of a Jewish homeland in Palestine, by settling and farming tracts of land there. However, the formal inception of the process that led to the founding of the modern state of Israel may be traced to the Balfour Declaration of 1917. This brief statement by the British government, which was soon to be granted the mandate by the League of Nations to govern Palestine in the aftermath of the First World War, expressed its support for a national home for the Jewish people in Palestine, " ... it being clearly understood that nothing shall be done which may prejudice the civil and religious rights of existing non-Jewish communities."

In the event, the state of Israel came into being not through any planned procedure devised by an outside agency, but in circumstances of extreme turmoil. Under the British mandate, tension and violence between Jews and Arabs had increased steadily throughout the 1920s and 1930s. Ever greater numbers of Jews arrived fleeing Nazi persecution, and after the Second World War, a pressing need arose for a resolution of the question of a homeland. As early as 1937, the British Peel Commission had concluded that Arab and Jewish interests could never be reconciled within a unitary state, and accordingly drew up a proposal for partition. This foresaw a large Arab state and a small Jewish state that covered the coastal plain in the north of the country and extended over to the Syrian border on the eastern shore of Lake Galilee. A strip running from the coast at Jaffa to Jerusalem would remain under British control.

By the time of the next partition proposals, the political and demographic landscape had undergone a seismic shift. In 1946, the Anglo-American Morrison-Grady committee put forward the idea of a binational state, a suggestion that failed to satisfy any of the parties involved. In particular, the *Yishuv*, the Jewish community in Palestine, took the view that only a Jewish sovereign state would be acceptable, and emphatically not a Jewish minority within an Arab state. Post-Holocaust, they felt an abhorrence toward any plan that smacked of further ghettoization. For their part, the Palestinians were totally opposed to any solution other than an Arab Palestine; they maintained this implacable ideological position right up to the early 1990s, when their leader Yasser Arafat finally recognized Israel's right to exist.

In 1947 the Jewish Agency Executive drafted their own plan, which laid claim to all Palestine except the West Bank—the area west of the Jordan River that included the large Arab towns of Jenin, Ramallah, and Nablus. While the partition proposal advanced by the UN Standing Committee on Palestine (UNSCOP) in November of that same year fell far short of the Jewish Agency's territorial wishes, it did guarantee the Jews their crucial sovereignty, and so was accepted. The UN plan granted large areas in the north, center, and south of the country (including the Gaza Strip) to Palestinian Arab communities, but was still rejected outright by their representatives. When David Ben-Gurion unilaterally announced the creation of the state of Israel on May 14, 1948, the fledgling state was immediately plunged into war with its Arab neighbors. The armistice settlement signed in 1949 left the victorious Israelis with as much territory as the Jewish Agency

RIGHT *"Palestine Index to Villages and Settlements" produced by Sami Hadawi in 1949. The legend indicates the UN partition plan of 1947, Arab State, Jewish State, the international zone of Jerusalem, armistice line of 1949, demilitiarized zones, "No-Man's Land," and "Land in Jewish Possession at 31 March 1945." The map was published by the Palestine Refugee Office in New York. At the armistice, Jewish forces occupied the mixed population cities of Tiberias, Safad, Haifa, and Jerusalem, 50 Arab-only towns and cities and about 500 Arab villages, plus about 100 Bedouin villages and localities.*

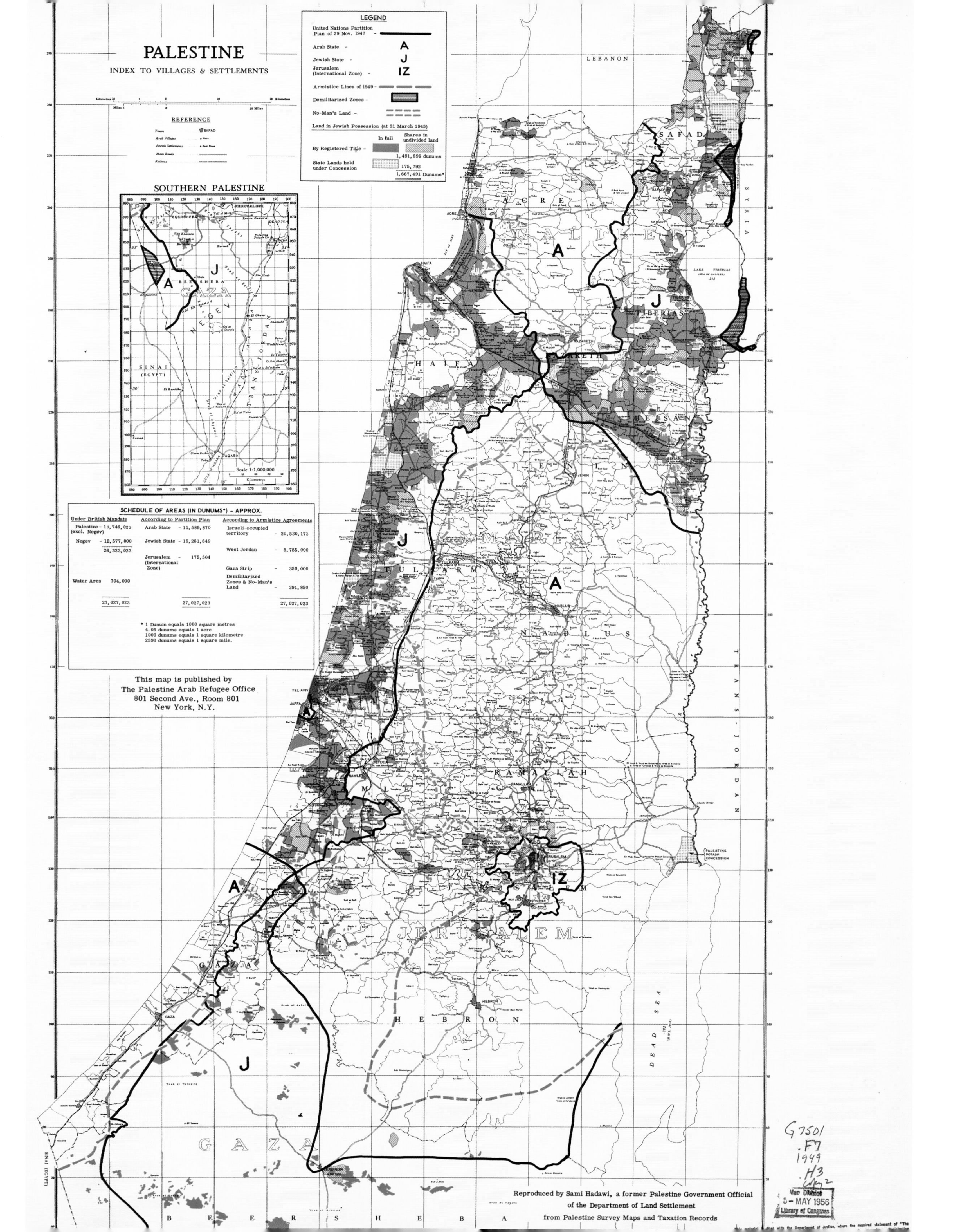
PALESTINE
INDEX TO VILLAGES & SETTLEMENTS
REFERENCE
LEGEND
United Nations Partition Plan of 29 Nov. 1947 -
Arab State - A
Jewish State - J
Jerusalem (International Zone) - IZ
Armistice Lines of 1949 -
Demilitarized Zones -
No-Man's Land -
Land in Jewish Possession (at 31 March 1945)
In full
Shares in undivided land
By Registered Title -
1,491,699 dunums
State Lands held under Concession
175,792
1,667,491 Dunums*
SOUTHERN PALESTINE
SINAI (EGYPT)
Scale 1:1,000,000
Kilometres
SCHEDULE OF AREAS (IN DUNUMS*) - APPROX.
Under British Mandate
Palestine - 13,746,023 (excl. Negev)
Negev - 12,577,000
26,323,023
Water Area 704,000
27,027,023
According to Partition Plan
Arab State - 11,589,870
Jewish State - 15,261,649
Jerusalem (International Zone) - 175,504
27,027,023
According to Armistice Agreements
Israeli-occupied territory - 20,530,173
West Jordan - 5,755,000
Gaza Strip - 350,000
Demilitarized Zones & No-Man's Land - 391,850
27,027,023
* 1 Dunum equals 1000 square metres
4.05 dunums equals 1 acre
1000 dunums equals 1 square kilometre
2590 dunums equals 1 square mile.
This map is published by
The Palestine Arab Refugee Office
801 Second Ave., Room 801
New York, N.Y.
LEBANON
SYRIA
TRANS-JORDAN
DEAD SEA
LAKE TIBERIAS
SAFAD
ACRE
GALILEE
TIBERIAS
HAIFA
BEISAN
JENIN
TULKARM
NABLUS
RAMALLAH
JERUSALEM
HEBRON
GAZA
BEERSHEBA
TEL AVIV
JAFFA
PALESTINE POTASH CONCESSION
Reproduced by Sami Hadawi, a former Palestine Government Official
of the Department of Land Settlement
from Palestine Survey Maps and Taxation Records

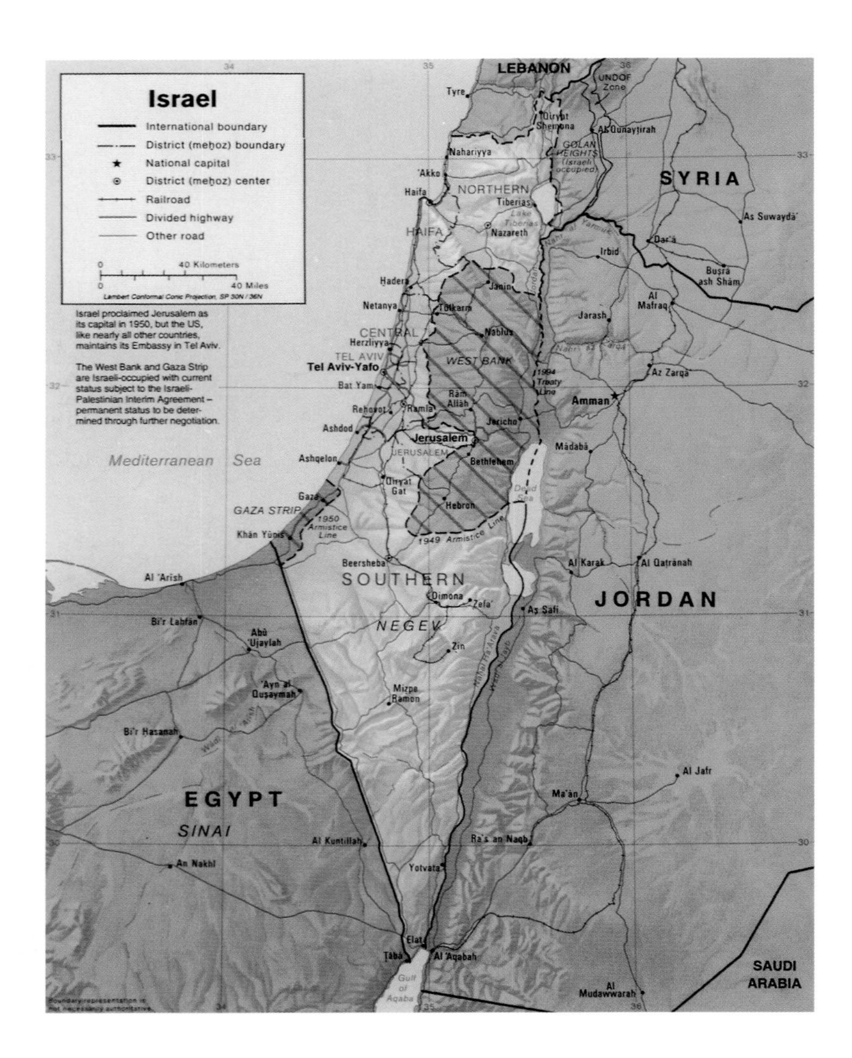

ABOVE AND RIGHT *CIA map of Israel, 2001 (above); "The West Bank and Gaza Strip are Israeli-occupied ... permanent status to be determined through further negotiation." The CIA map of the Gaza Strip in 2000 (right) distinguishes between "Israeli-controlled area" and (in yellow) "Israeli security control, Palestinian civil control." The dotted line is the security perimeter.*

proposal had envisaged. It is from this time that the so-called "Green Line" separating the Palestinian heartland of the West Bank from Israel proper dates. Under Jordanian jurisdiction, the West Bank fell under Israeli control in 1967, after a combined assault by Syria, Egypt, and Jordan had been repulsed in the Six-Day War. Curiously, although many atlases used today in Palestinian schools do not acknowledge the state of Israel, they refer back to the arbitrary borders established by the armistice of 1949 by clearly showing the course of the Green Line.

Israel reached its greatest extent after the astonishing gains made in the Six-Day War; having captured the whole of the Sinai Peninsula from Egypt, its troops now stood on the Eastern Bank of the Suez Canal. In the northeast, the strategically important Golan Heights were taken from Syria. The country again contracted in 1979-82, when, following the historic Camp David Accords with Egypt, Israeli forces made a phased withdrawal from the Sinai. A peace deal with Syria over the Golan Heights remains elusive. Undoubtedly the most significant change of recent times, however, has been the handing over in 1995 of administration of the Gaza Strip and parts of the West Bank (an area referred to by implacable Zionist settlers as Judea and Samaria) to a new Palestinian National Authority; at last, it appeared as though two states were on the brink of a relatively peaceful, if uneasy, coexistence.

However, relations between the embryonic Palestinian state and Israel are still in dangerous flux. On the one hand, Israel's push to build more homes in settlements in Palestinian areas, particularly to accommodate a large number of Jews leaving Russia after the collapse of the Soviet Union in 1991, has brought it into conflict with the United Nations and inflamed Arab opinion. On the other, a growing militancy among disillusioned Palestinians has manifested itself in a wave of suicide bomb attacks against Israeli civilians. The autonomous regions were reoccupied by Israeli forces to quell the violence.

RIGHT *Map of the route of the Security Fence issued by the Israeli Ministry of Defense. The accompanying notes state: "Every effort has been made to avoid including any Palestinian villages in the area of the Security Fence. The Security Fence does not annex territories to the State of Israel, nor will it change the status of the residents of these areas."*

It is in this context that a highly controversial new feature began to appear on the map of Israel from the early 2000s onward. The Security Fence or Separation Wall is a barrier designed to seal off the Palestinian areas of the West Bank to protect Israel from terrorist incursions. The largest infrastructure project in the country's history, its total length is estimated at around 410 miles (656 kilometers); it chiefly comprises a fence, although for 15 miles (24 kilometers), at places where snipers were once active, a 30-foot high concrete wall has been erected.

For its supporters, the fence is nothing more than a security barrier whose sole purpose is to save lives; they point to the dramatic drop in suicide attacks since completion of the first phase at the end of July 2003. Yet opponents—from both ends of the political spectrum—fear that the barrier represents a unilateral redrawing of Israel's borders. From the right wing, it is claimed that the route of the fence on or near the Green Line tacitly accepts the 1949 armistice line as the frontier (so preparing the Israeli public for a future wholesale handover of Judea and Samaria to the Palestinians, who would create a "Jew-free" state there). In contrast, those on the political left see the fence as the first step in corralling Palestinians into economically unviable enclaves. In raising the specter of "Palestinian Bantustans," their argument alludes to the notorious scheme instituted by the apartheid regime in South Africa to deprive Black Africans of South African citizenship by assigning them to barren "homelands" with only nominal independence. The Security Fence is a perfect example, some argue, of a border being built, a cartographic *fait accompli*.

Less emotive objections to the barrier point to the potential harm it will do to the economic, educational, and social cohesion of the community under the jurisdiction of the Palestinian National Authority. Some Palestinian farmers will lose their livelihoods through being unable to access their fields. In contrast, the fact that the government has already decided against building eastern sections of the fence, which would have enclosed the Palestinian areas, is seen as a final rebuttal of the "enclave" notion. The security barrier is just the latest manifestation of the issue of land. Given that the various changing maps of Israel since its founding only call to mind its history of reciprocal injustices and bloodshed, it is an irony that the latest US plan for resolution of the conflict is called the "Road Map for Peace."

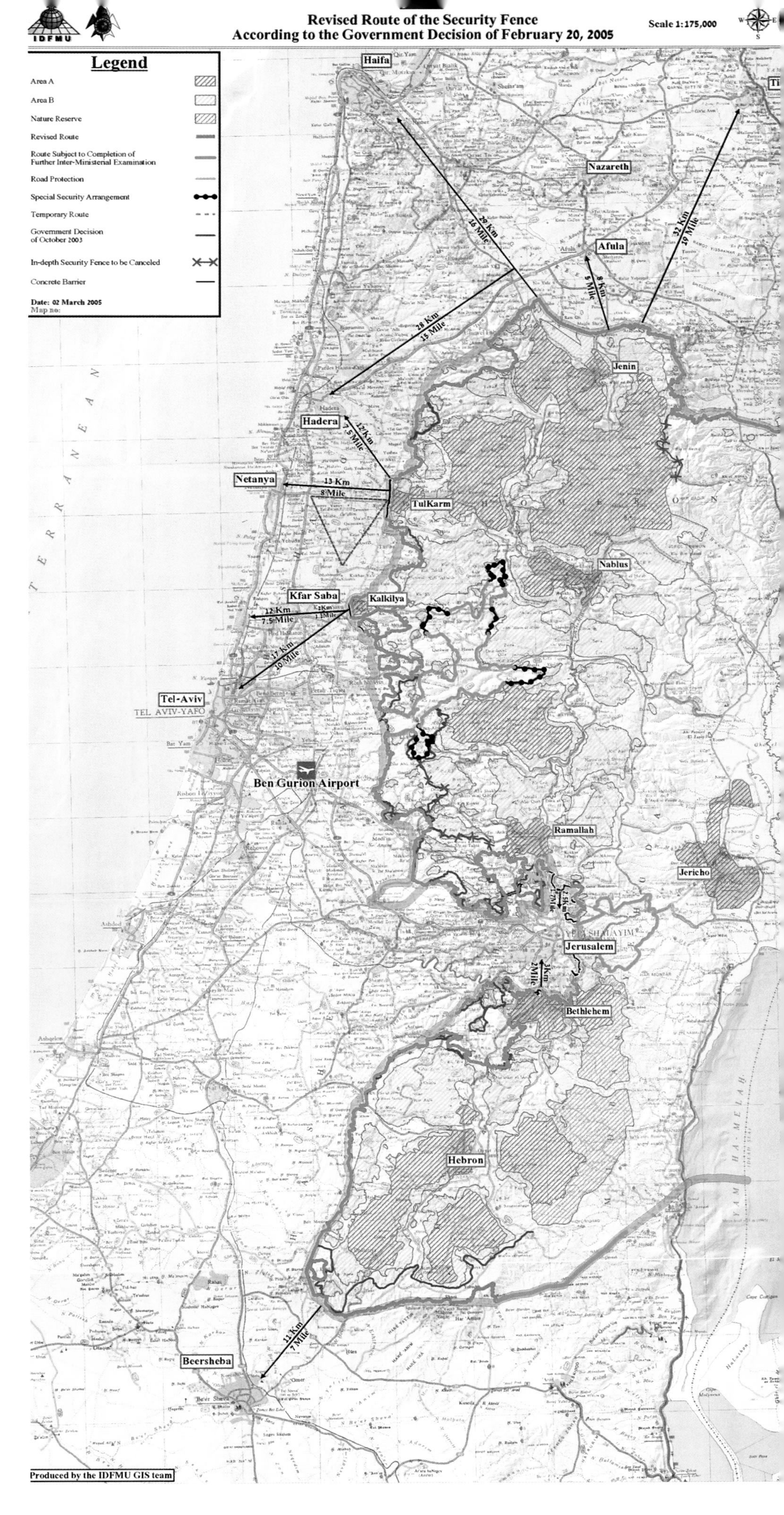

Septentrio

norvega

svetia

scotia

hyrlanda

Anglia

amborg

lubiech

alamagna

polonia

fiandra

collonia

EUROPA

paris

bretagna

Mar Atlanticu

ocidens

la rorella

bajona

prouenza

croatia

istria

dalma

hispania

catalogna

portogal

lisbona

granata

cadiz

corsica

Roma

sardegna

napoli

balearj insl

calabria

sicilia

barbaria

tunis

auster

Charte de navigatione de co
gnito orbe fin a tepi nri ~ cum
perenne gloria a Dio me fecit
fra Matheo de Chiara de arimino
ann. Dne M. D. XIX.

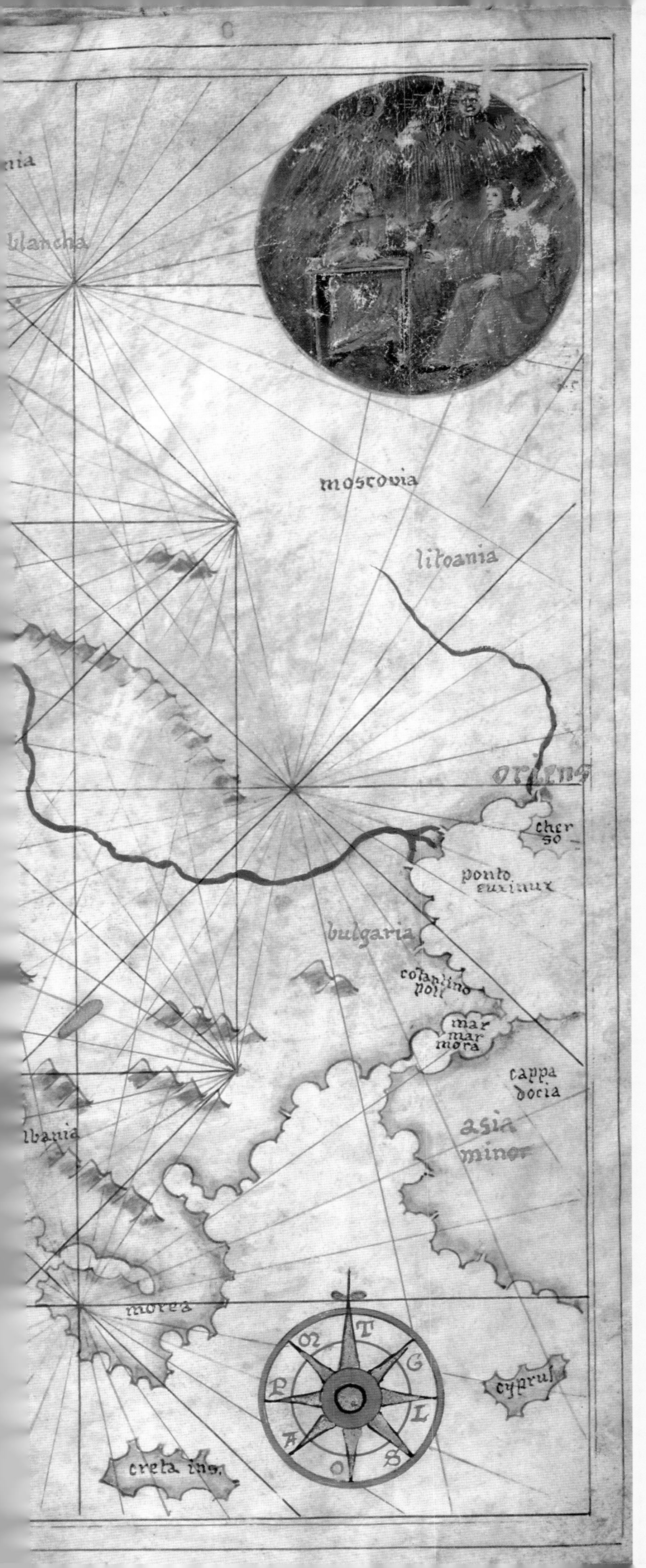

FANTASIES, FOLLIES, & FABRICATIONS

This portolan of Europe and the Mediterranean is one of four charts in an atlas apparently produced by Matheus de Chiara in 1519, as stated in ther bottom left corner. It is held in the Huntington Library, San Marino, California. The parchment is old and has worm holes and the black ink outlining the land masses is overtraced with ocher. But several things are wrong: the 32-line patterns of the rhumb lines are sometimes incomplete, rhumb lines do not radiate from the compass rose, and the handwriting appears to be of 19th or early 20th century origin. Crucially, the accompanying chart of North and South America shows far too much detail of the west coast of North America for such a date to be possible. This is a fake.

Race Memory Loss:

The Lost Continent of Atlantis

The belief in the lost continent of Atlantis is an ancient one. There is a widespread belief in a single, brilliant ancient civilization that inspired all Mankind through its achievements in the arts and sciences. The theory of the rise and disappearance of this civilization might provide answers to many mysteries: how did the Egyptians and Mayans develop identical pyramid building technologies? And why do Egyptians and Mayans (along with a score of other races) have identical legends about a lost civilization that vanished beneath the waves?

The earliest record of the legend of Atlantis comes to us in the 4th century B.C. by way of the Greek philosopher Plato in Timaeus. Plato tells us of the Athenian statesman Solon who traveled to Egypt. While visiting the temple of Sais in the Nile Delta, Solon spoke to a priest who told him the history of the lost civilization of Atlantis. According to Plato, the priest told Solon that Atlantis was once a great sea power, but was destroyed some 9,000 years before their time.

Plato's continent of Atlantis was described as being about the size of Spain and located in the western sea beyond the Pillars of Hercules (Straits of Gibraltar). Atlantis' power extended to all the nations of the world, but their desire for greater power seemed to know no bounds. Soon, it was believed the Sea Kings' pride and ambitions brought them to ruin. According to Plato's observations, Atlantis was constructed in concentric circles, connected by bridges over canals, with a massive circular temple at its the center. Plato's description of Atlantis seems to be expressed in terms of a musical allegory. The Kings of Atlantis eventually brought such discord that the foundations began to crack and collapse. In Plato's account, Zeus is called to pass judgement. We are not told of the judgement, but Atlantis vanishes, and appears to have sunk beneath the waves.

There have been many theories about the location (or former location) of Atlantis; from the Bahamas and the Bermuda triangle to the Canaries. However, the 17th century encyclopedist Athanasius Kircher drew a couple of maps of Atlantis. The one shown on page 222 is a map of the continent of Atlantis in the tenth millennium B.C. that takes up a large part of the North Atlantic. In his own time, Kircher believed the Azores and the Canaries were all that remained of Atlantis. Ever since the publication of *Twenty Thousand Leagues Under the Sea* in 1869, in which Jules Verne's fictional hero Captain Nemo discovered the ruined temples and palaces of Atlantis on the ocean floor; there have been hundreds of adventurers determined to make exactly such a discovery.

Curiously enough, one of the most committed believers in the existence of an historic Atlantis was the author J. R. R. Tolkien. From early childhood, Tolkien experienced one terrible recurring nightmare about a great towering wave looming up and flooding over the trees and green fields. What was most frightening to him was the feeling that it always felt more like a vivid memory than a bad dream.

It was not until he came to read the legend of Atlantis that Tolkien came to link this ancient disaster with his dark dream. Remarkably, because he never wished to alarm his children with this frightening dream, Tolkien did not learn until years later that his son Michael had somehow inherited the same recurring nightmare. It was this discovery that finally persuaded Tolkien that this dream of a great wave was actually an inherited "racial memory." J. R. R. Tolkien managed to live just long enough to have the destruction of an ancient Atlantis-like civilization proved; and his terrifying vision of a gigantic wave was confirmed as historically authentic.

Whatever the motivation, J. R. R. Tolkien went on to create or "reinvent" the most elaborate and complex history of Atlantis ever

RIGHT *Dr. Ulf Erlingsson, in his book* Atlantis From a Geographer's Perspective *argues that the empire of Atlantis is defined by the megalithic tomb distribution shown here, and Ireland is Atlantis. He returns to Plato's decription and finds equivalence in size and the fact that Ireland is a plain surrounded by mountains, as was Plato's Atlantis. Dr. Erlingsson is an expert on underwater mapping; but his theory evoked some strong feelings, especially in Ireland. His book's map has that delightful faux parchment faery land style—perhaps it didn't help his argument academically—though the book's subtitle is "Mapping the Faery Land."*

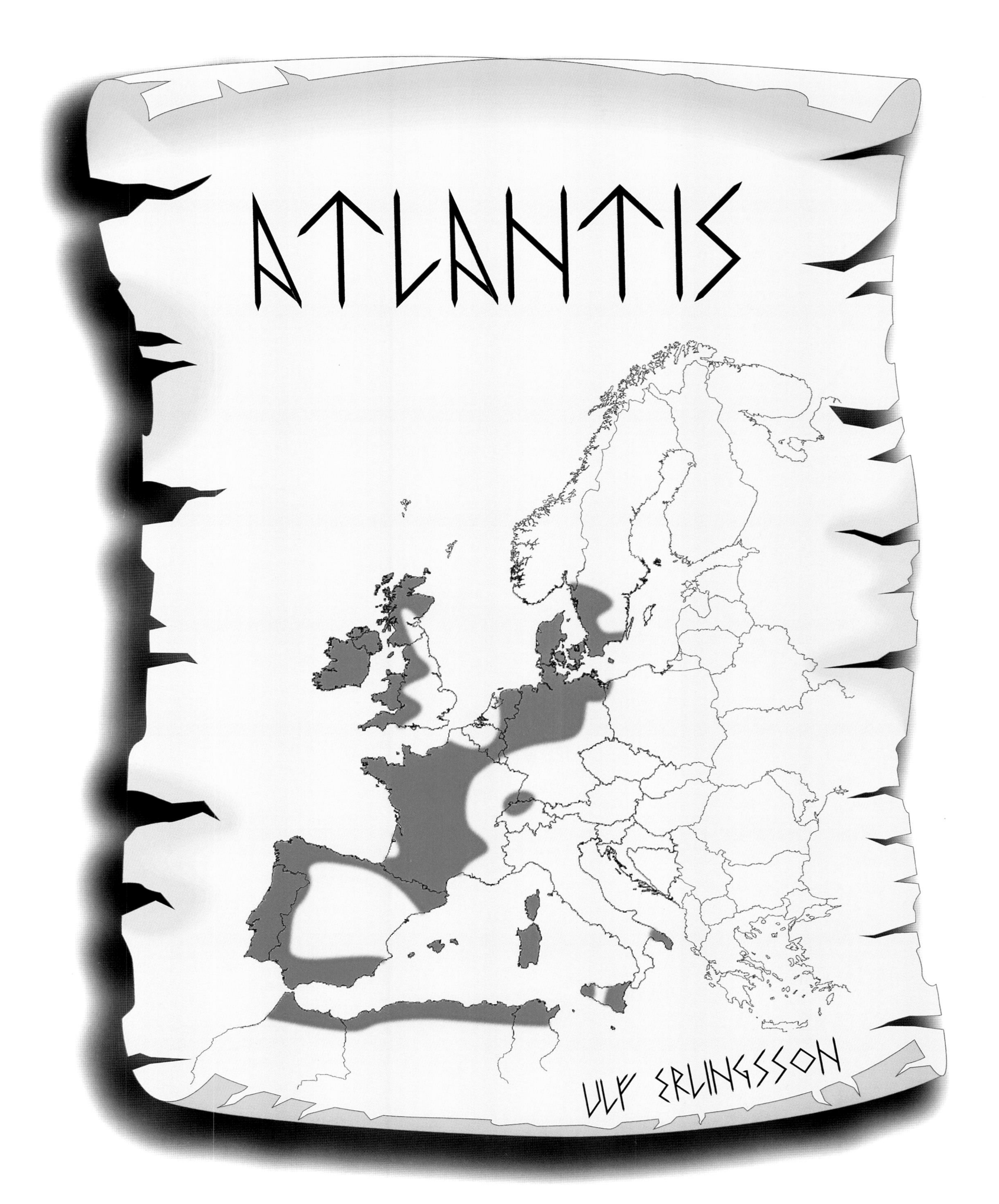
ATLANTIS
ULF ERLINGSSON

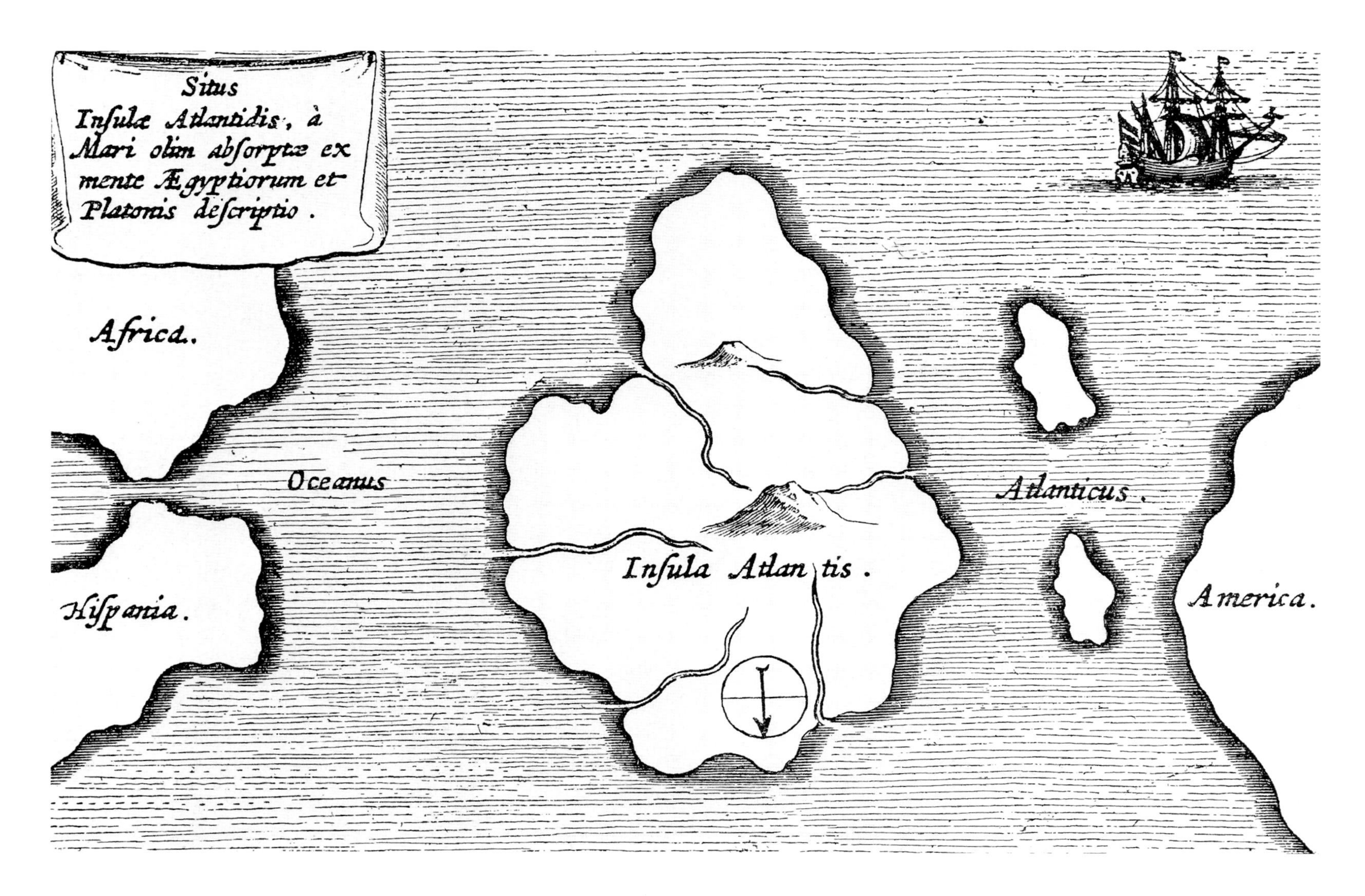

ABOVE *Map of "Insula Atlantis" engraved in Athanasius Kircher's* Mundus Subterraneus, *1665; Atlantis pretty well fills the North Atlantic. Kircher claimed to have based his map on Plato's description—as does Dr. Erlingsson.*

assembled. Tolkien compiled detailed chronologies covering 33 centuries of history as accounted for in the "Akallabeth" or "Downfall of Numenor." In case anyone had any doubts, we are informed that Numenor translates from Elvish to Atlante or Atlantis.

Tolkien's Atlantis was a roughly star-shaped island and was called Numenor or "Westerness." At its narrowest it measured 250 miles (400 kilometers) across, and double that from its furthest promontories. It was divided into six regions: one for each peninsula and one for its heartland, where stood the sacred mountain, Meneltarma, the "pillar of heaven." Its capital was Armenelos, the City of Kings, and it had three great port-cities. Numenor was a great star set in the middle of the Great Sea. To the east was the mortal world of Middle-earth; and to the west was the immortal world of the Undying Lands. Tolkien's Numenoreans become the most gifted and powerful race of mortals upon Middle-earth. The most noble of them live for several centuries; but they are still mortal. Eventually, they wish to become immortal and make war on the Gods. This brings about the destruction of Numenor that, like Atlantis, sinks into the Western Sea. This is a kind of second fall of man, as occurred during the time of Noah and the Great Flood.

In the mid-1960s, excavations in the Aegean Sea proved that a rich and powerful island kingdom had been destroyed by titanic volcanic eruption. This was followed by a massive tidal wave believed to be in excess of 300 feet (90 meters) in height. The wave swept through the entire Mediterranean basin from Anatolia to Gibraltar. The volcanic eruption was roughly equivalent to the force of six thousand hydrogen bombs, the greatest volcanic blast ever witnessed by humanity and almost beyond imagination to comprehend.

The island kingdom at the center of this event was Thera, one of the wealthiest and most powerful city-states in the ancient Minoan Empire. It was obliterated in a single cataclysmic eruption. The island's 20-

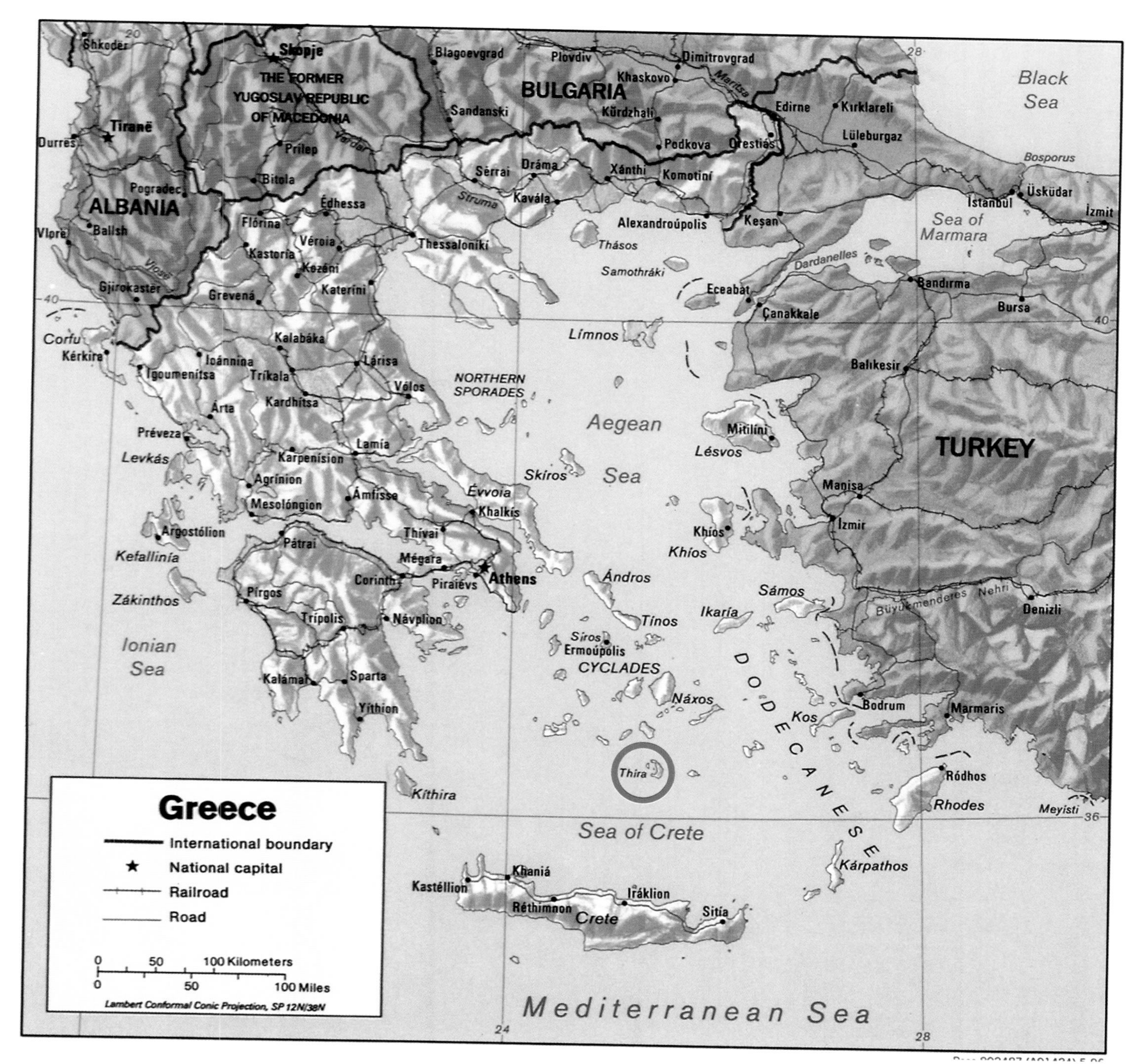

ABOVE *Map of the Aegean pinpointing the remains of Atlantis, if the massive volcanic eruption and destruction of a wealthy island kingdom here somewhere around 1500 B.C. (indicated by marine archeology) holds the key.*

square-mile (50-square-kilometer) landmass was blasted into the atmosphere. All that now survives of the island is a narrow crescent-shaped volcanic ridge a few miles long that constitutes the island now known as Santorini.

It seems undeniable that if Atlantis is based on any historic event in human history, it must be this volcanic blast that brought about the downfall of Thera. Furthermore, the destruction of this historic Atlantis occurred in the middle of the second millenium B.C.; or about 900 years (rather than an almost impossible 9,000 years) after the events reported by Plato. It seems very unlikely that an event of this magnitude would not be elevated to legendary status.

No one could possibly have survived the blast of the historic Atlantis. All of Thera's cities, fleets, and its entire population vanished beneath the sea. And it seems the entire Minoan Sea Empire's ports, harbors, towns, and coastal lands were swept away by the largest tidal wave ever to imperil the human race.

Rex Quondam Rexque Futurus:

King Arthur and the Isle of Avalon

At the end of Thomas Malory's "Le Morte d'Arthur," we are told of a terrible last battle of Camlann wherein nearly all the best of the knighthood of Britain is extinguished. There, the mortally wounded King Arthur commands the battle's last survivor, the loyal Sir Bedevere, to cast his sword Excalibur into the waters that lapped on the edge of the battleground. When the knight returns with the news that the king's sword was miraculously taken by the hand of a maid emerging from the lake, Arthur knew that his destiny was fulfilled.

... for I will into the vale of Avilion
to heal me of my grievous wound: and if thou hear never
more of me, pray for my soul.

Le Morte d'Arthur –Book XXI, Chapter V, Thomas Malory

It was at then, Malory tells us, in that twilight hour, that a great swan-like barge appeared upon the water bearing nine noble ladies all hooded and cloaked in black. With their healing hands these nine fair noble women lifted the wounded King and took him to their barge and sailed through the mists to the distant Isle of Avalon. And on that Isle of Avalon legends tell us that Arthur's mortal wounds were healed, and that he lives there still to this very day.

There are many traditions that proclaim that King Arthur upon Avalon continues to hold court as he did in Camelot. There, all the heroes of the world come to his table and feast and enjoy tournaments and practice all manner of sports and contests. From this place King Arthur watches over the world. For he is the guardian of the Britons and all the other people of his ancient realm. Prophecies foretell, in the time of their greatest need, King Arthur will return and be their salvation.

Although most famous for legends related to Arthurian romance, the Isle of Avalon had a far more ancient pedigree as an enchanted realm and refuge where all that was once mysterious and magical may still be found. Avalon, as the land of the nine Queens, the nine Muses, the nine Graces, and all the handmaidens who serve those invisible forces that inspire mortals, has taken on many forms over the ages.

The name Avalon, Celtic for "Isle of Apples," reveals something of its origins. There are scores of legends and tales of an earthly island paradise over the western sea where there is a tree on which grow "the golden apples of the sun and the silver flowers of the moon." Avalon's primary inspiration seems to come from Greek mythology, being modeled on the "Fortunate Isles" in the western sea beyond the Straits of Gibraltar. On these Isles the fabled Gardens of the Hesperides were to be found. The Hesperides were the nine daughters of Atlas the Titan on whose shoulders the skies rested. His wife was Hesperis the goddess of the west. ("Hesperis" is Greek for the "west.")

As the wedding gift of Zeus the King of the Gods to his Queen Hera, these miraculous gardens held many wonders. Most magnificent of all, however, was a tree that grew apples of gold from its branches; and had a spring of nectar flowing from beneath its roots. The golden apples provided infinite wealth, while the spring of nectar granted immortality to those who drank from it. These treasures embodied the universal wish for mortal newlyweds to have health and wealth forever. So precious were this tree and fountain, however, that Zeus kept them beyond the reach of mortals on an island west of the Pillars of Hercules (Straits of Gibralter). There, the sacred tree was placed under the care of the nine immortal daughters of Atlas and Hesperis. Around the base of the tree was coiled the great unsleeping serpent called Ladon, who was guardian of the golden apples and the fountain.

In Greek mythology, we find that one of the labors of Hercules was to pick three golden apples—the forbidden fruit—from this sacred tree in the Gardens of the Hesperides. In Norse mythology we have a similar quest involving the theft of Golden Apples from the garden of Idunn the Goddess of Youth. The consequence of this theft was the near

RIGHT *Some of the mooted locations of Avalon in the British Isles: 1 Tory Island; 2 Iona; 3 Isle of Man; 4 Anglesey; 5 Bardesy Island; 6 Glastonbury Tor; 7 Scilly Isles. Glastonbury Tor is a particularly attractive possibility, a dramatic and sacred place and no longer an island: thus a real place, but lost forever.*

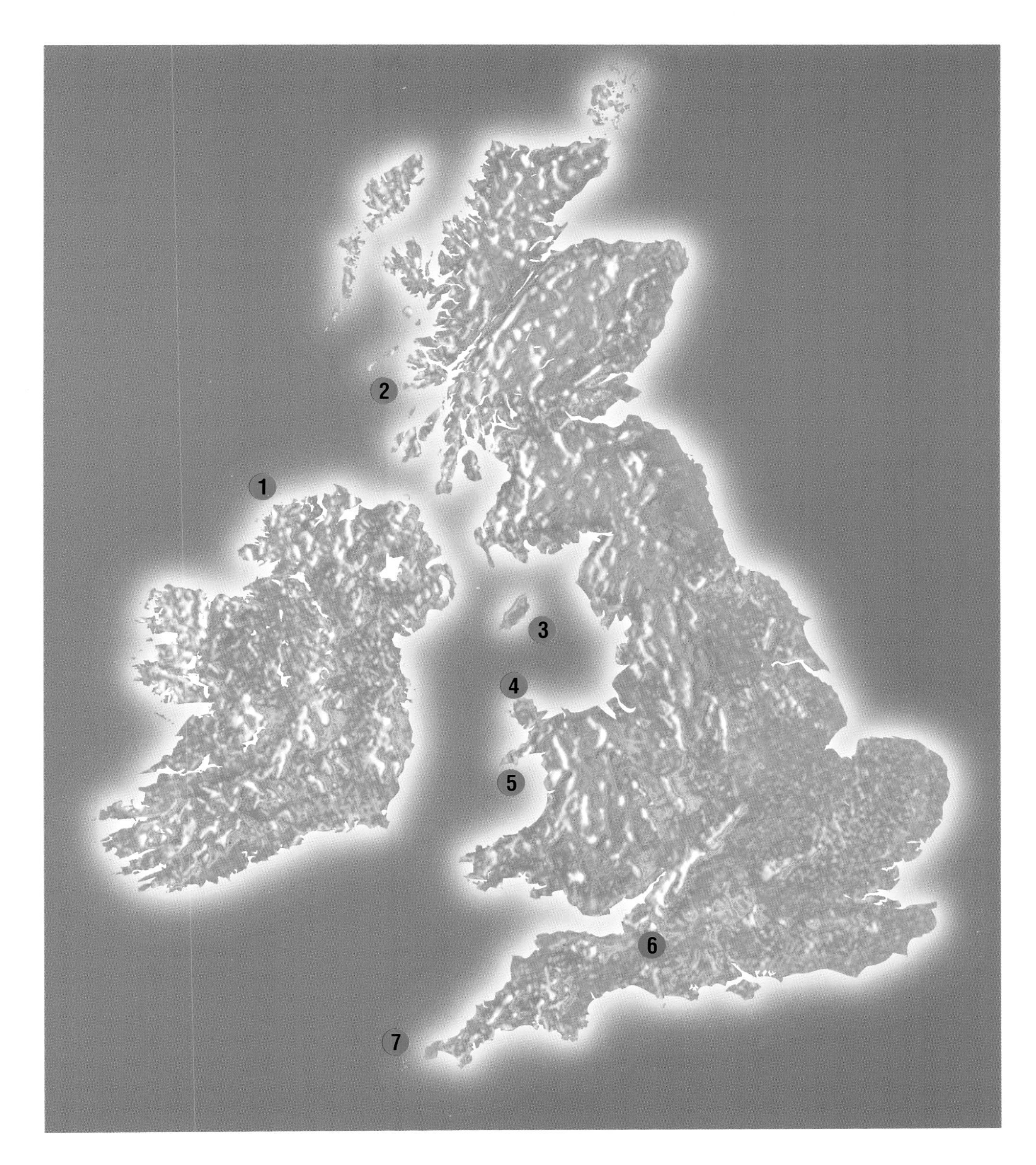

extinction of the Gods of Asgard whose immortality could only be sustained by these golden fruits.

In the Arthurian tradition, Geoffrey of Monmouth's early twelfth-century *Vita Merlini* ("Life of Merlin") offers one of the earliest descriptions of the Isle of Avalon, or *Insula Pomorum* ("Isle of Apples"). This was an isle of eternal summer warmth filled with self-propagating crops, and vines that sowed themselves. It echoes the medieval idea that upon the western sea there was still to be found an island that sustained itself forever as it was in the fabled golden age of the ancient world. Geoffrey tells us that on the Isle of Avalon all dwelled in peace, without illness, and lived for over 100 years. like Malory, Geoffrey claims that the wounded Arthur was taken there after the battle of Camlanus and

ABOVE *According to Geoffrey of Monmouth the dying Arthur sailed through the Straits of Gibraltar to be healed at Avalon. This would make one of the Canary Islands the likely destination. In his* Historia Regum Britanniae *(completed in 1138) he treats the Arthurian legend as a factual chronicle, in which the victorious king prepares to march on Rome before Mordred's treachery.*

ABOVE *West Africa and the Canary Islands: by Guillaume le Testu, 1556. Some traveled west not just in search of land and plunder, but in the hope of everlasting life, of discovering the lifegiving spring. (Le Testu's maps have been put forward as evidence that the Frenchman Paulmier de Gonneville was the first European to reach Australia, as early as 1503.)*

healed of his wounds. In the *Vita Merlini* Geoffrey even gives Avalon's precise geographic location: sailing south westward on prevailing winds from the Pillars of Hercules and through the straits of Gibraltar, the Isle of Avalon is to be found off the coast of Africa. By these directions one must conclude that Avalon is one of the Canary Islands, probably the largest, Grand Canary. "Macaronesia" is a modern collective name for several groups of islands in the North Atlantic Ocean near Europe and North Africa. The name comes from the Greek for "blessed islands," a term used by ancient geographers for islands to the west of the Straits of Gibraltar. "Macaronesia" takes in the archipelagos of the Canaries, the Azores, Cape Verde, Madeira, and the Ilhas Selvagens, or Savage Islands, that lie between the Canaries and Madeira. One could make an argument for any of the islands as being "blessed," and a potential Avalon. None of them was ever part of a continent, which makes their biogeography unique, and uniquely attractive; magical, in fact.

Grand Canary was only one opinion. There seems to be no shortage of suggested locations for Avalon. There is a long list of candidates: Anglesey, Tory Island, Isle of Man (described in an ancient Irish legend as "Ablach" which is Gaelic for "rich in apples"), Bardsey Island (the Welsh local favorite, home of Arthur, Merlin, and a magical cauldron that contained the long lost "Twelve Treasures of the Ancient Britons"), Iona Island (the Scot's "Isle of Dreams"), the Scilly Isles, and Avallon in

ABOVE *This undated woodcut shows Arthur sailing to Avalon. At the end of his poem Malory adds a final note: "Yet som men say in many partys of Inglonde that Kynge Arthur ys nat dede … that he shall com agayne, and he shall wynne the Holy Crosse."*

Burgundy. One of the most popular choices for Avalon is Glastonbury Tor, once an island surrounded by marsh water before the land was drained. It was claimed that there were once many apple trees there, hence its ancient name of "Inis Avalon."

During the 15th and 16th centuries, the belief in the existence of the Isle of Avalon resulted in many early explorers seriously going in search of its treasures on the newly discovered island of the Americas. In fact, the Spanish and Portuguese conquest of Central and South America resulted in the discovery of such vast quantities of gold among the Amerindian nations that many Europeans were inclined to believe that there might be some supernatural source for all this wealth. Certainly the expedition led by the Spanish explorer Ponce de Leon into the swamps and jungles of what is now Florida was in part motivated by a quest for infinite wealth and eternal life through the discovery of the fabled Fountain of Youth.

In the 19th century, although Tennyson's epic "Idylls of a King" revived interest in King Arthur in general, it was the American poet Longfellow and the English writer and artist William Morris who specifically revived interest in the legends relating to the Isle of Avalon. In the 20th century there have been many investigations and retellings of legends related to the Isle of Avalon: none perhaps more original than Marion Zimmer Bradley's *Mists of Avalon*, which gives us the story of Arthur from the perspective of Morgan Le Faye.

Baron de Lahontan:

New France or Fake America?

In the 17th century, Europeans' knowledge of the Great Lakes region of North America came from travelers and explorers, whose accounts formed the basis of maps of the region (see page 128). One such was French Baron Lahontan, who made maps and wrote one of the first travelogs. But, like some journalists who followed him, he never let the truth get in the way of a good story. In 1931 expert cartographer Louis Karpinski was moved to say of Lahontan's book and maps that "any real facts were only incidental or accidental."

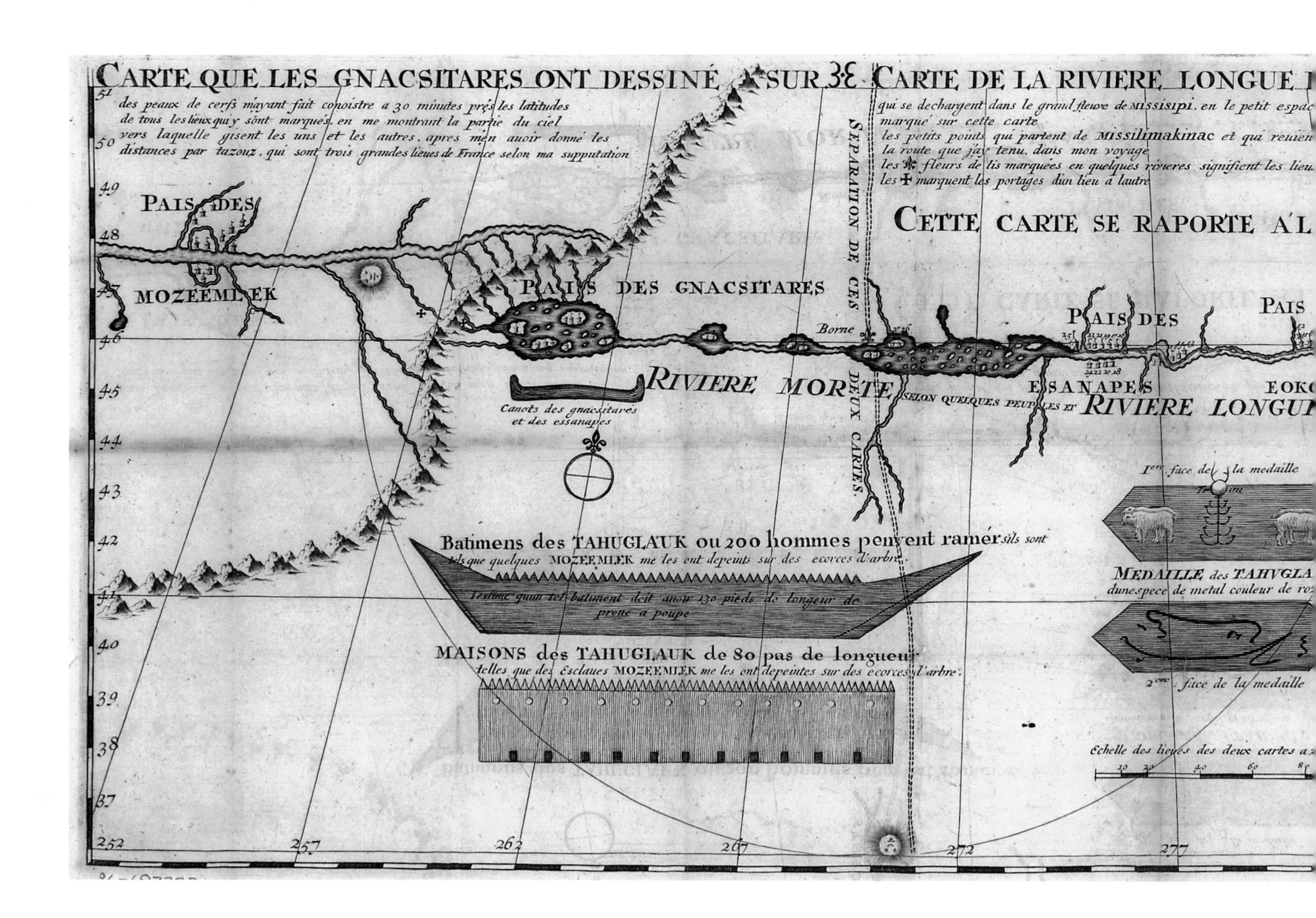

Louis-Armand de Lom d'Arce, impoverished third Baron Lahontan, was born in 1666 in La Hontan, in the French province of Béarn. He joined the army and was one of 200 troops that King Louis XIV sent to Québec, New France, in 1683 to help to put down the Iroquois Indians around the Great Lakes. After overwintering at Beaupré he wrote home that "In truth, the peasants here live much more comfortably than do many gentlemen in France". He then set off for Montréal, and left there in June 1684 with a scouting party to Fort Frontenac on Lake Ontario. Additional soldiers, local militia, and allied North American Indians swelled the force to 1,200, under the command of Governor Lefebvre de La Barre. Ravaged by fever, the army failed to suppress the Iroquois, and a disillusioned Lahontan returned to Ville-Marie (near Beaupré).

By the spring of 1685 Lahontan was at Forth Chambly and in the autumn he moved to Boucherville, where he stayed for nearly two years. With no arduous duties, he learned the Algonquin language and regularly went hunting, shooting anything with fur or feathers, including some of the Iroquois. In July he was given command of Fort Saint-Joseph (later to become Niles, Michigan) at the entrance to Lake Huron, which he and 100 men reached after getting around Niagara Falls, crossing Lake Erie and travelling up the Detroit River. After a hard winter and starved of supplies, he finally abandoned the fort.

Thus far the story is true, or as true as can be verified after more than 300 years. But then in September 1688 he set out on "another journey" with four or five "good Ottawa hunters," or so he said. According to his own account he arrived at Lake Michigan from the north, crossed the baie des Puants (Green Bay), canoed down Fox River and carried his canoes to the Wisconsin River. He then traveled west to the Mississippi and followed it as far as the "Long River" (all of this in the middle of winter). According to the Indians he met "the lower part of the river is adorned with six noble cities besides a hundred towns great and small" before the river emptied into a great salt lake. He set out on the equally laborious return journey in March 1689 and finally got back to the colony in July.

After returning to France for a year, Lahontan was again in Québec in late 1691, going back to France once more the following year to present to the French Minister for the Colonies his plans for forts and a fleet of ships on the Great Lakes. His plans rejected, he briefly became Lieutenant-Governor of Placentia, Newfoundland. But fearing arrest by the authorities for insubordination he deserted at the end of 1693 and took refuge in Holland. Using his diaries and notes, he then wrote his extremely popular travelogs, which were published in several volumes from 1703. His *Nouveaux Voyages* included references to previously unknown tribes with such unlikely names as Essanapes, Gnacsitares, Moozimlek, Nadouessioux and Panimobas ("sioux" being the only Indian-sounding syllable among them). He reported conversations with the local people in which they are romanticized as "noble savages." He even included a detailed map that purported to show his prolonged peregrinations around the Great Lakes and their associated rivers.

The conman was himself conned. Nicolas Gueudeville, an unfrocked monk, published in 1705 a new edition of the *Nouveaux Voyages*, with a much altered text. This brought further suspicion, though for many years after Lahontan's death in about 1715, indolent authors were still reciting his details of the marvels on the Long River.

BELOW Carte de la Riviere Longue; *and* Carte que les Gnacsitares ont dessiné sur des paux de cerfs. *The "Gnacsitare" tribe's deerskin map was supposed to show the area west of the Minnesota River.*

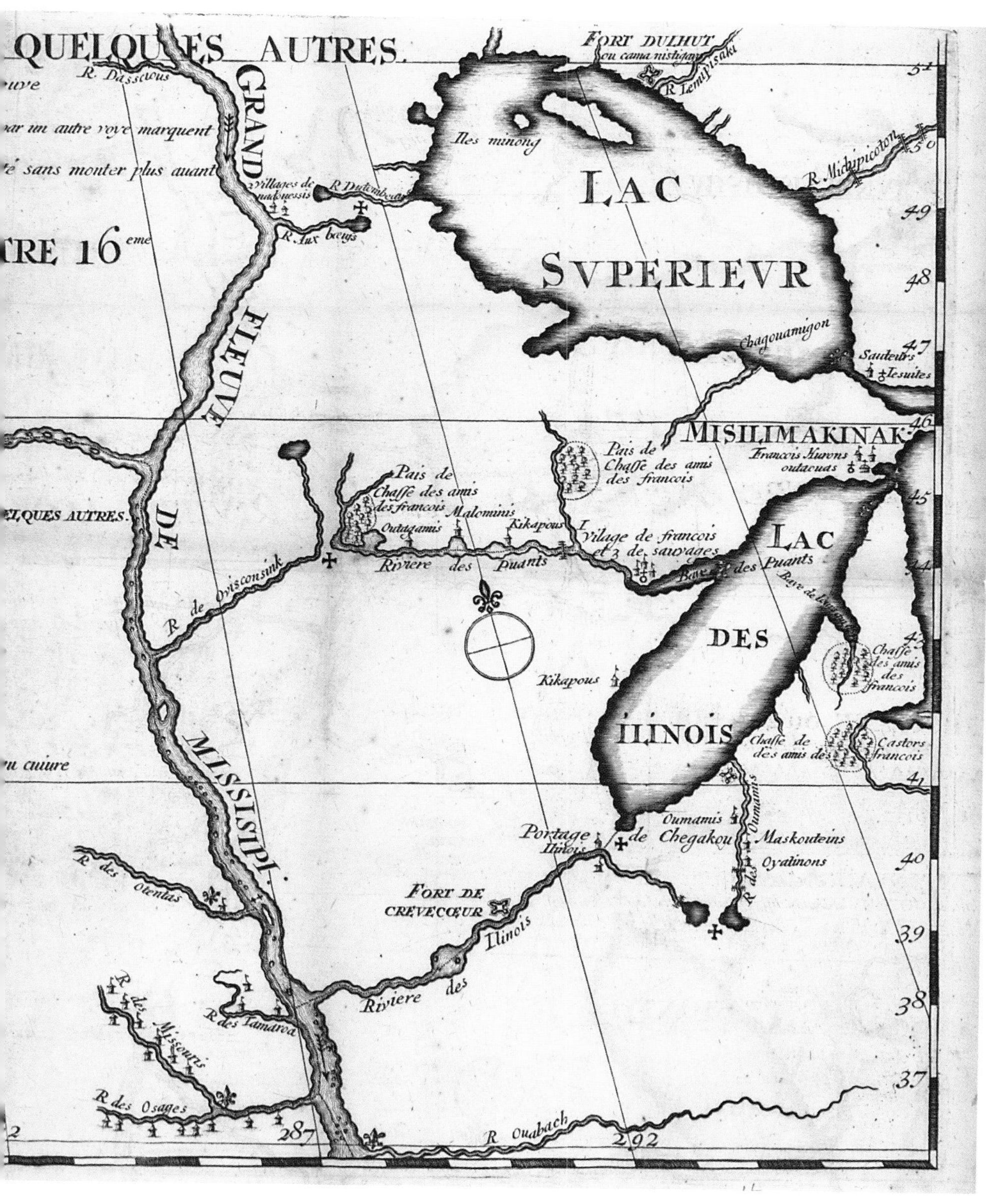

Etienne Cabet:

Utopia in Texas

In 1848 the February Revolution forced the abdication of the French King Louis-Philippe and established the Second Republic. Just before the shooting started, a group of socialist idealists left the inequalities of Europe to form a settlement in the United States. They were organized by—but not at first accompanied by—fledgling communist and would-be benevolent dictator of the world, Etienne Cabet.

Cabet was born in Dijon, France, on New Year's Day 1788, the son of a cooper. He trained as a lawyer and soon went into politics. His leftist views were too extreme for the authorities, and in 1835 he was arrested, tried and sentenced to exile. He spent four years in London, where he read *Utopia* by Thomas More and became friends with British industrialist and social reformer Robert Owen (1771-1858). Cabet wrote a fictional work called *Voyage to Icaria*, in which the English Lord Icar sets up a utopian communist state organized into communes. All property is shared, and everyone lives happily ever after.

Cabet returned to France in 1839 where his bestselling book struck a chord with the French working class, if not the bourgeoisie, as did his weekly newspaper *Le Populaire*. (Marx and Engels wrote to Citizen Cabet, asking for clarifications of the position of their Alliance of German Workers to be printed in the paper.) Cabet's "true Christianism" was deistic and rational, and his God was the god of equality and justice for the poor and oppressed. It yielded him considerable support among the workers. Icarian organizations sprang up around the country. Then in 1847 he announced that he was going to establish an egalitarian society at some undisclosed location, and appealed for volunteers to join him. The government of Texas agreed to let him have a million acres of land, to which Cabet would send an advanced guard to establish an Icarian presence. Each volunteer had to contribute the large sum of 600 francs for the privilege. Cabet's champion in Spain was the inventor Narcís Monturiol (1819-85), who in 1859 was to develop the modern twin-hulled submarine. He and his friends in Barcelona drummed up the 600 francs for a young Spanish doctor, Joan Rovira, to join the 69 emigrants (leaving behind his pregnant wife). On 2 February 1848 they sailed from Le Havre en route to New Orleans. Three weeks later the (stillborn) French February Revolution began.

When they arrived in New Orleans, the Icarians found that they had to trek hundreds of miles to their "free" land in northwestern Texas. They discovered that in order to keep the land, they had to build houses at a density of one house per 320 acres by 1 July. Of the required 3,125 houses they managed to throw up 32, entitling them to only 10,240 acres. The land would not support crops and, to

RIGHT AND INSET *What a tragically optimistic map this is, produced presumably at the behest of Cabet (above) and now in the collection of the New-York Historical Society. No boundaries to the expected million acres of paradise are shown in this "itinerary for the advanced guard." As author Matthew Stewart points out in his book,* Monturiol's Dream, *contemporaries agreed that "To set up a utopian ideal in the mind as an object of aspiration is noble … but to set it up in Texas is insane."*

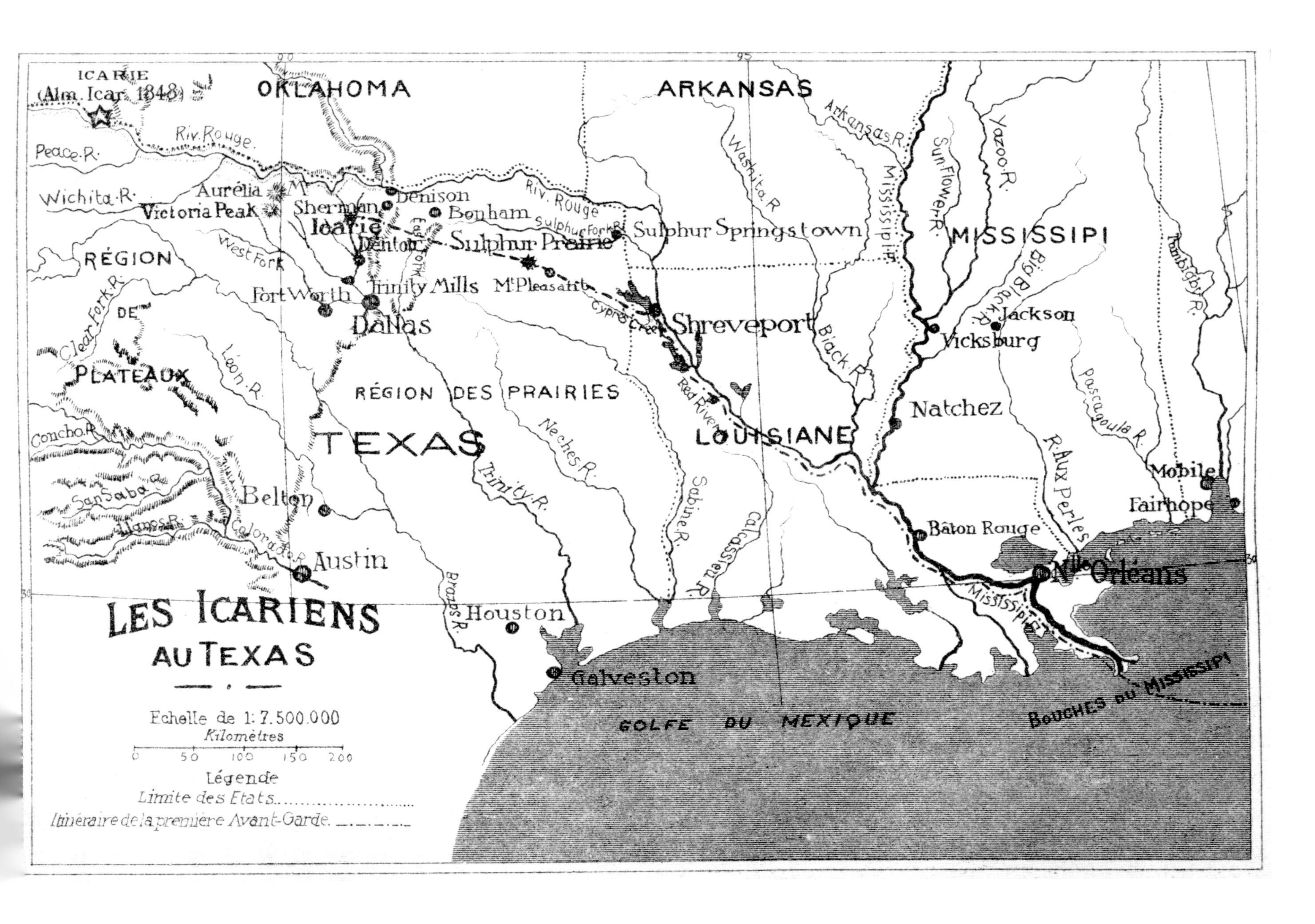

avoid a starving winter, the Icarians trekked back to New Orleans, losing several to injury and disease along the way.

In spring 1849 Cabet himself with a further 200 volunteers (including Rovira's wife and new baby) set sail for their promised land. In his absence Cabet was sentenced to two years imprisonment for fraud, as the sole controller of all the possessions of the Icarians. When they landed, Rovira had a showdown with Cabet about the mismanagement of the whole exercise. Later that year, in a fit of depression, Rovira shot himself. The expedition stayed in New Orleans for several months, while Cabet blamed their misfortunes on the advanced guard's leader, Adolph Gouhenant, who was thrown out of the movement (and went on to establish a series of successful saloons throughout Texas). Cabet then took his band of followers north up the Mississippi River to Nauvoo in Illinois, where he bought more suitable land recently vacated by the Mormons before they went to Utah.

In 1850 Cabet went back to France to reply to the charges of mismanagement brought by some local Icarians, but he was acquitted on appeal and returned to Nauvoo. In 1854 he became an American citizen. The elected board of the commune disagreed over Cabet's wanting to become sole arbiter/dictator once more; they took him to the state court, and failing in their action in 1856 expelled him from the community he had established. Cabet and a faithful few established a new commune in St. Louis, which they also called Icaria. Soon after the move he had a fatal stroke, and the Icarian movement died with him.

So Were Your Teachers Ever Funny?

Mocking National Stereotypes

Aleph is the first letter of the ancient Semitic—and later Hebrew—alphabet, based on an earlier Phoenician symbol representing an ox. Aleph was also the unlikely pseudonym of William Harvey, who in 1869 produced an atlas for children that he called Geographical Fun. Harvey's meritorious intention was to produce an educational European atlas that overcame the contemporary feeling among "young Scholars who commonly think that Globes and Maps but wearisome aids to knowledge."

He went on further to state that he hoped his unique atlas "may embue [sic] the mind with a healthful taste for foreign lands." Unlike his open-minded 17th-century namesake, who discovered the circulation of the blood, this William Harvey, who lived from 1796 to 1866, was a typical early Victorian teacher with firm ideas of what the various European nations were like. The maps in his 16-page atlas employ the outline of a country to create a caricature of what he regards as a typical inhabitant. To emphasize the point, each map is accompanied by a four-line verse summing up the national character. For example, Ireland is depicted as a female peasant with her baby strapped on her back. She carries a box with a baby's feeding bottle while the child clutches a herring in her hand. To emphasize the impoverished state of the Irish, the accompanying verse runs:

> And what shall typify the Emerald Isle?
> A Peasant, happy in her baby's smile?
> No fortune her's, though rich in native grace,
> Herrings, potatoes, and a joyous face.

So as long as you have fish and potatoes, what more do you want?

The twelve maps in the book begin, predictably enough, with England, represented by a sympathetic portrait of Queen Victoria. She sits in the guise of a helmeted Britannia, wielding a scepter and gripping a shield of the Union Flag. Next comes Scotland, depicted as a grimacing carrot-headed bagpiper with bare hairy knees. He peers over frameless spectacles as the wind blows up his kilt. The Scottish flag is shown as plain blue with a Union Flag in the left-hand upper quadrant. A goatee-bearded Owen Glendower—or Owain Glyndwr—represents Wales, complete with a golden crown, a cloak with a flying dragon, and what appear to be earrings as he sings "King Arthur's long, long pedigree". The Welsh flag is the same as the Union Flag.

After Ireland, described above, Aleph crosses the English Channel to vent some spleen on those ghastly foreigners. France comes under fire first, as the hook-nosed Empress of cooks with a mirror, or "flatt'ring glass," which declares to her that beauty, wealth, and power provide her imperial dower. A map of the Iberian Peninsula deals with Spain and Portugal together. Spain is a long-haired maiden in a flowing gown, holding a bunch of grapes that is also held by the friendly Portuguese bear next door. The bear is dressed in red trousers, a green tail-coat and a lace collar. Prominent at the bottom of the map is the Union Flag of Gibraltar. Italy is shown as a bearded revolutionary, the "Uncompromising friend of liberty!" He wears a sword and grips an upraised stick; with his peasant's hat still on his head, he nevertheless clutches the cap of liberty. On neighboring Sardinia, a tearful cardinal holds up a crucifix while hanging on to his rolled umbrella.

In these days before the unification of greater Germany, the nations of Prussia and Germany are each given maps of their own. Saxony, although mentioned on Prussia's southern border, does not get a map. On the map of Prussia a kneeling Wilhelm, Duke of Brunswick, hands a note to the Iron Chancellor, Otto von Bismarck, the "royal conscience keeper" (who became Prince Otto two years after the map was drawn). Wearing a pointed *pickelhaub* helmet and golden epaulettes on his characteristic long gray military greatcoat, Bismarck grips a musket as he stares over Poland looking for further conquests. In the event, it was

RIGHT *Not too many jokes in this one, no roast beef, warm beer, rowdiness, and colonial greed; just Queen Victoria, "Queen of Hearts." Though her "horn of plenty" helmet almost seems to have swallowed Scotland!*

Beautiful England,—on her Island throne,—
Grandly she rules,—with half the world her own;
From her vast empire the sun ne'er departs:
She reigns a Queen—Victoria, Queen of Hearts.

Austria that succumbed to his military might. With Germany, Aleph returns to a wasp-waisted female figure who takes delight at coming glories "shown by second sight" ... "Her joy expresses as a lady dancing."

Holland and Belgium are also female figures. Belgium, still retaining the region later annexed to Germany, is the taller of the two. She looks worriedly at the long-haired Dutch lady smoking a cheroot and hanging onto her bonnet straps. Holland is much smaller, reflecting the size of the country before the extensive land reclamation works had begun. The two represent "a land ... by perfect art made grand." Denmark's fragmented islands presented a problem. The main part is a female ice-skater who appears to be about to fall over backwards, while many of the islands are goblin-like creatures with distorted faces. The map is dedicated to Shakespeare's Prince and the Princess of Wales.

BELOW AND RIGHT *Germany as a lady dancing with delight (below) because she has looked into the future and sees great things ahead. Why this should be the case in 1869? The bagpipes and kilt are an easy target for a Sassenach (right); and the last two lines of the ditty have one of the worst forced rhymes in literary history. Russia (right) is a far more successful effort.*

The final map is devoted to Russia, illustrated by the classic Russian bear standing back to back with the white-bearded Tsar Alexander II dipping his toes in the Caspian Sea. The verse states:

> Peter, and Catherine, and Alexander,
> Mad Paul, and Nicolas, poor shadows wander
> Out in the cold; while Emperor A. the Second,
> In Eagles, Priests, and Bears supreme is reckoned.

How much should we censure Aleph? Well, it could be argued that none of the caricatures transgresses as far as the mildest Polack joke. Does the borrowed pedagogical authority that belongs to almost any kind of cartography make the joke more damaging, or less? As the title of his atlas says, it is meant to be "Geographical Fun: Being Humorous Outlines of Various Countries." The map drawings are credited to a 15-year-old young lady who apparently had the idea to amuse her brother who was confined to his bed by illness. Aleph hopes that "If these geographical puzzles excite the mirth of children; the amusement of a moment may lead to the profitable curiosity of youthful students."

GERMANY.

Lo! studious Germany, in her delight,
At coming glories, shewn by second sight,
And on her visioned future proudly glancing,
Her joy expresses by a lady dancing.

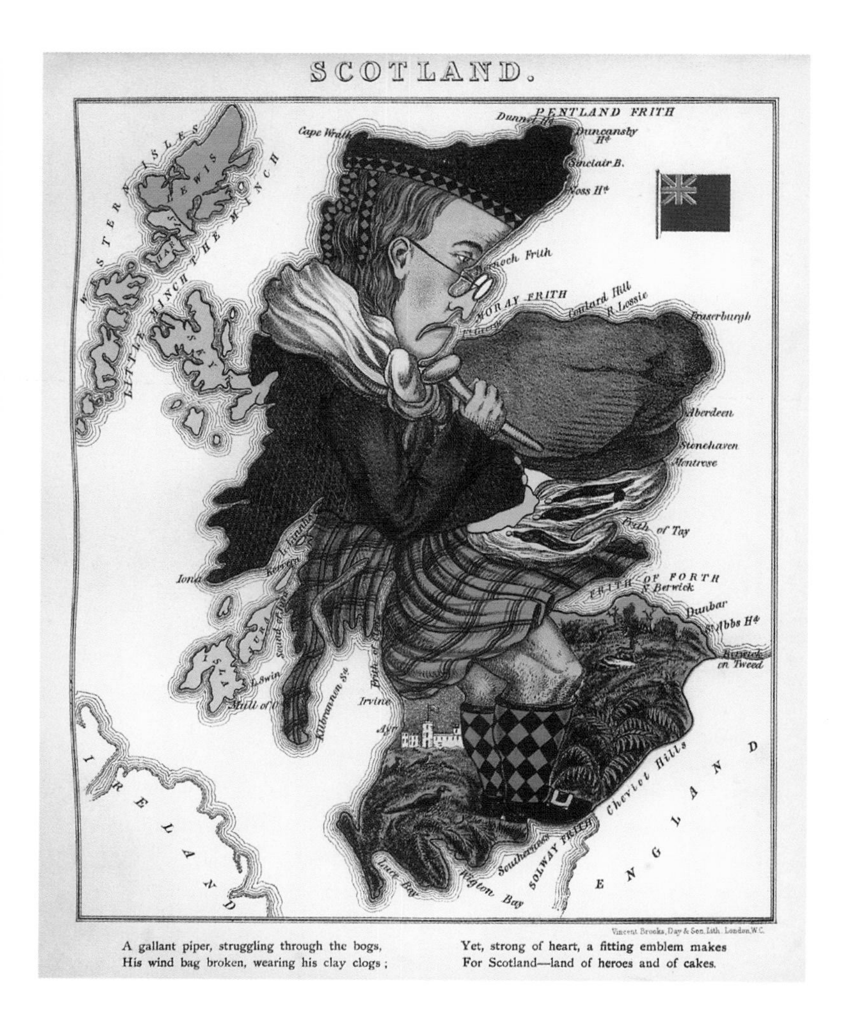
SCOTLAND.
A gallant piper, struggling through the bogs,
His wind bag broken, wearing his clay clogs;
Yet, strong of heart, a fitting emblem makes
For Scotland—land of heroes and of cakes.

Peter, and Catherine, and Alexander,
Mad Paul, and Nicholas, poor shadows wander
Out in the cold; while Emperor A. the Second
In Eagles, Priests, and Bears supreme is reckoned.

The National Socialist Blueprint:

Propaganda Maps

"The map is not the country." This is a truism that is almost always ignored. Maps shape the way we see the world. The map is not the country, but any map encloses and defines the country in a two-dimensional form. Maps—flat, oblong, even globes—impose a system of symbols and measurements. All maps have shortcomings as analogs for the real world. Nonetheless, most people tend to trust maps.

There is something about a map that makes us accept the informstion as presented. Once we are shown a map, we usually accept it as a set of images and symbols that objectively and accurately represent the world. We seem to forget that a map's perspective (both in the strict optical sense and in the more general application) is always selected by cartographers, or the person or government that commissioned them. Almost all maps have an agenda or bias; whether national, racial, cultural, political, philosophical, or financial.

During the Second World War, all sides extensively used maps for propaganda purposes. Worse, the same authorities that create propaganda maps are often themselves deluded by them. They come to believe their own propaganda. These maps frequently became instruments of self-delusion that have often resulted in military and political disaster.

German propaganda maps before the outbreak of the war showed Germany threatened on all sides and from within by evil powers, to create fear in the population and a rationale for acts of war. A classic example was the propaganda map shown opposite, top: "A Small State Threatens Germany." It shows how bomber aircraft from Czechoslovakia could reach all of Germany. Theoretically, yes; but the Czech air force did not possess any such aircraft.

This kind of propaganda mapping has been extensively employed by both the US and the Soviet Union throughout the Cold War period to justify overthrowing governments and financing wars throughout the Third World. Indeed, in the 21st century, the strategy of the 1930s Nazi-Czech propaganda map was exactly duplicated in maps in British newspapers to justify the invasion of Iraq. These maps showed how "weapons of mass destruction" from Iraq were capable of striking British bases in Cyprus within 45 minutes. Again, theoretically, this may have been true; but the Iraqis did not possess any such missiles, or any such warheads.

Maps that impose national, political, or racial identity on the geography of the world are *ipso facto* instruments of propaganda. Before contact with Europeans, to most nomadic native nations of the Americas the idea of ownership of the land made no more sense than the idea of ownership of the clouds or the wind. In any real sense how does any person or nation really "own" a mountain, or a stretch of tundra?

This use of maps as legal documents of ownership or control over the world is not a universal one. However, as shown elsewhere in this book, during the European Ages of Discovery cartography became the

ABOVE *"… It is ending! The war is nearing its end. The insanity that the enemy powers unleashed on humanity has gone beyond all bounds. The whole world feels only shame and disgust. The perverse coalition between plutocracy and bolshevism is collapsing!" Josef Goebbels, April 20th 1945, Hitler's 56th birthday.*

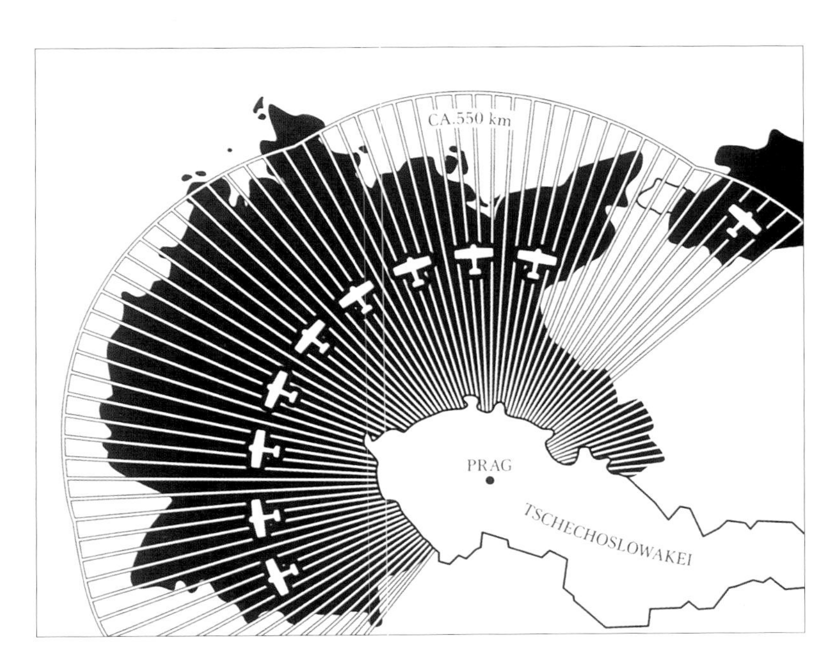

ABOVE Ein Kleinstaat Betroht Deutschland *(a small state threatens Germany); Nazi propaganda from the 1930s. Czechoslavakia had no bomber aircraft.*

convenient and accepted "legal" means by which European imperial powers were able to claim the land masses of the New World while ignoring their occupants. Maps were financed as a means of extending imperial control over the world not already claimed by other cartographers. Non-map-making nations had no legal rights or claims, even if those nations had occupied these 'new' lands for millennia. It seems this not only applied to native populations, but also to non-mapmaking Europeans.

This is illustrated by France's legal claim over what became Eastern Canada, established by Jacques Cartier's historic "discovery" in 1534. This concept of "discovery" is a curious one; especially as Cartier records the presence of over 1,000 Basque fishing vessels on his arrival in the Gulf of St. Lawrence! Basques were fishing cod off the shores of North America for decades before Cartier (and probably before Columbus); but as they provided no maps, they were given no legal claim, or credit for the "discovery" of the New World.

In the Americas, Africa, and Asia, all of the European colonial powers have used mapmakers as propagandists to exaggerate the extent

RIGHT *Supervised by Goebbels,* Signal *magazine was published throughout Europe in up to 20 languages during the War, and until Pearl Harbor, in the US. In 1943 the biweekly was selling three million copies. It was quite brilliantly designed. This image of the British using Europe as a buffer against Russia accompanies a* Signal *article entitled "The False Path of Pan-Europe."*

RIGHT *"A Survey of the Marvel of the Party Organization," printed in* Signal *magazine. The accompanying text to this bird's-eye map explains the ideal National Socialist organization of German society from the Block level (40 households each), to the Cell (200 households each), the Group (3,000 households), the District,* der Kreiss *(25,000), across 40 Provinces,* der Gau, *(500,000 households); a system to be exported to deserving races across the world. It helpfully indicates the salient points to look out for, including the Strength Through Joy seaside resort; a convalescent home for mothers; and a Hitler Youth political training college. See if you can find them all. This childish representation, with its baby palette and cartoon figures, is the most terrifying map in this book. Was it deliberately, knowingly infantile, or instinctively so?*

of territorial claims. This resulted in competitive "map wars" that often ludicrously exaggerated or distorted physical geography.

In the 20th century, propaganda maps became more sophisticated. Early in the century, maps of the world were color-coded to indicate the extent of control of European imperial powers. Most famously, the vast expanse of the British Empire was indicated by the color pink on most maps. As the pink landmasses covered about a quarter of the globe, these maps were impressive reminders of Britain's supremacy as a sea power and empire.

Nazi Germany was particularly adept at producing audacious propaganda maps. Before the war, Germany was portrayed as victimized and oppressed by other powers. But once war began, propaganda maps of Germany were redrawn to show how powerful and all-conquering the nation had suddenly become. Typographically and geographically, maps of Germany became bolder and larger. German cartographers ignored all common scales of measurement, and propaganda maps took on an epic dimension and style that was consciously imitative of the Roman Empire.

Early in the war, Nazi magazines published maps of major battles, and quite accurately portrayed the German war machine sweeping all other European forces before it. Each map and its commentary conveyed a sense of national destiny and historic significance comparable to the conquests achieved by the legions of Caesar. Later, when the tide of war had turned, the propaganda machine continued to map out victories despite massive losses on all fronts.

Nazi publications were adept at portraying "ideal" socially-engineered communities based on principles perceived to be for the good of the nation and the people. Nazi propaganda specialized in utopian visions: a future Germany of physically and intellectually superior human beings living in a perfectly organized society. The naiveté of the art (right) only adds to the chilling effect of this example.

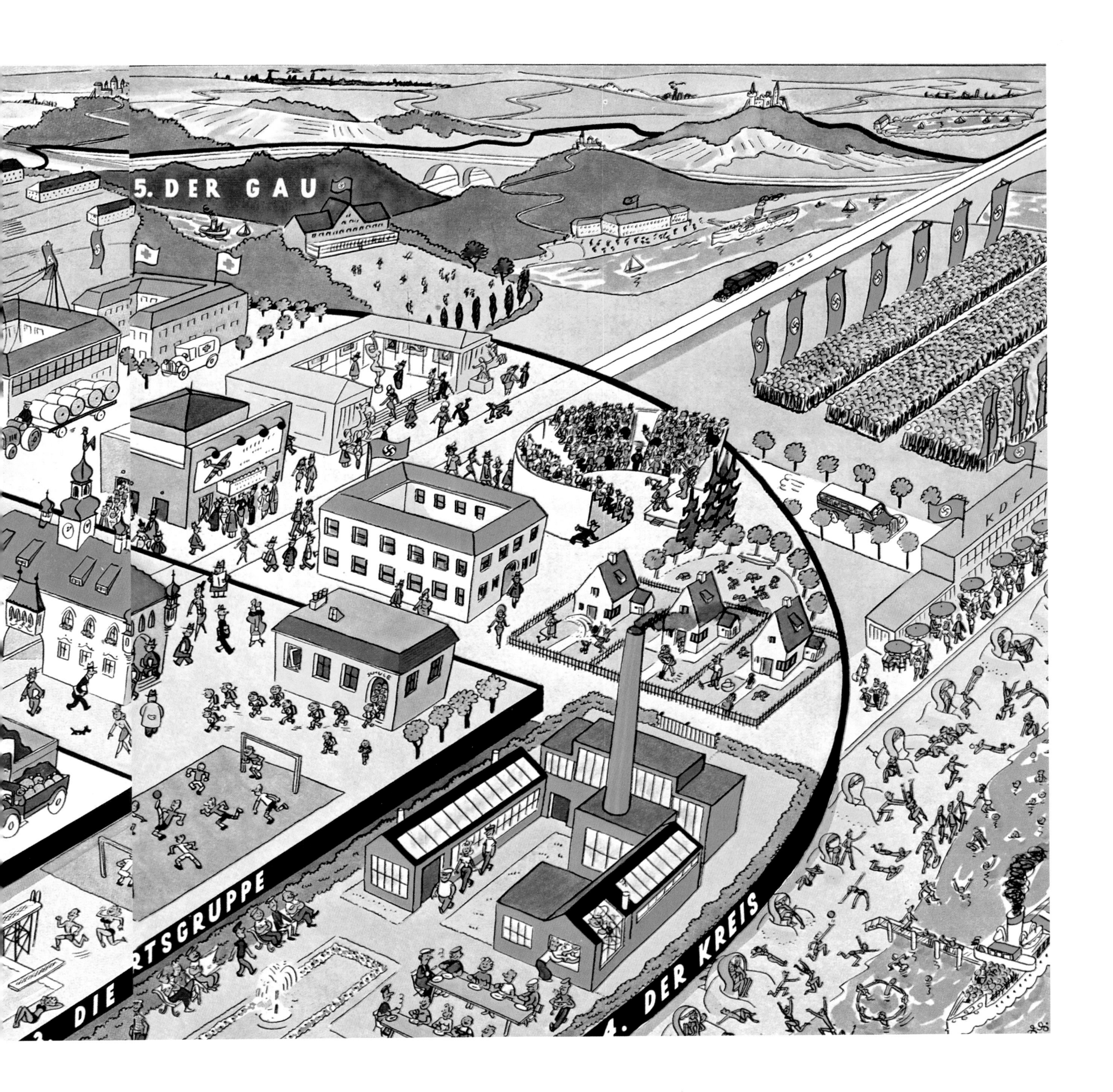

5. DER GAU
KDF
SCHULE
...RTSGRUPPE
DIE
4. DER KREIS

The Most Famous Fantasy Map:

Tolkien's Middle-Earth

All the epic adventures in J. R. R. Tolkien's two novels, **The Hobbit** and **The Lord of the Rings**, take place in a world now famously known as "Middle-earth." In a number of letters to readers and reviewers, Tolkien acknowledged that the location often confused people. Many assumed that Middle-earth was another planet. He found this a perplexing conclusion, because in his own mind he had not the least doubt about its locality.

Middle-earth was not an imaginary world to Tolkien. It was the world as perceived by his Anglo-Saxon ancestors. In the Old or Middle English language, the name for the world was "Middel-erd," a place that shared a great deal with "Midgard," the Old Norse name for their world (see page 28). As Tolkien once wrote in a letter to his publishers: "The name is a modern form of midden-erd/middel-erd, an ancient name for the *oikoumene*, the abiding place of Men, the objectively real world, in use specifically opposed to imaginary worlds (as Fairyland) or unseen worlds (as Heaven or Hell)."

Tolkien has always been clear about the geographic location of "The Shire," the Hobbit homeland. The Hobbits are the ancestral inhabitants of the green and pleasant lands of the pre-industrial English Shires of the rural Midlands. As Tolkien frequently pointed out, The Shire could be nowhere else than this ancient heartland of England: "After all the book is English, and by an Englishman ..."

Even more instructively, in answer to journalists' questions, Tolkien has given a larger scale "real world" geographic location for his epic lands: "The action of the story takes place in North-west of Middle-earth, equivalent in latitude to the coastline of Europe and the north shore of the Mediterranean." Hobbiton and the Elf refuge of Rivendell were to be found on the same latitude as Oxford is today; while the White Tower of Gondor was to be found some 600 miles south on the same latitude as the Italian city of Florence. At the end of the Ring War, Tolkien explained elsewhere, the Reunited Kingdom of the Dunedain covered a land mass geographically comparable to the Holy Roman Empire of Charlemagne; that is, roughly, the area now defined as Western Europe.

The real trick to understanding Tolkien's world is not where, but when. He explained this many times in letters to readers: "The theatre of my tale is this earth, the one in which we now live, but the historic period is imaginary." That imaginary time is the heroic mythic age that ended just as the age of recorded history began. It was an age when the first humans shared the world with dragons, elves, dwarves, and giants.

Tolkien's placing of Middle-earth in the dimension of "Myth Time" is consistent with a mystical understanding of the evolution of world civilizations. It has much in common with the theories of Mircea Eliade, the 20th century's most influential historian of religions and mythologies. According to Eliade, myths are tales about the creation and ordering of the world in a timeless, sacred realm. Rituals and recitations re-enact archetypal actions of a god or ancestor *in illo tempore* (before history began). Tolkien's Middle-earth was an ideal, heroic world of archetypes: the world where all the dreams and nightmares of our civilization have their origin.

Although the epic romance of *The Lord of the Rings* mapped out the vast and complex realm of Middle-earth, it was not until the publication of *The Silmarillion* that readers had any idea of this world's true dimensions in both time and space.

The Silmarillion is a compendium of Elvish history and mythology (which has been supplemented by a dozen other posthumous drafts and notes on Middle-earth) which begins with the creation of the cosmos by means of a divine composition, known as the"'Great Music." Within this music of manifest bright spheres of air and light was the World. This was a flat World with a single massive

RIGHT *Beleriand was in the extreme north-west of Middle-earth. A series of wonderful kingdoms was built there, until at the end of the first Age of the Sun and the War of Wrath, it sank beneath the waves. Does it have a specific geographical equivalent? Does it lie beneath the North Sea or the Atlantic?*

UNDYING LANDS
HELCARAXË (GRINDING ICE)
IRON MOUNTAINS
Tol Sirion
Belegost
Nogrod
Doriath
Gondolin
Nargothrond
Narog River
Sirion River
Menegroth
Gelion River
BLUE MOUNTAINS
BAY OF BELAR
MIDDLE EARTH
BELERIAND
BELEGAER (WESTERN SEA)

Железные холмы
Эсгарот
Одинокая гора
Долгое озеро
Дейл
р. Быстротечная
Дворец короля эльфов
Чёрный лес
скала Каррок
Дом Беорна
Туманные горы
Последний домашний приют
Дикий край
С
Ривенделл
Хоббитон
З
В

LEFT *Middle-earth is surely the most mapped fantasy world ever; which is as it should be, because the sense of place is what inspired Tolkien. This is a 1976 engraving by Mikhail Belomlinsky for the Russian translation of "The Hobbit." Its ancient pictograms and starkness are wholly apposite.*

continent. Initially inhabited by angelic powers known as Valar and Maiar; these were akin to pagan gods and demi-gods that shape the mountains and stars and tame the earth's fires and control its floods. After these angelic powers, other legendary races and species drawn out of Old English bestiaries begin to appear: Elves, Dwarves, Ents, Woses, Wargs, Orcs, Balrogs, Trolls, and Dragons.

Although Darwinian evolution has no place in Tolkien's world, he does seem to be an enthusiastic fan of Lyle's theory of geological evolution. Too enthusiastic, in fact: he puts the theory of continental drift on fast forward. One super continent (over periods of tens of thousands of years) breaks up and shifts into various parts and dimensions—speeding up the movement of landmasses and drifting continents by several hundred million years.

Remarkably, we are 30,000 years into the history of Tolkien's world before the human race actually appears on Middle-earth. Then after nearly four more millennia there is the cataclysm known as the Change of the World, wherein the lands of mortals and immortals are ever after separated. Middle-earth comes to exist within a globed world; while the Undying Lands of the Valar are set in another dimension.

By the time of "The Lord of the Rings" Tolkien's world is 37,000 years old before it has evolved into the shapes shown in his own map of Middle-earth. Considering the fact that the actual quest of the Ring is completed in just 90 days, 37 millennia of historical scene setting seems somewhat excessive! Nor is that the end of it. After the War of the Ring, Tolkien himself estimates that there is another six millennia to account for before Middle-earth's chronicles leads on "eventually and inevitably to ordinary history." Working backwards from our own time, this would place the the War of the Ring at somewhere between 4000-5000 B.C.; and the creation of his world at about 41,000 B.C.

Why all this evolution and mapping out of Middle-earth through time? The answer is that Tolkien was not just creating a map for a couple of entertaining adventure stories. From his earliest years, J.R.R. Tolkien had a mission in life: to create or—as he would see it—recover the lost mythology of the Anglo-Saxon peoples. Important as the epic tale of "The Lord of the Rings" was to him, it was almost a secondary concern. Tolkien's greatest passion was focused on the creation of an entire mythological world for the English people.

Tolkien's creation of Middle-earth is his attempt to recreate a world

ABOVE *The Witch-king of Morgul, the evil Lord of the Wringwraiths, "cannot be slain by the hand of man." Tolkien deliberately echoes the legend of the Scottish King Macbeth. In this, as elsewhere, he is reclaiming, rebuilding, lost archetypes.*

of archetypes that have survived in the racial unconscious of the English people. The enormity of Tolkien's undertaking is staggering. It would be as if Homer, before writing the Iliad and the Odyssey, had first to invent the whole of Greek mythology and history. What is most remarkable is that Tolkien actually achieved his ambitions to an extraordinary degree. His Hobbits of Middle-earth are now as much part of the English heritage as Leprechauns are to the Irish, Gnomes to the Germans, and Trolls to the Scandinavians. All that the English are and know comes from this world. All the great events in English history are prefigured in archetypal form in this ancient mythic world of Middle-earth.

The Play of Human Emotions:

The Vinland Map

The Yale Beinecke Rare Book and Manuscripts curator probably resisted the simultaneous temptation to genuflect and whoop with joy as he watched the check being signed. In 1958 $1 million was hardly loose change. But what a treasure it was about to secure! Thirty years later the anonymous benefactor came forward, out of the shadows and into the academic firefight, to defend his purchase. Was the map genuine, or a brilliant forgery?

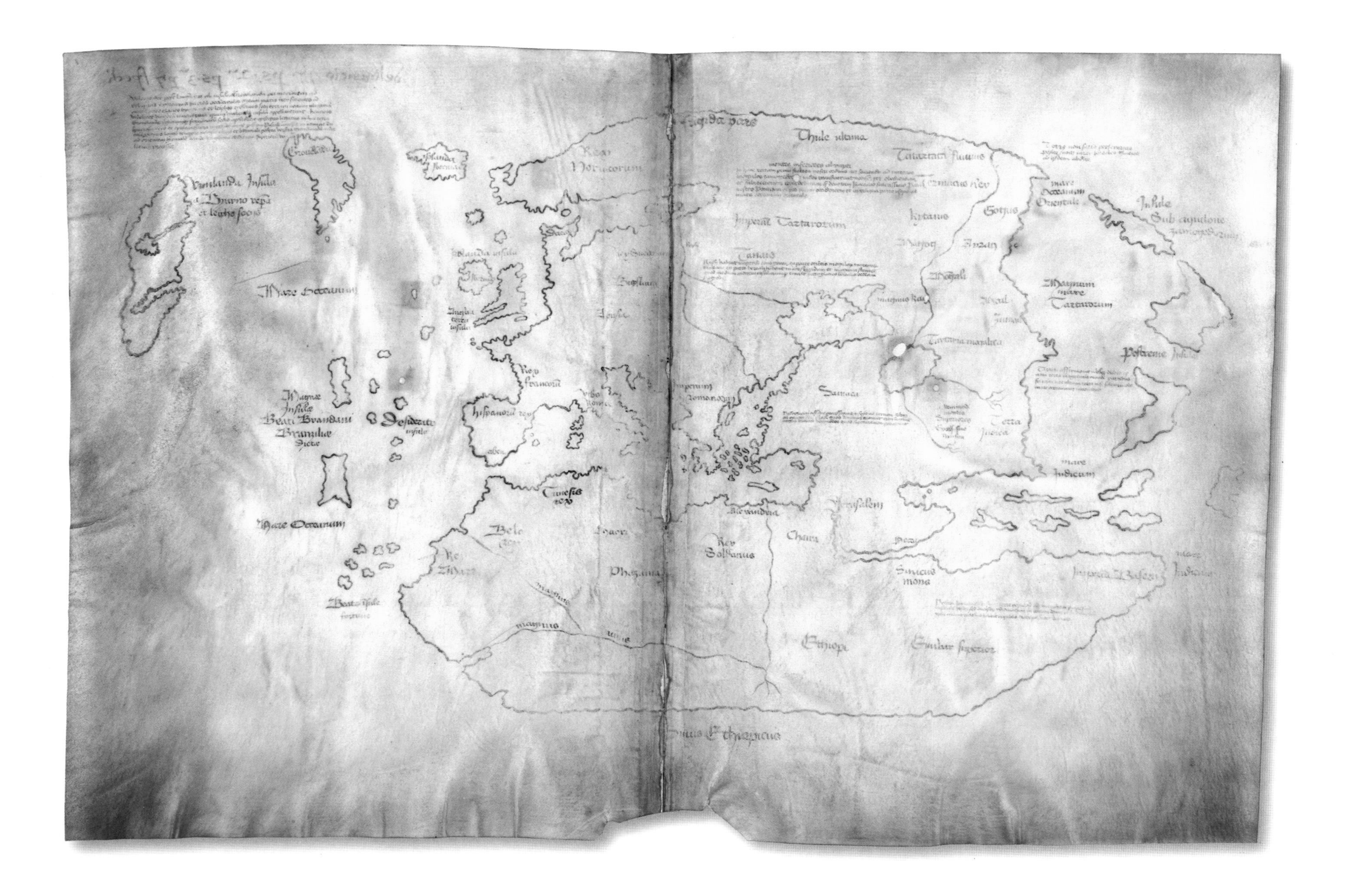

The map was first discovered by a bookseller in Barcelona in a bona fide manuscript entitled *Tartar Relation*, which recounted an expedition to central Asia in the 13th century. In 1957 a New Haven dealer took a chance and bought it for $3,500 and showed it to the curator of maps at Yale University, Alexander Vietor, who secured Yale's right of first refusal. The anonymous benefactor stepped in and in 1965—pointedly the day before Columbus Day—Yale not only announced the existence of the map to the world but also released an accompanying book, *The Vinland Map and the Tartar Relation*.

Why all the fuss over a map, why the million dollars and the book fanfare? (The latter was an even greater risk than the former, to Yale.) The Vinland Map appears on a single sheet of vellum, which has the consistency of a very thick sheet of tracing paper. It measures 11 x 16 inches (279 x 406 millimeters) and folds down the middle. When folded in half, the map fits perfectly into the book of manuscripts where researchers theorized it had existed for hundreds of years. The map showed Europe, North Africa, Greenland, and part of the northeast coast of North America. It contained statements in Latin, one of which is located beside a large island. Its coastline is unmistakably part of Canada's Labrador and Newfoundland. The text translates as:

> "The companions Bjarni and Leif Ericksonn discovered a new land, extremely fertile and even having vines, the which island they named Vinland."

The map was reckoned to date from about 1440: so here was proof of the first European landfall in the New World, 50 years before Columbus. It is important to remember that the discovery of the map predates any archeological evidence for the Viking's temporary stay. Here, as *Time* magazine trumpeted, was "By far the most important cartographic discovery of the century."

Note the lead-time between purchase and presentation: more than six years. Yale had not been overhasty, partly at the behest of the purchaser, Paul Mellon, a long-time Yale benefactor and not a fool, who had demanded in-depth research before publication.

LEFT *The paleographer Paul Saenger asserts that "the Nordic details, ranging from the presence of Vinland to the astonishingly accurate representations of Greenland and Iceland ... unsupported by any medieval Nordic cartographical tradition ... strongly suggest fraud." A German Jesuit priest (and cartographer) Josef Fischer, is a suspect. Some say he faked the map in reaction to Nazi aggression toward the Catholic Church in the 1930s, to suggest that the church arrives in the New World with the Vikings (with whom the Nazis identified).*

Doubts surfaced fairly quickly. At a conference held at the Smithsonian Institution in 1966, several scholars cried foul. Meanwhile Gerald Crone, map curator at the Royal Geographical Society carried on a vigorous skeptical correspondence in the London *Times*. It is not difficult to imagine the response from Yale. As Wilcomb Washburn, Director of the American Studies Program at the Smithsonian would point out fully 30 years later—at a second conference debating the same subject— "No map has been a subject of greater controversy than the Vinland Map. The play of human emotions effects the consideration of truth in history."

In 1972, the results of chemical analysis of the map's ink raised doubts about its authenticity. Chemist Walter McCrone removed and analyzed portions of the map's ink. He concluded that the ink contained a significant amount of titanium anatase in it, a material scientists thought was invented after 1920. So in 1974, researchers declared the Vinland Map a forgery, shattering scholarly reputations and years of research. That appeared to be that. It wasn't of course.

In 1985, Yale's Beinecke Library secretly lent UC's Thomas Cahill the map for four days. Cahill and his colleagues performed a different series of chemical tests on the ink. Cahill used PIXIE—particle-induced x-ray emission tests—to determine the substances of the ink as a whole and not just a few fragments. He found only a minute presence of the titanium anatase, which scientists have since discovered occurs naturally. So a 15th-century scholar's ink could contain this substance. Then in 1995 chemist Garman Harbottle of the Brookhaven National Laboratory at the University of Arizona was permitted to trim off a three-inch sliver of the vellum itself for Carbon-14 dating. He could confidently assign a date of 1434, give or take 11 years. So what? was Walter McCrone's reaction: it's "still a 20th-century fake."

In 2002 his position seemed to be strengthened by two chemists, Robin Clark and Katherine Brown of University College London. They used a technique called laser Raman microprobe spectrometry. In this technique, a laser beam is directed at an object and a small portion of the light scatters off the molecules as radiation with different colors. Every material has a unique scattering spectrum that acts as a fingerprint, allowing scientists to identify it. They concluded that the forger had overlaid a black line over a yellow one to mimic the effect of aging. This time, so what? was Thomas Cahill's reaction: "They're drawing a lot of conclusions from a very limited data set."

The argument continues today. To date, scientific analysis cannot resolve it. Perhaps it is not so preposterous for the non-expert, (the reader of this book perhaps) to simply look at the map, consider what it is supposed to be, and guess.

A Different Point of View:

Propaganda Projections

As this book demonstrates, no map projection is perfect. It is just not possible to accurately portray a curved surface on a flat piece of paper. Some projections are better than others, but one cartographer had his dubious ideas accepted not through merit but through lobbying and canny marketing. Another—not taking things quite so seriously—did not make such an impact, but nevertheless became something of a hero on home soil.

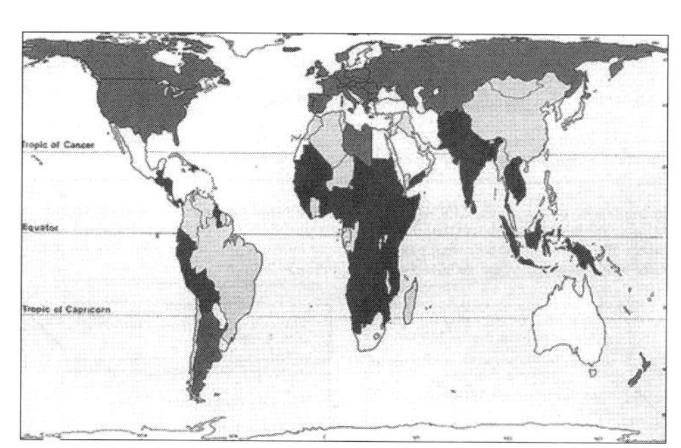

The central character in the first story of map manipulation is Arno Peters. He was born in Berlin in 1916 and studied history before becoming a film-maker in Hitler's Nazi Germany of the 1930s. He was awarded his doctorate in 1942 for a thesis on political propaganda. He became a journalist and acquired an interest in cartography as a possible method of expressing political ideals. He continued in this work while living in Bremen, where he died in 2002.

In 1973 Peters called a press conference at which he denigrated the commonly used Mercator world map and introduced his "new" equal-area projection. The Mercator map—which employs a type of cylindrical projection—distorts the northern latitudes occupied by North American and Eurasian countries, making them appear much larger than they actually are. It has survived for so long because a line drawn on the map preserves true direction, for making or taking bearings, for example.

Peters claimed that it would be "fairer" to the underdeveloped nations of the world if the standard parallels (lines of longitude) were shifted southwards by about 45 degrees, while maintaining parallel lines of latitude and longitude on a rectangular graticule. The result is a more-or-less equal-area map, which one prominent modern cartographer described as resembling "wet, ragged long winter underwear hung out to dry on the Arctic Circle." Another expert commented that Peters projection was "a scam capitalizing on the cartographic ignorance of most people" surviving "in a climate of political correctness where it is inappropriate to criticize anyone who claims to criticize the status quo."

Peters' equal-area approach was far from original. It was first proposed in 1722 by Swiss-born German mathematician Johann Lambert (1728-77), whose map had perforce to abandon correct angular relationships. The projection was put into practical form in 1855 by Scottish cleric, cartographer, and mathematician James Gall (1808-95), who published it in the *Scottish Geographical Magazine* and called it the "orthographic equal-area projection." The Lambert azimuthal and Albers conic projections also preserve land areas in their correct proportions. The modern Goode projection opens out the globe into four roughly elliptical shapes on which the continents have almost the same forms and proportionate areas as when the Earth is viewed from space.

But Peters' own propaganda paid off and the media seized on this "first non-racial map." By 1983 the world map appeared in an English edition. Meanwhile, Peters had drummed up endorsement from various high-profile international organizations, including Christian Aid, the World Council of Churches, UNESCO (The United Nations Educational, Scientific, and Cultural Organization) and UNICEF (the United Nations Children's Fund). Few Christian left-wing, non-governmental charitable organizations escaped Peters' catchment area.

INSET *Peters orthographic, equal-area map is now commonly called the Gall-Peters projection in recognition of its earlier presentation by James Gall, in 1855. The oft-quoted problem with the Mercator projection that Peters was attacking is that, for example, it shows Africa to be about the size of Greenland, when it is in fact 13 times larger. This is so; just as the Peters world map shows Africa to be twice as long north-south as it is east-west, when it is in reality roughly the same.*

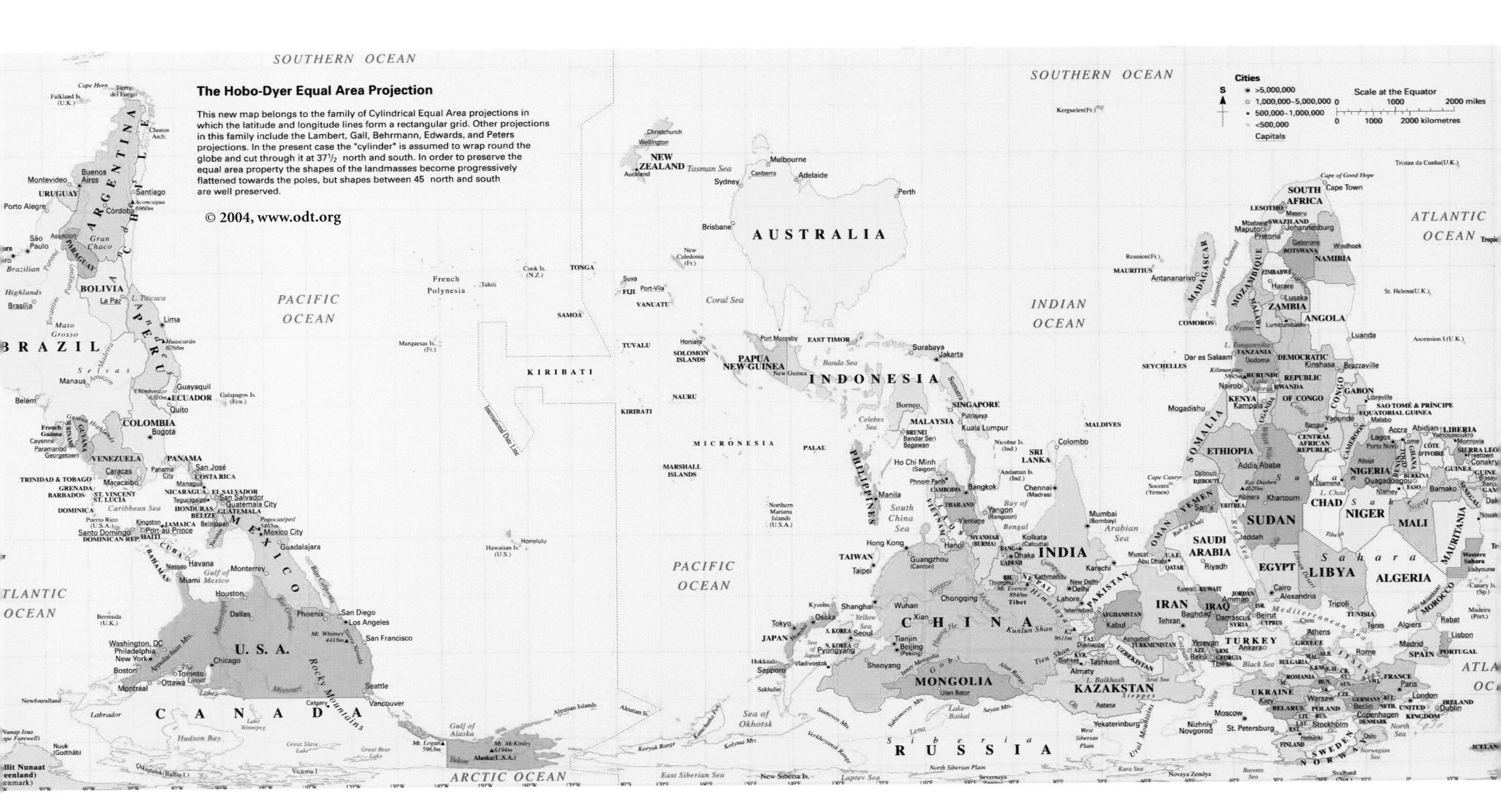
The Hobo-Dyer Equal Area Projection
This new map belongs to the family of Cylindrical Equal Area projections in which the latitude and longitude lines form a rectangular grid. Other projections in this family include the Lambert, Gall, Behrmann, Edwards, and Peters projections. In the present case the "cylinder" is assumed to wrap round the globe and cut through it at 37½ north and south. In order to preserve the equal area property the shapes of the landmasses become progressively flattened towards the poles, but shapes between 45 north and south are well preserved.
© 2004, www.odt.org
SOUTHERN OCEAN
SOUTHERN OCEAN
PACIFIC OCEAN
PACIFIC OCEAN
INDIAN OCEAN
ATLANTIC OCEAN
ARCTIC OCEAN
AUSTRALIA
INDONESIA
CHINA
RUSSIA
INDIA
MONGOLIA
KAZAKSTAN
U. S. A.
CANADA
MEXICO
BRAZIL
ARGENTINA
SUDAN
LIBYA
ALGERIA
Cities
Scale at the Equator

He also had the backing of the West German government. None listened to those who pointed out that the Peters projection is not original and it is not a particularly good map, the desire for equal-area denying the competing claims of direction and distance. A spokesperson at the US Geological Survey maintains that the Peter's map "isn't any better than similar maps that have been in use for 400 years." But more than a thousand educational sites on the Internet still provide links to the Peters propaganda.

Perhaps this is too harsh on Peters. After all, his father was imprisoned by the Nazis—both his parents were social activists, and some of this certainly rubbed off. In 1989 seven North American geographic organizations put foward a resolution urging all map publishers not to use rectangular maps *at all*, on the grounds that, among others, "world maps have a powerful and lasting effect on peoples' impressions of the shapes and sizes of lands and seas … Such maps promote serious, erroneous conceptions by severely distorting large sections of the world." Undeniable: but does a visual reminder that Greenland is smaller than Africa help to ease world poverty?

ABOVE *This is not the original McArthur Universal Corrective World Map discussed below—but you get the idea, and this is more interesting. The Hobo-Dyer equal-area map was specially commissioned by a map publisher and is a modification of the 1910 Behmann projection. The "cylinder" of this south-oriented cylindrical projection cuts through the globe at 37½° North and South; and Australia, of course, is on top of the world. Russia is big, as it should be, but it's taken a pounding. It could be argued that the Pacific is a waste of space.*

A far less dour attack on cartographic norms came from Stuart McArthur of Melbourne, Australia, who drew his first South-Up map when he was 12 years old, in 1970. His geography teacher told him to redo the assigment. Age 21, he tried again; and to date "McArthur's Universal Corrective Map of the World" has sold some 350,000 copies. The original map declaims: "At last, the first move has been made—the first step in the long overdue crusade to elevate our glorious but neglected nation from the gloomy depths of anonymity in the world power struggle to its rightful position—towering over its northern neighbours, reigning splendidly at the helm of the universe."

Junk History?:

The Pizzigano "New World" map

In 2002, a British writer and "amateur" historian, Gavin Menzies, published the book **1421–The Year China Discovered the World**, which put forward the intriguing idea that a Chinese expeditionary fleet under the renowned eunuch Admiral Zheng He (1371-1433) had circumnavigated the globe and explored the Americas fully 70 years before Christopher Columbus made landfall on Hispaniola.

The starting point for this piece of archival detective work was a map of 1424 by an obscure Venetian cartographer. Does the map in question truly support Menzies' thesis? One fact is indisputable—the authenticity of the 1424 portolan chart, painted on vellum by a mapmaker called Zuane Pizzigano, is not in question. Held by the James Ford Bell Library at the University of Minnesota, it shows the British Isles, the coastline of mainland Europe around France and the Iberian peninsula, and part of the western seaboard of Africa. Far out in the Atlantic, four variously sized islands named Antilia, Satanazes, Saya, and Yamana have been drawn, and boldly colored. Menzies identifies the first two (and largest) of these islands as Puerto Rico and Guadeloupe, and speculates that the information for the map must have come from a Venetian traveler named Niccolò da Conti, who was in the major trading port of Calicut on the Malabar Coast in southwestern India at the same time as the Chinese treasure fleets in 1421, and may well have spoken there to the official historian of the expedition, Ma Huan. The fact that no Chinese map exists documenting the voyage and its new discoveries is attributed to the wholesale destruction of records after 1424, when the emperor Zhu Di died and China entered a period of isolation, suspending all naval expeditions abroad. Menzies claims that the inclusion of these islands on a chart predating Columbus' voyage must necessarily derive from a real feat of navigation, rather than just hearsay. It is a main plank of his argument, and should be tested for its soundness. Pizzigano clearly labels the larger of the two islands "Antilia," the first known use of this term. A study of the map by the Portuguese historian Armando Cortesão gives the etymology of the name as a combination of anti ("before") and ilha ("island"); by extension, "an island lying before a continent." Thereafter, the name appears on a map by the Genoese cartographer Battista Beccario in 1435 and, perhaps most famously, on the Nuremberg terrestrial globe made by Martin Behaim, a German navigator and geographer working as an advisor on navigation to King John II of Portugal, shortly before Columbus set sail in 1492.

ABOVE *This early 20th-century junk in Hong Kong harbor is not much smaller than Christopher Columbus' 85-foot (26-meter)* Santa Maria. *Admiral Zheng's flagship was 400 feet (122 meters), one of the largest wooden ships ever. The sails would have looked similar, recognizably "eastern"—but there were nine masts. The Admiral erected a tablet in the province of Fujian in 1432: "We have traversed more than 100,000 li (50,000 kilometers) of immense waterspaces and beheld in the ocean huge waves like mountains rising in the sky, and we have set eyes on barbarian regions far away hidden in a blue transparency of light vapors."*

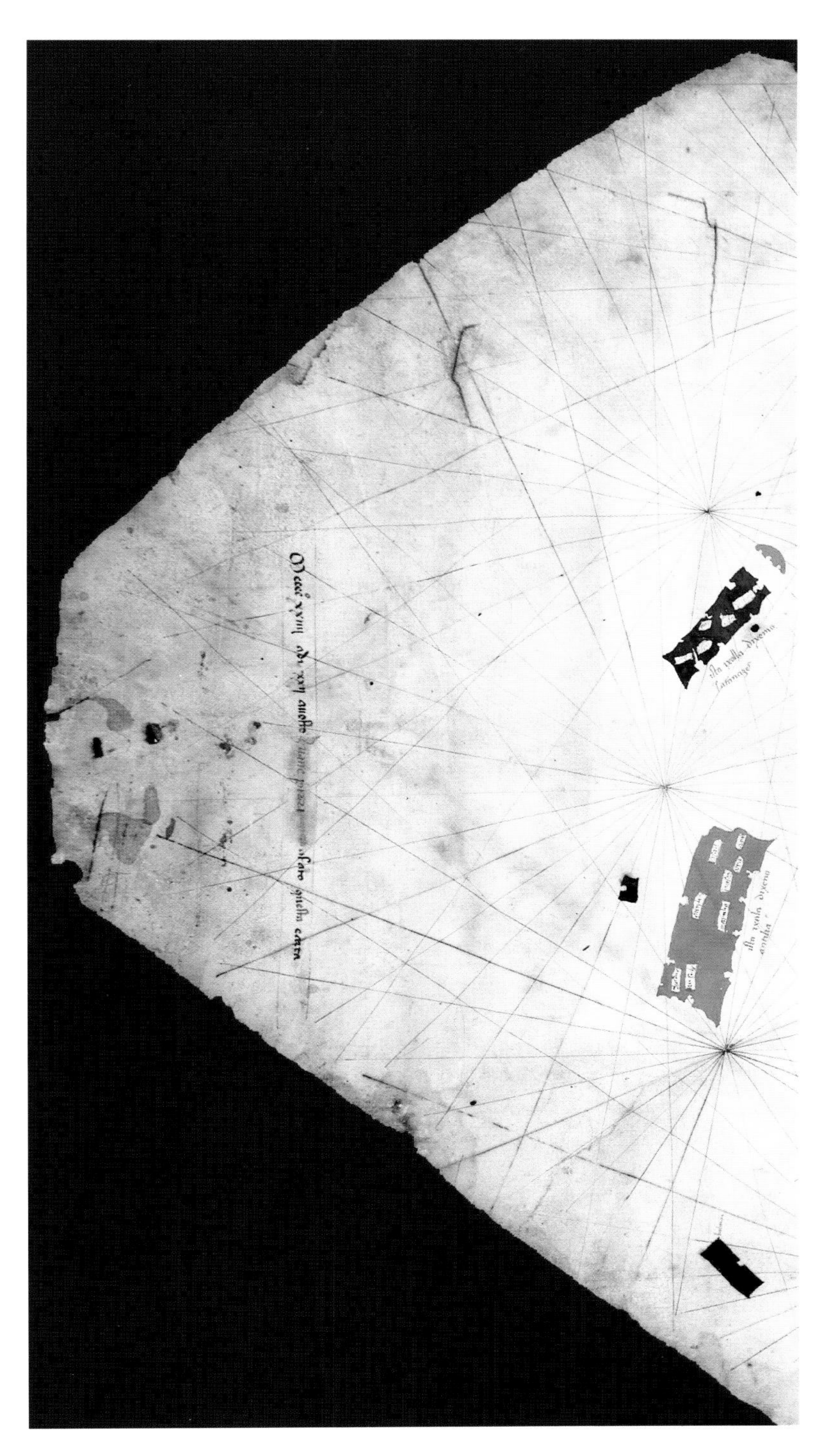

ABOVE *Pizzigano's 1424 map shows the coastlines of Portugal, Spain, and Africa, plus, Menzies argues, "with the help of some complex translation," Puerto Rico and Guadeloupe. The two key islands are shown in the closeup above. Menzies compared the outline of the red island, Antillia, with Puerto Rico.. For two of the bays, Guayanilla and San Juan, the similarity is quite convincing, and Puerto Rico itself is distinctly rhomboid. But elsewhere the match is not good. Others have argued that the islands represent Newfoundland, Florida, or Formosa. One unshakable supporter of the Menzies thesis is the Chinese government.*

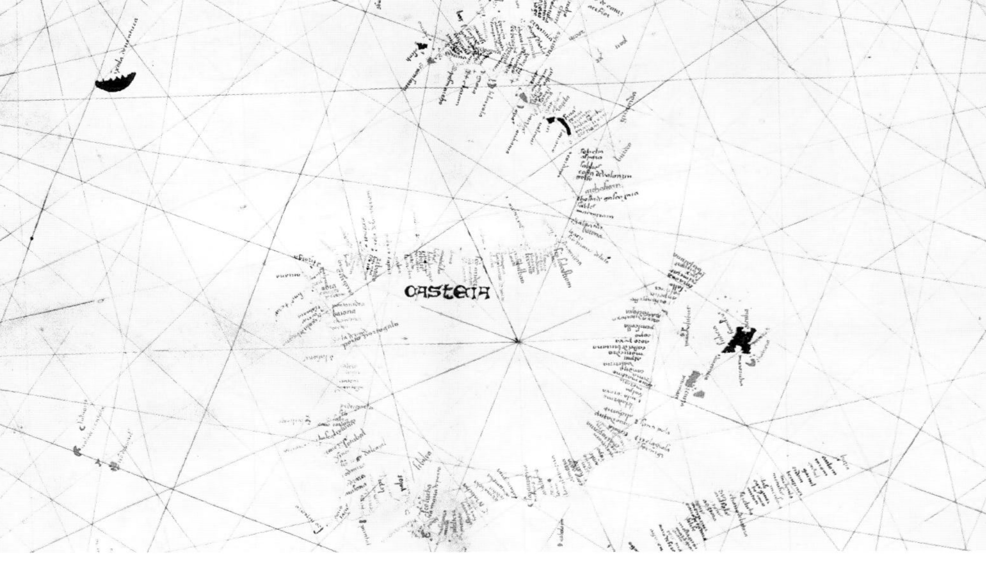

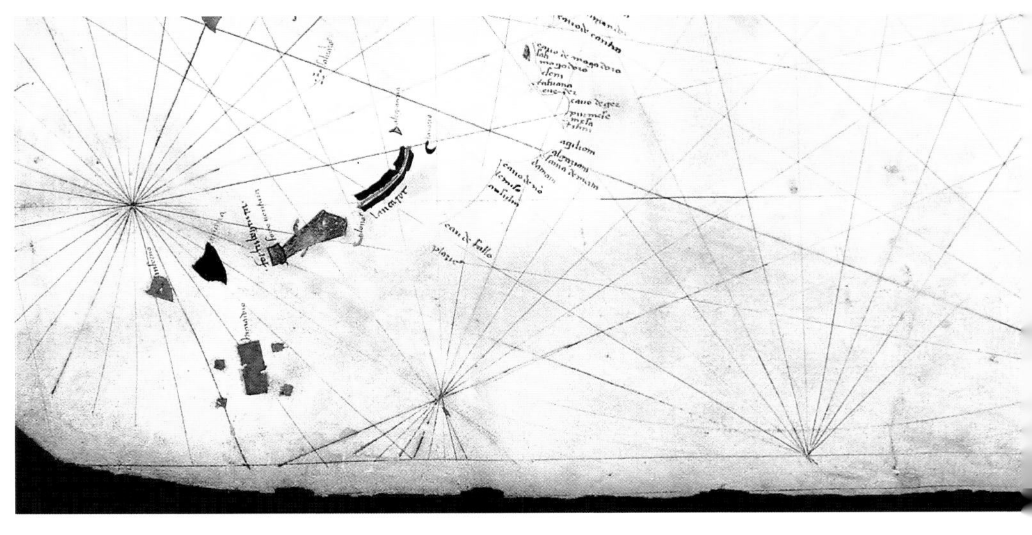

One key point about Portolan charts such as that by Pizzigano is that they were compendia of the accumulated wisdom—be it accurate or apocryphal—of previous navigators and mariners. The very name Portolan, meaning "reaching a haven or safe port," suggests that their makers included every piece of information that might possibly enable a seafarer to make land without mishap. One such chart was produced in 1367 by the same Venetian family of mapmakers to which our cartographer is thought to belong; this decorated Portolan by the brothers Francisco and Domenico Pizzigano (sometimes referred to as "Pope Urban's Map," from the name of its patron Urban V) is especially notable for showing a group of large islands far out in the Western Atlantic. Although it makes no mention of "Antilia," this map clearly demonstrates that the whole notion of exotic lands far to the west of the Azores was already common currency long before the 1424 map. It may well be the case, then, that the island of "Antilia" (or "Antillia") was a legend that, over time, assumed corporeal reality in the minds of cartographers simply as a result of having been depicted time and again on maps throughout the 15th century.

The earliest reference to the islands may be in the writings of the 1st century A.D. Greek essayist Plutarch, who refers to four "isles of the Demons," supposedly situated in the Atlantic Ocean; to be sure, this designation tallies neatly with Pizzigano's "Satanazas" ("Devil's Island"). But by the 15th century, an altogether different myth was in circulation. In 1474, a Florentine scientist called Paolo Toscanelli wrote a speculative letter to the king of Portugal, concerning a western sea passage to India, which asserts: "And from the Island of Antillia, which you call the 'Island of the Seven Cities' to the very famous island of Cipango is 2,500 miles." The key to this allusion is supplied by Behaim's later globe. On it, Antillia is described in the following way:

> "In the year 734 A.D., when all of Hispania was captured by the infidels from Africa [i.e. the Moorish invaders from Tangier, who actually invaded Spain in 711], the above island of Antilia, called the Seven Cities, was colonized by an Archbishop from Oporto, Portugal, together with six bishops and other Christian men and women who fled Hispania by ship with cattle, goods, and belongings. In 1414 a ship from Hispania sailed close to it."

Behaim's description places Antillia very firmly in the realms of Christian "wish-fulfilment" fables, which had their origins in the meteoric and militant rise of Islam and the loss of large territories to the new faith. Another, more famous legend of this kind is that of Prester John, the mythical Christian priest-warrior who was believed to inhabit a Central Asian realm somewhere east of Armenia and Persia, and who, it was rumored, would come to rescue Jerusalem and Christian Europe from the Muslim threat at the time of the Crusades.

RIGHT *It is fitting that the last image should be a 14th-century manuscript illumination depicting Marco Polo's journey through Siam from* The Book of Marvels. *As Zheng He sailed west, so Polo traveled East, to the court of Kublai Khan, 150 years earlier. The account of his journeys captured the Medieval imagination and is a mix of accurate reportage and fantasy that influenced cartography for centuries. We know that Columbus took Marco Polo's book with him; some think he also took Chinese maps of the New World.*

So durable are legends that, even after 16th-century navigators had dispelled the notion of Antillia (the name being reassigned to the island chain that makes up the West Indies), the myth of the Seven Cities (*Sete Cidades*) persisted, resurfacing in a new guise as the "Seven Cities of Cibola," a region of rich deposits of gold and silver allegedly located in North America just north of the Gulf of Mexico.

Meantime, Antillia continued to be shown—with remarkable consistency in its location, some 700 nautical miles west of the Azores—on maps throughout the 15th and early 16th centuries. Other constant features of its depiction were its area, around two-thirds the size of Portugal, and its almost perfect rectangular shape. Menzies is persuasive on the latter point, giving a thorough account of the correspondences between features of the coastline of the island on Pizzigano's map and those of Puerto Rico. However, Satanazes is a similar shape, and the evidence linking it to the decidedly non-rectangular Guadeloupe is less than compelling. Are we not dealing, rather, with a schematic outline that was applied indiscriminately to several notional lands?

Menzies' thesis has attracted criticism on the grounds that much of its reasoning is based on insufficiently rigorous research. For example, his claim that the Chinese expedition introduced giant ground sloths (mylodons) to Australia from South America founders on the fact that these creatures were long extinct by the 15th century. Similarly, it appears that the initial premise of his argument concerning Pizzigano's map is fatally flawed; the mere presence of Western Atlantic islands on pre-Columbian charts such as this in no way presents incontrovertible evidence of early exploration there. The opinion of the curators of the James Ford Bell Library must be germane:

> "Because of the details shown on the islands, with names of places given on some of them, it is very tempting to believe that they are more than the results of legends or myths. But

the names on maps cannot be identified with modern places or even the origin of names determined with certainty. Too often the reading is made to satisfy some theory about 'firsts'."

The Chinese government are not unhappy about the theory. As the *China Daily* pointed out in July 2004:

"Question for the ages: Who circumnavigated the globe 87 years before Italian explorer Cristopher Columbus (1451-1506) and 114 years before Portuguese explorer Ferdinand Magellan (1480-1521)? ... To prepare to mark the 600th anniversary of Zheng's first voyage next year, the country has set up a directorate headed by Minister of Communications Zhang Chunxian to stage a massive celebration."

It is certainly unfair to include Menzies' theory in a chapter entitled "Fantasies, Follies, and Fabrications." There is no deceit here, and some of the criticism from the history academic "establishment" smacked a little of *de haut en bas*. So, Mr Menzies, if you come across this book we hope you will be happy with this; your map story is the last in this work because it touches upon exploration, politics, interpretation—several of the themes that make cartography a proper study of mankind.

Bibliography and Websites

Bibliography

Alder, Ken *The Measure of All Things: The Seven-Year Odyssey and Hidden Error that Transformed the World*, Simon & Schuster, 2002
Black, Jeremy *Maps and History; Constructing Images of the Past*, Yale University Press, 1997
Black, Jeremy *Maps and Politics*, University of Chicago Press, 1997
Blake, John *The Sea Chart: The Illustrated History of Nautical Maps and Navigational Charts*, Conway Maritime Press, 2004
Brown, Lloyd A. *The Story of Maps*, Dover Publications, 1980
Day, David *Tolkien: The Illustrated Encyclopedia*, Mitchell Beazley 1993
Erlingsson, Ulf *Atlantis from a Geographer's Perspective; Mapping the Faery Land*, Lindorm Publishing, 2004
Harvey, Miles *The Island of Lost Maps; a True Story of Cartographic Crime*, Broadway, 2001
Haywood J. et al., *Atlas of World History*, Barnes & Noble, 2001
Hearn, Chester G. *Tracks in the Sea: Matthew Fontaine Maury and the Mapping of the Oceans*, Ragged Mountain Press, 2002
Monmonier, Mark *How to Lie with Maps*, University of Chicago Press, 1991
Monmonier, Mark *Cartographies of Danger*, University of Chicago Press, 1998
Monmonier, Mark *Rhumb Lines and Map Wars: A Social History of the Mercator Projection*, University of Chicago Press, 2004
Thrower, Norman J. W. *Maps and Civilization; Cartography in Culture and Society*, University of Chicago Press, 2nd edition 1999
Robinson, Arthur H. et al., *Elements of Cartography*, Wiley, 6th edition 1995
Sobel, Dava *Longitude: The True Story of a Lone Genius who Solved the Greatest Scientific Problem of His Time*, Penguin 1996
Schwartz, Seymour L. and Ehrenberg, Ralph E. *The Mapping of America*, Wellfleet, 2001
Wilford, John N. *The Mapmakers*, Knopf, 2000
Winchester, Simon *The Map That Changed the World: William Smith and the Birth of Modern Geology*, Perennial, 2002
Whitfield, Peter *New Found Lands: Maps in the History of Exploration*, Routledge, 1998
Woodward, David (ed) *Art and Cartography*, University of Chicago Press, 1987
Woodward, David and Harley, J. B. (eds) *The History of Cartography: Cartography in Prehistoric, Ancient and Medieval Europe and the Mediterranean*, Vol. 1, University of Chicago Press, 1987
Woodward, David and Harley, J. B. (eds) *The History of Cartography, Volume 2, Book 1: Cartography in the Traditional Islamic and South Asian Societies*, University of Chicago Press, 1992
Woodward, David and Harley, J. B. (eds) *The History of Cartography, Volume 2, Book 2: Cartography in the Traditional East and Southeast Asian Societies*, University of Chicago Press, 1995

Websites

www.auslig.gov.au/
National mapping division of Australia.

http://earthtrends.wri.org/
World Resources Institute national and world thematic mapping; see page 86.

http://bell.lib.umn.edu/hist/
Historical maps from the James Ford Bell Library, University of Minnesota, organized alphabetically by mapmaker, from Dutchman Carel Allard (1648-1709) to Italian Bolognini Zaltieri (fl. 1550-1580).

www.lib.utexas.edu/maps/
Historical and modern mapping collection, including CIA maps.

http://www.maphistory.info/
Straightforward search engine.

http://memory.loc.gov//ammem/gmdhtml/
Library of Congress Geography and Map Division map collections. Maps from ca 1320-present day, with focus on the US, but by no means limited to American maps.

http://www.nmm.ac.uk/collections/explore/index.cfm/category/charts
View a selection of the 100,000 maps and charts of all the seas and oceans in the collection of the National Maritime Museum, Greenwich; search by region, title, maker, and century.

http://oddens.geog.uu.nl/index.html
The best cartographic link to other sites.

www.ordsvy.gov.uk/
Site of one of the oldest mapping institutions, the Ordnance Survey.

http://www.pmel.noaa.gov/tsunami/indo_1204.html
Mapping and research following the December 2004 tsunami disaster under the auspices of the US National Oceanic and Atmosphere Administration (NOAA); see page 84.

www.un.org/Depts/Cartographic/english/htmain.htm
Country profiles by one of the most prolific mapmaking agencies in the world, the United Nations.

http://www.usno.navy.mil/library/
The US Naval Observatory Library; a collection of rare maps and complete atlases, with high resolution imagery.

Index

Figures in *italic* indicate captions.

Picture Acknowledgments

R=Right, L=Left, C= Center, T=Top, B=Bottom.

© **Ancient Art & Architecture Collection Ltd:** 16, 19.
© **Alan Lee,** from *Tolkien's Ring* (Pavilion Books, 1999): 243.
© **The Art Archive** / Biblioteca Estense Modena / Dagli Orti: 16. / © The Art Archive / Bodleian Library Oxford / The Bodleian Library: 24. / © The Art Archive / National Library Cairo / Dagli Orti: 25.
Beinecke Rare Book and Manuscript Library, Yale University: 244.
© **Bridgeman Art Library:** 32. / © The Stapleton Collection/Bridgeman Art Library: 35. / © Musée de la Poste, Paris, France, Archives Charmet/Bridgeman Art Library: 43. / Courtesy of the Warden and Scholars of New College, Oxford/Bridgeman Art Library: 103. / © Royal Geographical Society, London, UK/Bridgeman Art Library: 226R. / © illustration from *The Hobbit* by J. R. R. Tolkien (1892-1973), edition published 1976 (engraving), Belomlinsky, Mikhail (contemporary artist), private collection/Bridgeman Art Library: 242.
© **British Crown Copyright:** 149.
By permission of the British Library: 61.
© **Bodleian Library, Oxford:** 183.
© **Chrysalis Image Library:** 99T, 99B, 134, 168, 186T, 236, 237L, 237R, 238-239, 246.
© **Chester Hearn Collection:** 62, 63, 65B.
© **Central Intelligence Agency:** 88TR, 88BR.
© **CORBIS:** 126, 213. / © Michael Maslan Historic Photographs/CORBIS: 15. / © Yann Arthus-Bertrand/CORBIS: 23. / © Charles & Josette Lenars/CORBIS: 30. / © Gianni Dagli Orti/CORBIS: 39, 40. / © Bettmann/CORBIS: 56, 197, 200, 222, 227. / © Royalty-Free/Corbis: 14, 47, 49, 72, 73, 93, 95, 96, 106, 110, 114-115, 119, 123, 131, 137, 196-197, 206. / © Stapleton Collection/CORBIS: 58, 121. / © Reza; Webistan/CORBIS: 89. / Archivo Iconografico, S.A./CORBIS: 111. / ©The Andy Warhol Foundation for the Visual Arts/Corbis: 189. / © David Turnley/CORBIS: 207.
© **Digital Vision:** 4, 6, 7, 10, 45, 46, 50, 54, 107, 108, 109, 113, 116, 117, 108, 109, 113, 116, 117, 118, 120, 129, 204, 205, 225, 226L.
Dixson Library/State Library of New South Wales: 141.
© **American Map Corporation – Release AMC062805:** 77.
© **Heritage image partnership / The British Library:** 142.
Reproduced by permission of the Huntington Library, San Marino, California: 218.
© **Imperial War Museum:** 176, 177, 178.
© **The James Ford Bell Library, University of Minnesota:** 249.
Library of Congress Geography and Map Division: 2, 8, 11, 22, 26, 27, 57, 65T, 69, 70, 88BL, 90, 94, 100-101, 105, 125, 127, 130, 138, 143, 146-147, 155, 156T, 157, 190, 193, 201, 203, 209, 210, 211, 215, 216L, 216R, 223, 228, 233, 234L, 234R, 235; **Library of Congress, Geography and Map Division / Preparation Route source: Frank Muhly, Lewis and Clark Trail Heritage Foundation, Philadelphia Chapter**: 145; **Library of Congress, Prints & Photographs Division: [HABS, RI,3-NEWP,3-1]** 31, **[LC-USZC2-3365]** 64, **[LC-USZC4-4544]** 146, **[LC-USZC4-724]** 156R, **[LC-USZ62-118818]** 248.
© **Ludington Limited:** 79.
© **Martin Marix Evans:** 160, 161, 163T, 163B, 164, 165, 167, 169, 171, 172, 173, 175, 179, 181, 186B.
© **NASA:** 1, 3, 81, 82, 83.
Courtesy of The National Archives of the UK: (C0700 NY 23) 13, **(FO 931/1900)** 36, **(CO 700 AMERICAN COLONIES GENERAL 1)** 133, **(CO700 AMERICAN COLONIES GENERAL 2)** 134-135, **(W078/1006)** 150, **(MPI/1450 (4)** 185T, **(MPI 1/450 (3)** 185B, **(MPI 1303)** 194-195.
Collection of the New-York Historical Society: 230, 231.
© **Octopus Publishing Group/Sally Davies:** 241.
© **Photo Scala, Florence:** 153.
From the NORSE MYTHS by Kevin Crossley-Holland, © 1980 by Kevin Crossley-Holland. Used by permission of Pantheon Books, a division of Random House Inc.: 29.
© **Science Museum/Science & Society Picture Library:** 53, 199.
© **Rex Features:** 76.
© **2003, The State of Israel / MOD:** 217.
The Tank Museum Collection, Bovington: 180.
Trustees of the Corsham Estate: 159.
© **Ulf Erlingsson, 2004.** *Atlantis From a Geographer's Perspective: Mapping the Fairy Land* (Lindorm.com): 221.
© **US Department of Defense:** 187.
© **U.S. National Oceanic & Atmospheric Administration (NOAA) and U.S. National Tsunami Hazard Mitigation Program:** 85.
© **US Naval Observatory Library:** 33.
© **Wellcome Library, London:** 67, 75.
© **World Resources Institute:** Data from China Human Development Report 1999, provided by CIESN, boundaries by ESRI, map reproduced from http://earthtrends.wri.org, a project at the World Resources Institute: 87TL. / "Poverty in Bangladesh: Building on Progress," World Bank Poverty Assessment 2002, map reproduced from http://earthtrends.wri.org, a project at the World Resources Institute 87TR. / Global map provided by CIESIN, anthropometric data by Demographic and Health Survey. Map reproduced from http://earthtrends.wri.org, a project at the World Resources Institute. 87BL, 87BR, 88TL.
© **ODT, Inc.** For maps and other related teaching materials contact: ODT, Inc., PO Box 134, Amherst MA 01004 USA; (800-736-1293; Fax: 413-549-3503; E-mail: odtstore@odt.org. Web: **www.odt.org:** 247.
© **Front cover CORBIS.**

With thanks to Randall Bytwerk, Professor of Communication Arts and Sciences at Calvin College in Grand Rapids, Michigan, for the Goebbels quotation on page 236. http://www.calvin.edu/academic/cas/gpa/goebmain.htm

With thanks to James Methuen-Campbell for permission to reproduce the map on page 159.

Contributors

In addition to the co-authors introduced on the back flap of the book jacket, the following have kindly contributed to this title:

Book packager, editor and author Marcus Cowper is a US history specialist and so has chosen to write about such topics as the North American pioneers John Smith and John White, Sanson and de Champlain cartography, and the Anglo-French "Map Wars."

David Day, who has written here on some of the fantasy maps and on ancient cosmologies, has written no less than six books on the works of J. R. R. Tolkien. He is also author of *The Search for King Arthur* (Facts on File, 1996).

Chet Hearn, who has contributed on the work of American Civil War cartographer Jedediah Hotchkiss and oceanographer Maury, is the author of more than 20 books, including *Tracks in the Sea: Matthew Fontaine Maury and the Mapping of the Oceans* (Ragged Mountain Press, 2002) and *Civil War: Virginia* (Salamander Books, 2005).

Gillian Hutchinson, who has contributed analysis of portolans and British Admiralty sea-charts, is Curator of the History of Cartography at the National Maritime Museum, Greenwich.

Author, translator, and editor Peter Lewis has brought his historical academic background to bear on Renaissance and 17th-century cartography: including Cellarius, Saxton, Apian, Kaerius, de Wit, and the Pizzigano map.

Editor's note: There are in fact 156 maps in this book; but some are reproduced small, and some are representative of similar cartographic themes; it is hoped and intended that 100 maps are of sufficient size and difference to be examined more closely.